Artificial Muscles

Artificial Muscles

Applications of Advanced Polymeric Nanocomposites

Second Edition

Mohsen Shahinpoor

CRC Press

Taylor & Francis Group
Boca Raton London New York

CRC Press is an imprint of the
Taylor & Francis Group, an **informa** business

Second edition published 2022
by CRC Press
6000 Broken Sound Parkway NW, Suite 300, Boca Raton, FL 33487-2742

and by CRC Press
2 Park Square, Milton Park, Abingdon, Oxon, OX14 4RN

© 2022 Taylor & Francis Group, LLC

First edition published by CRC Press 2007

CRC Press is an imprint of Taylor & Francis Group, LLC

Library of Congress Cataloging-in-Publication Data

Names: Shahinpoor, Mohsen, author.
Title: Artificial muscles : applications of advanced polymeric nanocomposites / Mohsen Shahinpoor, Department of Mechanical Engineering, College of Engineering, Graduate School of Biomedical Science and Engineering, University of Maine, Maine, USA.
Description: Second edition. | Boca Raton : CRC Press, 2022. | Includes bibliographical references and index.
Identifiers: LCCN 2021021421 | ISBN 9780367857905 (hbk) | ISBN 9781032107646 (pbk) | ISBN 9781003015239 (ebk)
Subjects: LCSH: Nanotechnology. | Muscles. | Artificial organs. | Biomimetic polymers. | MESH: Polymers--chemistry. | Artificial Organs. | Biomimetic Materials. | Electrophysiology--methods. | Muscles--physiology. | Nanostructures.
Classification: LCC TP248.25.N35 S53 2022 | DDC 660.6--dc23
LC record available at https://lccn.loc.gov/2021021421

ISBN: 978-0-367-85790-5 (hbk)
ISBN: 978-1-032-10764-6 (pbk)
ISBN: 978-1-003-01523-9 (ebk)

DOI: 10.1201/9781003015239

Typeset in Times LT Std
by KnowledgeWorks Global Ltd.

Contents

List of Figures

List of Tables

Acronyms

AMPS	2-Acrylamido-2-methyl-1-propanesulfonic acid monomer
CAM	composite artificial muscles
DLVO	combination double-layer forces and Van der Waals forces in ionic gels
HEMA	2-Hydroxyethyl methacrylate
IEM	ion-exchange membrane
IEM-Pt	ion-exchange membrane platinum
IEMMC	ion-exchange membrane metal composite
IEMPC	ion-exchange membrane platinum composite
IPCC	ionic polymeric conductor composite
IPCNC	ionic polymeric conductor nanocomposite
IPMC	ionic polymeric metal composite
IPMNC	ionic polymer-metal nanocomposite
MBAA	N,N'-Methylene-bis-acrylamide
PAAM	polyacrylic acid bis-acrylamide
PAM	polyacrylamide
PAMPS	poly(2-acrylamido-2-methyl-1-propanesulfonic acid)
PAN	polyacrylonitrile
PANi	polyaniline
PPy	polypyrrol

Symbols

a	maximum swelled distance from the center of the gel
B	reduced electric potential
b	polyion separation distance
b_i	spacing between ith row of polyion segments
C	total ion-exchange capacity, ml/g dry membrane; also, gel cylindrical sample half radial thickness
C_p	power coefficient
C_T	thrust coefficient
C_g	specific capacitance of the gel, F/g
C_{mi}	Fourier coefficients
C_{oi}	Fourier coefficients
c_i	concentration of the ith species
D	dielectric constant of the liquid phase; also, drag force
D_{eff}	effective diffusivity coefficient in cm^2/s
D_0	diameter of cylindrical polymer sample
ΔE	driving force pumping ions; also, gradient of electric field
e	electron charge, 1.602×10^{-19} C
E	Young's modulus
F	total free energy; also, force performing mechanical work; also, force produced in the gel; also, mean coulomb attraction/repulsion forces associated with R^*
F_0	force during isometric contraction
ΔF	free energy decrease as a result of contact between two polymer networks
F_e	free energy due to work done against electric field
F_g	free energy due to gel deformation
f	frequency, Hz; also, number of ionized segments out of N0 total; also, friction coefficient between polymer network and liquid medium
G	shear modulus
g	local gravity acceleration, 9.81 m/s^2
H	total amount of hydrogen ions (including undissociated)
h	hydrogen ion concentration
$\bar{h}$	hydrogen ion concentration inside the membrane
I	electric current, A
i	current density of the gel
K	dissociation constant; also, modified Bessel function
k	proportionality constant; also, Boltzmann constant, 1.381×10^{23} J/K; also, bulk elastic modulus of the gel polymeric component
L	gel length when fully swollen
L_0	length of cylindrical polymer sample
l	uniaxial elongation of muscle fiber
N_0	number of freely jointed segments of polymer
n	number density of counter-ions; also, number of polyions
n_i	number density of ions for the ith species
P	pressure term, N/m^2; also, hydrostatic pressure term
pH	a measure of acidity or alkalinity of solution
Q	electric charge, C; also, charge per unit mass or specific charge of the gel, C/g
q	electric charge, C; also, degree of swelling
q_n	quantity of mobile ions

R	electric potential field per unit charge
R_+	electric potential field corresponding to positive charges
R_-	electric potential field corresponding to negative charges
R^*	total electric field due to all strands of polymer network
R_g	specific resistance of the gel, Ω/g
Re	Reynolds number
r	hydrodynamic frictional coefficient; also, radius of ionic gel sample
r^*	mean radius of ionic gel sample
r_i	inner radius of elemental cylinder in gel fiber; also, cylindrical polar coordinate
r_0	outer radius of elemental cylinder in gel fiber
S	total amount of salt cation; also, entropy in thermodynamic context; also, wetted surface area in fluid mechanics
S_w	linear swelling ratio
s	concentration of salt cation
$\bar{s}$	concentration of salt cation inside the membrane
T	absolute temperature, K; also, thrust force, N
t	time, s
t^*	thickness of gel cylindrical sample
U	total internal energy, J; also, steady-state velocity term
u_0	percent change of sample dimension at final state
u_{ij}	displacement vector of the gel elemental volume from its position when the swelling process has gone to completion
V	water volume; also, volume in thermodynamic sense; also, velocity of contraction
V_P	volume of the dry polymer sample
V_t	total volume
$\bar{V}$	water volume inside the membrane
v	voltage across the thickness of the gel, V
W	dry weight of the muscle membrane or gel
w_0	water content of the muscle membrane, ml/g
X	total concentration (dissociated and undissociated) of weak acid groups in the membrane
Z	distance of an element from free end of the gel; also, number of ionizable groups
Z_i	cylindrical polar coordinate
ΔZ_0	elemental disk thickness before applied electricity
z	valance number
z_i	valance of the ith species
α	degree of dissociation; also, degree of ionization; also, factor increasing D_0 after application of electricity
β	factor increasing ΔZ_0 after application of electricity; also, chemical stress term
β_m	positive root of Bessel function of order zero
Γ	modified Bessel function
Δ	thickness term
δ_{ij}	mechanical strain
ε	dielectric constant; also, electric field energy; also, average electric charge
η	fish propulsion efficiency
θ	cylindrical polar coordinate angle
Φ	contraction rate of the gel
ϕ	volume fraction of the polymer network
ϕ_0	concentration of polymer (no interaction between segments or reference states)
κ	inverse of Debye length or effective thickness of the ionic layer surrounding the charge sites of the individual fibrils; also, phenomenological coefficient

λ	dimensionless parameter relating to β
μ	shear modulus; also, average mobility of the medium in the gel; also, solvent viscosity
μ_h, μ_s	hydrogen-ion and salt-cation mobility
μ_i	chemical potential (energy) for the ith species
ξ	positive root of Bessel function of order zero
ρ	density of liquid solvent
ρ^*	charge density
σ_{ij}	tensile stress, N/m^2
τ	reduced temperature
ν	number of polymers cross-linked in the network; also, velocity of contraction of the gel; also, three-dimensional liquid velocity vector
ψ	local electrostatic potential

About the Author

Mohsen Shahinpoor is a Professor and Director at the University of Maine College Of Engineering, Department of Mechanical Engineering. He is also a professor in the Graduate School of Biomedical Science and Engineering at the University of Maine. He also serves as Director of two Engineering Minor Programs, one in Robotics and another in Biomedical Engineering. He also has affiliations with the Civil and Environmental Engineering Department (stimuli-responsive nanomaterials), Electrical and Computer Engineering Department (advanced robotics and MEMs) and Chemical and Bioengineering Department (Biomedical Engineering).

He has also served as Richard C. Hill Professor and Chair of the Mechanical Engineering Department for 7 years. He is the Director of the Advanced Robotics Laboratory, the Director of the Smart Materials, Artificial Muscles and Tissue Manufacturing Laboratory and the Director of the Biomedical Engineering and Robotic Surgery Laboratory. He is a Fellow of the National Academy of Inventors.

He is a Fellow of the American Society of Mechanical Engineers (ASME), Institute of Physics (IOP), Royal Society of Chemistry (RSC) and International Association for Advanced Materials (FIAAMs).

He is active in research and development in biomimetics, soft, flexible robotics, robotic surgery, smart wearable robotic braces, smart materials, electroactive polymers and ionic polymer-metal composite (IPMCs) soft actuators, self-powered energy harvesters and sensors.

Prologue

In this book, thorough reviews of existing knowledge connected with electroactive polymers (EAPs) and, in particular, ionic polymeric conductor nanocomposites (IPMCs, IPCNCs) are presented. Ionic polymeric metal nanocomposites (IPMCs) are biomimetic stimuli-responsive, multifunctional distributed nanosensors, nanoactuators, nanotransducers, nanorobots, artificial muscles and electrically controllable intelligent polymeric network structures. The book further discusses how biomimetic actuator materials like IPMCs, in general, have been developed based on biological muscle operation and how the latter are limited in their performance compared with biological muscles.

Where possible, comparisons have been made with biological muscles and applications in noiseless, biomimetic marine propulsion and unmanned aerial vehicles (UAVs) and flapping-wing systems using such electroactive polymeric materials. Furthermore, the book introduces and discusses in detail methods of fabrication and manufacturing of several electrically and chemically active ionic polymeric sensors, actuators and artificial muscles, such as polyacrylonitrile (PAN), poly(2-acrylamido-2-methyl-1-propanesulfonic) acid (PAMPS) and polyacrylic-acid-*bis*-acrylamide (PAAM), as well as a new class of electrically active composite muscles such as IPCNCs or IPMNCs. These discoveries have resulted in seven US patents regarding their fabrication and application capabilities as distributed biomimetic nanoactuators, nanosensors, nanotransducers, nanorobots and artificial muscles.

In this book, various methods of IPMNC manufacturing and fabrication are reported. In addition, manufacturing and characterization of PAN muscles are discussed. Conversion of chemical activation to electrical activation of artificial muscles using chemical plating techniques is described. Furthermore, other methodologies, such as physical/chemical vapor deposition methods or physical loading of a conductor phase into near boundary of such materials, are briefly discussed. The technologies associated with pH-activated muscles like PAN fibers have also been detailed. Experimental methods are described to characterize contraction, expansion and bending of various actuators using isometric, isoionic and isotonic characterization methods.

Several apparatuses for modeling and testing the various artificial muscles have been described to show the viability of application of chemoactive as well as electroactive muscles. Furthermore, fabrication methods of PAN fiber muscles in different configurations (such as spring-loaded fiber bundles, biceps, triceps, ribbon-type muscles and segmented fiber bundles) to make a variety of biomimetic nanosensors and nanoactuators have been reported here.

Theories, modeling and numerical simulations associated with ionic polymeric artificial muscles' electrodynamics and chemodynamics have been discussed, analyzed and modeled for the manufactured material. The book concludes with an extensive chapter on all current industrial and medical applications of IPMNCs as distributed biomimetic nanosensors, nanoactuators, nanotransducers, nanorobots and artificial muscles.

Mohsen Shahinpoor

1 Introduction to Ionic Polymers, Ionic Gels and Stimuli-Responsive Materials and Artificial Muscles

1.1 INTRODUCTION

Many scientists and researchers' focus has been to achieve efficiencies as high as 50% in the direct conversion of chemical to mechanical energy as occurs in biological muscles. In comparison, most internal combustion engines and steam turbines have lower efficiency at best. First, it would be useful to briefly review electroactive polymers (EAPs) in general and other stimuli-responsive and multifunctional materials and artificial muscles. For a brief description of biological muscles, see Appendix A of the first edition, Shahinpoor, Kim and Mojarrad, (1992).

1.2 A BRIEF HISTORY OF ELECTROACTIVE POLYMERS (EAPs) AND ARTIFICIAL/SYNTHETIC MUSCLES

Roentgen (1880) appears to have been the first to make an EAP. He used a rubber band that could change its shape by being charged or discharged. Later, Sacerdote (1899) formulated the strain response to electric field activation in polymers. In 1925, Eguchi (1925) reported the discovery of a piezoelectric polymer that he called an electret. He found that when molten carnauba wax, rosin and beeswax were solidified by cooling in the presence of a direct current (D.C.) electric field, the resulting material was piezoelectric. Generally, electrical excitation is only one type of stimulator that can induce elastic deformation in polymers. The polyvinylidene fluoride (PVDF) crystalline solids possess microscopic regions with dipole charges such that under mechanical stress, the internal arrangements of embedded electrical dipoles generate a voltage across the material boundaries. Conversely, an applied voltage to the material changes the embedded internal dipole charges' orientation and generates a deformation or strain in the solid. Zhang and colleagues (1998) observed a substantial piezoelectric activity in PVF2-TrFE as early as 1998.

Other activation mechanisms include chemical (Katchalsky, 1949; Kuhn et al., 1950; Steinberg et al., 1966; Shahinpoor, 1992; Otero et al., 1995; Shahinpoor et al., 1997a, 1997b), thermal (Tobushi et al., 1992; Kishi et al., 1993; Li et al., 1999), pneumatic (Shahinpoor et al., 2001), optical (van der Veen and Prins, 1971) and magnetic (Zrinyi et al., 1997) mechanisms.

Chemically stimulated polymers were discovered more than half a century ago when it was shown that collagen filaments could reversibly contract or expand when dipped in acidic or alkaline solutions, respectively (Katchalsky, 1949). This early work pioneered synthetic polymers that mimic biological muscles (Steinberg et al., 1966; Shahinpoor et al., 1998a, 1998b). However, electrical stimulation has remained the best means of EAP material actuation and sensing. Shahinpoor and Mojarrad (1994, 1996, 1997a, 1997b, 1997c, 1997d, 2000) were among the pioneers in making electrically active – in sensing and actuation – ionic polymer conductor nanocomposites (IPCNCs) and ionic polymer-metal nanocomposites (IPMNCs).

DOI: 10.1201/9781003015239-1

The most significant progress in EAP materials' development has occurred in the last ten years, where useful materials that can induce strains exceeding 100% have emerged (Perline et al., 1998). Generally, EAPs can be divided into two major categories of electronic and ionic polymers based on their activation mechanism. The electronic polymers (electrostrictive, electrostatic, piezoelectric and ferroelectric) require high activation fields (~200 MV/m) close to the dielectric breakdown field levels in polymers.

The electronic EAP materials have a higher mechanical energy density and can be operated in the air with no significant constraints. Ionic EAP materials (gels, polymer-metal composites, conductive polymers (CPs) and carbon nanotubes and graphene), on the other hand, require drive voltages as low as a few volts and small fields of about a few thousand volts per meter (~2–4 kV/m). The induced displacement of the electronic and ionic EAPs can be designed geometrically to bend, stretch, roll, twirl, twist or contract.

Next, we will describe some specific EAPs.

1.2.1 ELECTRICALLY CONDUCTIVE AND PHOTONIC POLYMERS

McGehee et al. (1999) described it had taken almost 20 years to develop synthetic metals or CPs such as polyaniline. These studies' success has led to the emergence of CPs that exhibit electronic properties approaching the levels in metals and semiconductors (Cao et al., 1991) and offering the processing advantages and mechanical properties of polymers. High-performance devices have been fabricated, including light-emitting diodes, light-emitting electrochemical cells, photodiodes and lasers. The brightness of some light-emitting diodes based on photonic polymers is as high as that of a fluorescent lamp. The performance of some photonic polymers has been improved such that their performance is comparable to or better than their inorganic counterparts. Compared to some conventional charge-coupled devices (CCDs), EAP-based systems are expected to respond about ten times faster, provide larger pixel arrays at low power and effectively operate at low light levels. There are currently a large number of conducting polymers used industrially or medically. Well-known conducting polymers are polypyrrole (PPy), polyaniline, polythiophene, polyphenylene vinylene, polyacetylene etc., which can be manufactured by chemical or electrochemical oxidation and reduction (REDOX) procedures.

Early work on CPs reported that polyacetylene's conductivity increased millions of times when it was oxidized by "**doping**" with iodine vapor. The molecular structures of CP possess both single and double chemical bonds to enhance charge transfer.

The pioneering work of Professors Heeger (2001), MacDiarmid (2001) and Shirakawa (2001) in these areas of CPs earned them the Nobel Prize in chemistry in 2000.

1.2.2 MAGNETICALLY ACTIVATED POLYMER GELS AND POLYMERS

Magnetically activated gels, so-called ferrogels, are chemically cross-linked polymer networks that are swollen by the presence of a magnetic field (Zrinyi et al., 1999). Zrinyi et al. (1999) has been a pioneer of this technology. Such a gel is a colloidal dispersion of monodomain magnetic nanoparticles. Such magnetic ferrogel materials deform in the presence of a spatially nonuniform magnetic field because their embedded nanoparticles move in reaction to the field. The ferrogel material can be activated to bend, elongate, deform, curve or contract repeatedly. It has a response time of less than 100 ms, which is independent of the particle size. Note that magnetic gels are softer and more stretchable, and maneuverable in the magnetic field. An intro to the development of ferrogels was a classic paper by Rosenzweig in 1985 on ferrohydrodynamic. A colloidal ferrofluid, or a magnetic fluid, is a colloidal dispersion of monodomain magnetic particles. The monodomain magnetic particles are about 10–15 nm in size and are superparamagnetic in which magnetization can randomly flip direction under the influence of temperature. Magnetic gels or ferrogels belong to the general family of magnetostrictive materials producing strain when exposed to a magnetic field. Magnetic gels belong to the family of hydrogels, polymeric gels and generally polyelectrolyte gels. They are highly swollen molecular network which is cross-linked and they create a hydrophilic solid.

1.2.3 ELECTRONIC EAPs/FERROELECTRIC POLYMERS

The conducting crystal or dielectric material exhibits spontaneous electric polarization. These are based on the phenomenon of piezoelectricity, which is found only in noncentrosymmetric materials such as PVDF. These polymers are partly crystalline, with an inactive amorphous phase, and have Young's modulus of elasticity of about 1–10 GPa. This relatively high elastic modulus offers high mechanical energy density. A large applied A.C. field (–200 MV/m) can induce electrostrictive (nonlinear) strains greater than 1%. Sen et al. (1984) investigated the effect of mixing heavy plasticizers (–65 wt%) of ferroelectric polymers hoping to achieve large strains in reasonably applied fields. However, the plasticizer is also amorphous and inactive, resulting in decreased Young's modulus, permittivity and electrostrictive strains.

Recently, Zhang et al. (1998) introduced defects into the crystalline structure using electron irradiation to dramatically reduce the dielectric loss in a PVDF trifluoroethylene or P(VDF-TrFE) copolymer. This copolymerization permits A.C. switching with much less generated heat. The electric-field-induced change between nonpolar and polar regions is responsible for the large electrostriction observed in this polymer. As large as 4% of electrostrictive strains can be achieved with low-frequency driving fields having amplitudes of about 150 V/μm.

As with ceramic ferroelectrics, electrostriction can be considered the origin of piezoelectricity in ferroelectric polymers (Furukawa and Seo, 1990). Unlike electrostriction, piezoelectricity is a linear effect, where the material will be strained when voltage is applied, and a voltage signal will be induced when stress is applied. Thus, they can be used as sensors, transducers and actuators. Depoling due to excessive loading, heating, cooling or electric driving is a problem with these materials.

1.2.4 ELECTRETS AND PIEZOELECTRIC POLYMERS

Electrets, which were discovered in 1925, are materials that retain their electric polarization after being subjected to a strong electric field (Eguchi, 1925). Piezoelectric behavior in polymers also appears in electrets, which are essential materials that consist of a geometrical combination of a hard and soft phase (Sessler and Hillenbrand, 1999). The positive and negative charges within the material are permanently displaced along and against the field's direction, respectively, making a polarized material with a net zero charge. Piezoelectric materials are piezoceramics like lead-zirconate-titanate (PZT) and piezopolymers like PVDF. The piezoelectric effect describes a reversible electrodynamic relation in some solid crystalline structures with embedded dipoles. These crystalline solids possess microscopic regions containing dipole charges such that under applied mechanical stress, they change their internal arrangements of embedded electrical dipoles generating a voltage subsequently across the material boundaries. Conversely, an applied voltage to the solid crystalline changes the orientation of the embedded internal dipole charges and generates a deformation or strain in the solid.

1.2.5 DIELECTRIC ELASTOMER EAPs

Polymers with a low elastic stiffness modulus and a high dielectric constant can be packaged with interdigitated electrodes to generate large actuation strains by subjecting them to an electric field. This dielectric elastomer EAP can be represented by a parallel plate capacitor (Perline et al., 1998). The induced strain is proportional to the square of the electric field, multiplied by the dielectric constant, and is inversely proportional to the elastic modulus. Dielectric elastomer EAP actuators require large electric fields (100 V/μm) and can induce significant levels of strain (10–200%).

Recently, Perline and colleagues (2000) introduced a new class of polymers that exhibits an extremely high strain response. These acrylic-based elastomers have produced large strains of more than 200% but suffer from the fact that they require large electric fields in the range of hundreds of megavolts per meter. If rubbery elastomers like a silicone rubber sheet are sandwiched between two compliant electrodes, any imposed electric field induces electrostatic forces (attraction) between the electrodes. Thus, the rubber sheet in between them can be compressed by the electrostatic forces, which then cause the rubbery sheet to expand sideways due to the Poisson's ratio effect and, thus, actuation results. Röntgen in 1880 demonstrated this actuation by using two glasses as dielectrics, and once the opposing surfaces of these glasses were charged, small thickness changes were observed by Röntgen. Later electrostatically induced pressures acting to compress dielectrics became known as the "Maxwell stress". It was, however, Perline, Kornbluh and Joseph in 1998 who introduced dielectric elastomer technology with compliant electrodes. They concluded that by deliberately choosing polymers with relatively low moduli of elasticity, the field-induced strain response due to Maxwell stress could be large. Polymer actuators designed to exploit Maxwell stress in this manner became known as "Dielectric Elastomers or DEs".

1.2.6 Liquid Crystal Elastomer (LCE) Materials

Liquid crystal elastomers (LCEs) were pioneered at Albert-Ludwig Universitat in Freiburg, Germany (Finkelmann et al., 1981). These materials can be used to form an EAP actuator by inducing isotropic-nematic phase transition due to temperature increase via joule heating. LCEs are composite materials that consist of monodomain nematic LCEs and CPs that are distributed within their network structure (Shahinpoor, 2000d; Finkelmann and Shahinpoor, 2002; Ratna et al., 2002). LCEs have been made electroactive by creating a composite material that consists of monodomain nematic LCEs and a conductive phase such as graphite or conducting polymers that are distributed within their network structure. The actuation mechanism of these materials involves phase transition between nematic and isotropic phases for less than a second. The reverse process is slower, taking about 10 s, and it requires cooling to cause expansion of the elastomer to its original length. The mechanical properties of LCE materials can be controlled and optimized by effective selection of the liquid crystalline phase, density of cross-linking, the flexibility of the polymer backbone, coupling between the backbone and liquid crystal group and coupling between the liquid crystal group and the external stimuli. These materials can serve as robotic actuators by inducing nematic-isotropic phase transition in them by temperature increase. This thermal actuation causes them to shrink, as described in a thorough review of these intelligent multifunctional materials by Brand and Finkelmann (1998). The actuation mechanism of these materials involves phase transition between nematic (cholesteric, smectic) and isotropic phases over less than a second. The reverse process is slower, taking about 10 s, and requires cooling the LCE back to its initial temperature as it expands back to its original size.

1.2.7 Ionic EAPs/Ionic Polymer Gels (IPGs)

Polymer gels can be synthesized to produce strong actuators with the potential of matching the force and energy density of biological muscles. These materials (e.g., polyacrylonitrile [PAN]) are generally activated by a chemical reaction, changing from an acid to an alkaline environment and causing the gel to become dense or swollen, respectively. This reaction can be stimulated electrically, as was shown by Shahinpoor and Mojarrad (1994, 1996, 1997a, 1997b, 1997c, 1997d, 2000).

Current efforts are directed toward the development of thin layers and more robust electrode techniques. Progress was recently reported by researchers using a mix of conductive and PAN fibers at the University of New Mexico (Schreyer et al., 2000). Osada and Ross-Murphy describe the mechanism responsible for the chemomechanical behavior of ionic gels under electrical excitation (1993). A model for hydrogel behavior as a contractile EAP is described in Gong et al. (1994a, 1994b). A significant amount of research and development has been conducted at the Hokkaido

University, Japan, and applications using ionic gel polymers have been explored. These include the electrically induced bending of gels (Osada and Hasebe, 1985; Osada et al., 1992) and electrically induced reversible volume change of gel particles (Osada and Kishi, 1989).

1.2.8 NONIONIC POLYMER GELS/EAPs

Nonionic polymer gels containing a dielectric solvent can be made to swell under a D.C. electric field with a significant strain. Hirai and his coworkers (1995, 1999) at Shinshu University in Japan have created bending and crawling nonionic EAPs using a poly(vinyl alcohol) gel with dimethyl sulfoxide. A 10- × 3- × 2-mm actuator gel was subjected to an electrical field and exhibited bending at angles greater than 90° within 60 ms. This phenomenon is attributed to charge injection into the gel and the flow of solvated charges that induce an asymmetric pressure distribution. Another non-ionic gel is poly(vinyl chloride) (PVC), which is generally inactive when subjected to electric fields. However, if PVC is plasticized with dioctyl phthalate (DOP), a typical plasticizer, it can maintain its shape and behave as a nonionic elastic gel.

1.2.9 IONIC POLYMER-METAL COMPOSITES (IPMCs)

The ionic polymer-metal composite (IPMC) is an EAP that bends, rolls and twists in response to a small electrical field (5–10 V/mm) due to the mobility of cations in the polymer network. In 1992, the IPMC was realized based on a chemical-plating technique developed by Merlet, Pinneri and coworkers in France and Kawami and Takanake in Japan in the 1980s. The first working actuators were built by Oguro and colleagues (1992) in Japan and Shahinpoor (1992) and Mojarrad (2001) in the United States. The first working sensors of this kind were fabricated in the United States by Shahinpoor (1992) and Sadeghipour et al. (1992).

The operation as actuators is the reverse process of the charge storage mechanism associated with fuel cells (Heitner-Wirguin, 1996; Kim et al., 1998, 2000). A relatively low electric field is required (five orders of magnitude smaller than the fields required for PVDF-TrFE and dielectric elastomers to stimulate bending in IPMCs), where the base polymer provides channels for the mobility of positive ions in a fixed network of negative ions on interconnected clusters. To electrode the polymer films chemically, metal ions (platinum, gold, palladium or others) are dispersed throughout the hydrophilic regions of the polymer surface and are subsequently reduced to the corresponding zero-valence metal atoms. These methodologies and characteristics will be fully explained in this chapter. Ionic polymeric networks contain conjugated ions that can be redistributed by an imposed electric field and consequently act as distributed nanoactuators, nanosensors and energy harvesters. The chapter briefly presents the manufacturing methodologies.

A linear irreversible thermodynamics model of the underlying actuation and sensing mechanisms is presented. These mechanisms are based on two driving forces: an electric field $\mathbf{E}$ and a solvent pressure gradient $\Delta \mathbf{p}$ and two fluxes, electric current density $\mathbf{J}$ and the ionic+plasticizer flux $\mathbf{Q}$. Gel- and chitosan-based conductor composites have also been considered electrically active composite smart materials.

1.2.10 CONDUCTIVE POLYMERS (CPs) OR SYNTHETIC METALS

CPs operate under an electric field by the reversible counter-ion insertion and expulsion during REDOX cycling (Gandhi et al., 1995; Otero et al., 1995). Oxidation and reduction occur at the electrodes, inducing a considerable volume change due to mainly the exchange of ions with an electrolyte. When a voltage is applied between the electrodes, oxidation occurs at the anode and reduction at the cathode. The presence of a liquid electrolyte containing conjugated ions or a solid polyelectrolyte medium in close proximity to CPs, such as PPy, is often necessary to cause charge migration into and out of the CP.

Ions (H^+) migrate between the electrolyte and the electrodes to balance the electric charge. The addition of the ions causes swelling of the polymer, and, conversely, their removal results in shrinkage. As a result, the sandwich assembly bends. CP actuators generally require small electric fields in the range of 1–5 V/μm; the speed increases, with the voltage having relatively high mechanical energy densities of over 20 J/cm³, but with low efficiencies at the level of 1% or less. In recent years, several CPs have been reported – for example, PPy, polyethylene dioxythiophene, poly(p-phenylene vinylene)s, polyanilines and polythiophenes. The operation of CPs as actuators at the single-molecule level is currently being studied, taking advantage of the intrinsic electroactive property of individual polymer chains.

The initial attempts to create artificial muscles date back to the pioneering work of Kuhn et al. (1950) and Katchalsky et al. (1949) in connection with pH-activated muscles (or simply pH muscles) (Steinberg et al., 1966; Shahinpoor, 1992, 1993; Otero et al., 1995). Note also the other means of activating artificial muscles, such as thermal (Tobushi et al., 1992; Kishi et al., 1993; Li et al., 1999), pneumatic (Shahinpoor et al., 2001), optical (van der Veen and Prins, 1971) and magnetic (Zrinyi et al., 1997) means.

Chemically stimulated polymers were discovered more than half a century ago when it was shown that collagen filaments could reversibly contract or expand when dipped in acidic or alkaline solutions, respectively (Katchalsky, 1949). This early work pioneered synthetic polymers that mimic biological muscles (Steinberg et al., 1966; Shahinpoor et al., 1998).

However, actuation and sensing have remained the best means to describe artificial muscle materials. Shahinpoor and Mojarrad (1994, 1996, 1997a, 1997b, 1997c, 1997d, 2000) were among the pioneers of making electrically active – in sensing and actuation – IPCNCs and IPMNCs (see also Mehran Mojarrad's PhD dissertation, 2001). Zhang and colleagues (1998) observed a substantial piezoelectric activity in PVF2-TrFE as early as 1998.

The most progress in the development of artificial muscle materials has occurred in the last ten years, where active materials that can induce strains exceeding 100% have emerged (Perline et al., 1998). All the preceding categories of EAPs also fall under the category of artificial and/or synthetic muscles.

Next, we will describe some more specific artificial muscle materials.

1.2.11 SHAPE-MEMORY ALLOYS (SMAS) AND SHAPE-MEMORY POLYMERS (SMPS)

The history of shape-memory alloy (SMA) and shape-memory polymer (SMP) artificial muscles is extensive and will not be reported here. SMPs belong to the family of shape-memory materials (SMMs), which can deform into a predetermined shape under some imposed specific fields and stimuli such as temperature, electric or magnetic and strain and stress. The pioneering works of Liang and Rogers (1990, 1992) and Liang et al. (1997) in developing SMA and SMP actuators are noteworthy. In their work, a load of a spring-biased SMA actuator was modeled as a dead weight. However, many practical applications involve varying loads, such as the cases of SMA rotatory joint actuators (Shahinpoor, 1995d; Wang and Shahinpoor, 1997a, 1997b, 1997c). A general design methodology of various types of bias-force SMA actuators has been investigated by Shahinpoor and Wang (1995) (see also Guoping Wang's PhD dissertation, 1998).

These transformations are essentially due to the elastic energy stored in SMMs during the initial deformation. As a member of SMMs, SMPs are stimuli-responsive polymers. SMPs frequently use either heat or laser light energy as a stimulant to change shape. The thermally induced shape-memory effect can be observed by irradiation with infrared light, exposure to alternating magnetic fields, imposing an electric field or immersion in water.

1.2.12 METAL HYDRIDE ARTIFICIAL MUSCLE SYSTEMS

The binary combination of hydrogen and a metal or metal alloy can absorb large amounts of hydrogen via surface chemisorption and subsequent hydriding reactions. This phenomenon can be used to fabricate artificial muscle systems as described in Shahinpoor and Kim (2001d), U.S. Patent 6,405,532 (Kim et al., 2002; Lloyd et al., 2002; Shahinpoor, 2002c; Shahinpoor and Kim, 2002f). Useful characteristics of metal hydrides as artificial muscles are their large uptake/discharge capacity of hydrogen, safe operation (hydrogen desorption is a highly endothermic process), rapid kinetics and environmentally benign characteristics. Metal hydrides are traditionally used for hydrogen storage and thermal devices.

1.2.13 ELECTRORHEOLOGICAL FLUIDS (ERFs) SMART MATERIALS

Electrorheological fluids (ERFs) are suspensions consisting of dielectric particles of size 0.1–100 m and dielectric base fluid. ERFs belong to a class of smart materials capable of changing from a liquid phase to a much more viscous liquid and then to an almost solid phase in the presence of an electric field. Since the dielectric constant of suspension particles differs from the dielectric constant of the base fluid, the external electric field polarizes particles. These polarized particles interact and form chain-like or even lattice-like organized structures. Simultaneously, the rheological properties of the suspension change effectively; for example, the effective viscosity increases dramatically. ER suspensions also have a magnetic analog consisting of ferromagnetic particles and the base liquid. Because the viscosity of the ER liquid can be controlled with the electric field strength, the viscosity of magnetorheological fluid (MRF) is sensitive to the magnetic field.

The response time of ERFs is of the order of 1–10 ms. In principle, this enables the use of these liquids in such applications as electrically controlled clutches, valves and active damping devices. The use of ERFs as artificial muscles has not been reported anywhere. However, several publications in the pertinent literature (Gandhi et al., 1989a, 1989b; Furusha and Sakaguchi, 1999) concern the applications of ERF to robotics and as intelligent materials and composites. Some other relevant references are Huang et al. (2003) and Dwyer-Joyce et al. (1996). The solid phase of an ERF has mechanical properties similar to a solid like a gel. They can perform the phase change from liquid to thicker liquid like honey and then almost solid or reversely from almost solid transform to a thick liquid and then the thin liquid in few milliseconds. The effect is called "**Winslow effect**" after its discoverer, Willis M. Winslow, who was awarded a U.S. patent on the effect in 1947 and published an article on the effect in 1949. Note that the change is not just a change in fluid viscosity but also a change in the emerging solid-like properties. The effect is an electric field-dependent yield shear stress, such as what happens in a Bingham plastic (a type of viscoelastic material like thick honey or wax), with a shear stress yield point dependent on the electric field strength. The ERF once in a yielding shearing mode behaves like a Newtonian fluid when there is no yield shear stress and stress is directly proportional to shear rate γ.

1.2.14 MAGNETORHEOLOGICAL FLUIDS (MRFs) SMART MATERIALS

MRFs are mostly suspensions of micron-sized, magnetizable particles in the oil. MRFs are similar to ERFs, but they are 20–50 times more reliable. They can also be operated directly from low-voltage power supplies and are far less sensitive to contaminants and extremes in temperature. Under normal conditions, an MRF is a free-flowing liquid with a consistency similar to that of motor oil. Exposure to a magnetic field, however, can transform the fluid into a near-solid in milliseconds. Just as quickly, the fluid can be returned to its liquid state with the removal of the field. The degree of change in an MRF is proportional to the magnitude of the applied magnetic field. When subjected to the field, MRFs develop yield strength and behave as Bingham solids. The change can appear as a substantial change in effective viscosity like that in ERFs.

Applications include automotive primary suspensions, truck seat systems, control-by-wire/tactile feedback, pneumatic control, seismic mitigation and prosthetics. These applications are more than just a demonstration of MRF functionality. Each represents a commercially field-proven MR system that embodies all of the necessary refinements required to make it fully functional, reliable, cost-effective and long-lived. Some relevant references are Jolly et al. (1996), Jolly and Carlson (2000), Carlson (1999a, 1999b), Carlson and Weiss (1994) and Weiss et al. (1994). MRFs are suspensions of micron-sized magnetic particles, such as carbonyl iron powder, in a liquid host. The host medium is usually a type of oil with some additives to minimize particle sedimentation and particle wear and tear. When the MRF suspension is placed in a magnetic field, the suspended colloidal particles reconfigure to form chains in the direction of the magnetic flux and make the solution more solid-like than liquid.

1.2.15 Magnetic Shape Memory (MSM) Smart Materials

The magnetically controlled SMM is a new way to produce motion and force. The magnetic shape memory (MSM) mechanism was suggested by Ullakko (1996) and O'Handley (1998) and was demonstrated for a Ni–Mn–Ga alloy as early as 1996. Magnetic SMAs (MSMAs), also referred to as ferromagnetic SMAs (FSMAs), are an exciting extension of the class of SMMs. The MSM effect has demonstrated that certain SMMs that are also ferromagnetic can show substantial dimensional changes (6%) under the application of a magnetic field. These strains occur within the low-temperature (martensitic) phase.

Ferromagnetic shape memory (FSM) materials are a new class of active materials that combine the properties of ferromagnetism with those of a diffusionless, reversible, martensitic transformation. Materials such as SMA Ni_2MnGa, which has a cubic Heusler structure in the high-temperature austenitic phase and undergoes a cubic-to-tetragonal martensitic transformation, clearly exhibit MSM effects. The FSM effect refers to the reversible field-induced austenite-to-martensite transformation or the rearrangement of martensitic variants by an applied magnetic field, leading to an overall change of shape.

Typically, contractile/expansive deformation of the order of 6% is routinely observed in these materials. See also Lavrov et al. (2002). The magnetic control of the shape-memory effect would lead to a more rapid response of the actuator than the thermal control. The magnetic field controls the reorientation of the twin variants as the stress controls the twin variants in standard SMAs. The MSM effect has demonstrated that certain SMMs that are also ferromagnetic can show substantial dimensional changes (6–10%) under the application of a magnetic field. These strains occur within the low-temperature (martensitic) phase.

1.2.16 Giant Magnetostrictive Materials (GMMs)

Magnetostrictive materials allow the interchange of mechanical and magnetic energies that, for example, produce strains in the magnetic field. These strains are called magnetostriction λ. Magnetostriction in ferromagnetic materials causes them to change their shape when subjected to a dynamic magnetic field. The effect was first identified in 1842 by James Joule when observing a sample of nickel. This effect is known as the Joule effect. At a fundamental level, the change in dimensions results from the interactive coupling between an applied magnetic field and the magnetization and magnetic moments of the material's domains or magnetic dipoles. The most advanced magnetostrictive materials (called giant magnetostrictive materials, GMMs), such as commercially available TERFENOL-D, among other materials, exhibit $\lambda = 1,000$ ppm (or about 0.1% strain) in $H = 80$ kA/m. They are alloys composed of iron (Fe), dysprosium (Dy) and terbium (Tb): $Tb_xDy_{1-x}Fe_y$. In the applied magnetic field (H), the domains move or rotate so that the magnetic poles align, causing a dimensional change (d). Note that magnetostriction produces mechanical deformation in nearly all ferromagnetic materials.

1.2.17 Fibrous Ionic Polyacrylonitrile Gel (PAN, PANG) as Artificial Muscles

PAN fibers in an active form (PAN or PAN gel modified by annealing/cross-linking and partial hydrolysis) elongate and contract when immersed in pH solutions (caustic and acidic solutions, respectively). Active PAN fibers can also contract and expand in polyelectrolyte solutions when electrically and ionically activated with cations and anions, respectively. The change in length for these pH-activated fibers is typically greater than 100%. However, more than 900% contraction/expansion of PAN nanofibers (less than 1 μm in diameter) has been observed in the laboratories. PAN muscles present great potential as artificial muscles for linear actuation. The basic unit of commercially available PAN fiber (Mitsubishi Rayon Co., Japan, Orlon or artificial silk) is a "single strand". One PAN strand consists of approximately 2,000 filaments. The typical diameter of each filament is approximately 10 μm in the raw state and 30 μm in a fully elongated state (gel). The advantages of PAN fiber, among other electrolyte gels, are that PAN fiber gel has excellent mechanical properties, which can be compared to those of mammalian biological muscles. The substantial volume change of PAN fiber gel also allows the reduction in the size of the gel, which is an essential factor in determining response time. PAN fibers can convert chemical energy directly into mechanical motion. Based on the ion diffusion theory, the response time of swelling will be proportional to the square of the gel fiber diameter. The surface/volume ratio also affects the response time. Note that such pH-induced contraction expansion of modified PAN fibers can also be induced electrically in a chemical cell by electrolysis and production of H^+ and OH^- ions.

1.2.18 Piezoresistive Materials as Smart Sensors

Piezoresistivity is a property of specific materials such as metals and semiconductors for which the materials' electrical resistance changes due to mechanical pressure, force, acceleration, strain and stress. Note that metals do not exhibit piezoresistivity as they do not have a bandgap. The resistance of strained metal samples changes due to dimensional changes – this may not be considered piezoresistivity. The unit of piezoresistivity is ohm-meter or symbolically $\Omega - m$. Metals and semiconducting materials exhibit such a property. The piezoresistive effect in semiconductors is generally several orders of magnitudes more significant than the geometrical effect. This effect is present in some semiconductors like germanium, amorphous silicon, polycrystalline silicon, silicon carbide, among others. Semiconductor strain gauges can be designed, built and operated in various smart sensor applications and as microelectromechanical (MEMs) or nanoelectromechanical (NEMs) devices and systems.

1.2.19 Electrostrictive Materials as Smart Actuators and Sensors

Electrostriction is the electromechanical coupling in all electrical-nonconductors (dielectric materials). Under the application of an electric field, these materials show deformation, strain and stress. Note that all electrostrictive materials exhibit the second-order nonlinear coupling between the elastic strains or stresses. The dielectric terms that couple the strain tensor with the applied electric field is a nonlinear product of the vectors of the electric fields. For a single uniaxial strain (deformation), the induced strain (deformation) is directly proportional to the square of the applied electric field (voltage). The domains of the material get polarized, and opposite sides of these domains become differently charged and attract each other, reducing material thickness in the direction of the applied field, increasing thickness in the orthogonal directions due to the Poisson's ratio. The resulting strain tensor S_{ij} ($i, j = 1, 2\ 3$) is proportional to the product of the polarization vectors P_k. For simple one-dimensional domains, the strain is proportional to the square of the applied electric field.

1.2.20 Hydrogels and Nanogels as Smart Actuators and Sensors

Hydrogels and nanogels can be used for particularly many applications and are increasingly developed also because of their possible biocompatibility. Such smart materials can bind or release drugs, pollutants, catalysts upon interaction with external effectors and swell or shrink under the influence of different pH, various chemical compounds, temperature or light. Hydrogels have a water content typically between 80 and 99%, which can change by external stimuli; this is the basis of many applications. Natural sources of hydrogels are agarose, chitosan, methylcellulose or hyaluronic acid. Smart nanogels are the most exciting innovations emerging in the field of nanomedicine and biomedical applications. The recent advancements in the applications of biomaterials, nanogels have emerged as a novel candidate for drug delivery, biosensing, imaging, tissue engineering. Nanogels owe the synergistic properties of interpenetrating networks and the nanoscale properties such as small size high surface to volume ratio. In brief, this chapter discusses the natural and synthetic polymers deployed for the synthesis of nanogels.

1.2.21 NEW STIMULI-RESPONSIVE SMART MATERIALS ACTUATORS AND SENSORS

1.2.21.1 Giant Magnetoresistive (GMR) Materials

Magnetoresistance is defined as the property of a material to change its electrical conductivity or its inverse electrical resistance when an external magnetic field is applied to it. William Thomson (Lord Kelvin), in 1851, discovered that when pieces of iron or nickel are placed within an external magnetic field, the electrical resistance increases when the current is in the same direction as the magnetic force which is aligned with the magnetic N-S vector, and decreases when the current is perpendicular to the direction of the magnetic force. He noted that the magnitude of changes in conductivity or resistivity was greater with nickel than iron. This magnetoresistance effect is referred to as anisotropic magnetoresistance (AMR). Lord Kelvin was unable to reduce the electrical resistance of any metal by more than about 5%. This effect is commonly called ordinary magnetoresistance (OMR) effect to differentiate it from the more recent discoveries of giant magnetoresistance (GMR), colossal magnetoresistance (CMR), tunnel magnetoresistance (TMR) and extraordinary magnetoresistance (EMR). Giant magnetoresistive materials generally possess alternating layers of ferromagnetic and nonmagnetic but conductive layers such as iron-chromium and cobalt-copper.

1.2.21.2 Mechanochromic Smart Materials and Mechanical Metamaterials

This chapter reviews two recent families of smart materials, namely mechanochromic materials and mechanical metamaterials, respectively. Mechanochromic materials change their optical properties and in particular photoluminescence characteristic if subjected to mechanical loading or interactions with their environment. Chemical and physical molecular changes across various length scales and the rearrangement of molecular chemical bonds to modifications in molecular arrangements in the nanometers regime generally trigger the mechanochromic characteristics. Metamaterials define materials that are not ordinarily produced in nature. Note that "meta" means "beyond" and metamaterials have properties that go beyond conventional materials. Metamaterials are nanocomposite materials made of a periodically repeated micro or nanounits of metals, alloys and plastics that exhibit properties different from the natural properties of the participating materials. In the following sections, these families of mechanochromic and metamaterials are further described.

1.2.21.3 Smart Nanogels for Biomedical Applications

Smart nanogels are one of the most important innovations emerging in the field of nanomedicine and biomedical applications. The recent advancements in the applications of biomaterials, nanogels have emerged as a novel candidate for drug delivery, biosensing, imaging, tissue engineering and targeted delivery of bioactive. The present chapter gives a basic understanding of the hydrogels and introduces the nanoparticle form of hydrogel known as "nanogels". Nanogels owe the synergistic properties of interpenetrating networks as well as the nanoscale properties such as small size and high surface to volume ratio. These hybrid materials owe high drug loading, capable of crossing strong barriers as well as are highly biocompatible. In brief, this chapter describes the basic synthesis methodology, characterization techniques of nanogels. It also discusses the natural and synthetic polymers deployed for the synthesis of nanogels. Moreover, it highlights the important literature reported for the biomedical applications of nanogels.

1.2.21.4 Janus Particles as Smart Materials

In ancient Roman times, *Janus* was the god who had two faces (beginnings and endings) (Figure 1.1(a)). In modern science, we have adopted the term to describe particles with two distinct and usually contrasting sides. These particles have the resemblance of the Taijitu symbol in ancient Asian philosophy (Figure 1.1(b)), where Yin and Yang (dark and bright) were used to describing seemingly opposite forces. It is believed that these two basic elements give rise to complicated change and transition in the whole world. In the same sense, Janus particles are defined by their duality, which can take on a variety of forms and create a wide range of new materials with the simple Janus motif. The possibilities for properties that can be assigned to each half of the Janus particles are vast (for example, hydrophobicity and charge) and are limited only by the fabrication capabilities of their creators.

Furthermore, the particle geometry is not limited to a sphere with two equal hemispheres; Janus rods, dumbbells and sheets are structures that exhibit unique properties in their own right, and the ratio of two surfaces does not necessarily need to be 1:1. As a result of these features, Janus particles can self-assemble into many unique structures. Also, when one surface can respond to an external field, Janus particles become smart materials that are capable of adapting their assembly structures to the environmental changes. This chapter will provide an overview of the properties and applications of Janus particles as smart materials. To do this effectively, we must first go over some of the common strategies for fabricating Janus materials, since fabrication is still the primary limit for which combinations of properties can be arranged. Then we will delve into the properties that emerge from these combinations, citing several examples of the self-assembly structures demonstrated by Janus particles. Finally, we will look at some potential applications for Janus systems.

1.2.21.4.1 Self-healing Materials

It is a family of self-healing materials. The self-healing characteristics of these materials and in particular biomaterials and the concepts of the self-healing processes in nature and biology are already well known by scientific communities. One can start by describing their impact and occurrence in nature, in plants, in animals and human beings. These understandings of self-healing processes in biology and nature are particularly more advanced in terms of dermatology and skin repair by scar tissues, and they have further led to the most recent industrial applications and scientific discoveries. This chapter will introduce, describe briefly and explain a wide range of self-healing smart materials. These materials will have internal structural abilities and characteristics that enable them to automatically repair damage to themselves with almost no external intervention or diagnosis. It is well recognized that using various materials over time will degrade them due to several phenomena such as fatigue failure, environmental degradation or damages such as cracks, fracture and creep

FIGURE 1.1 (a) Janus God image engraved on an ancient Roman coin; (b) Taoist Taijitu symbol used on the Flag of South Korea.

incurred during operation. In general, internal cracks are difficult to detect, and manual intervention may be necessary. The advantage of self-healing materials is that they can treat material degradation by initiating a repair mechanism that responds to the incurred damage or degradation. Smart materials and structures also play important roles in self-healing materials because they are multifunctional and are capable of handling various environmental conditions.

1.3   A BRIEF HISTORY OF ELECTROMOTIVE POLYMERS

Kuhn and Kunzle (Kuhn et al., 1948) of Basle University in Switzerland and Katchalsky (1949) of Weizmann Institute of Science in Israel were among the first scientists to discover the shape change of ionized polymers such as a polyacid or polybase when stimulated by various pH solutions. We call this class of ionic polymers pH-activated ionic polymer gels (or merely pH muscles).

A mechanically stressed foil of polyacrylic acid (PAA) containing glycerol and H_2SO_4 was heated to make a contractile filament. The ionization of the resulting polymeric acid caused a change of molecular shape in an aqueous solution. Stretch of the polymer molecules was attributed to electrostatic repulsive forces between carboxylic ions present in solution. The contraction of the polymer molecule was attributed to the Brownian motion resulting from the neutralization of charged groups. In their experiment, Kuhn and Kunzle and Katchalsky used copolymer of methacrylic acid with divinylbenzene. They used an intermittent wash cycle with distilled water to remove salt formed between the application of acid and alkali. They proposed that equilibrium swelling of the polymeric acid gels was caused by:

1. solution tendency of polymeric molecules and osmotic pressure of the cations of the alkali bond by the gel
2. rubber-like contraction tendency of the stretched polymer molecules

They observed that the swelling capacity of the gel decreased with increasing degrees of cross-linking of the polymer. The degree of ionization depended on the charge distribution of the polymer chain. For about 50% neutralization of these charge groups, maximum swelling could be achieved.

Later, Hamlen and coworkers (1965), three research scientists at General Electric Company, were able to stimulate copolymer of polyvinyl alcohol (PVA)-PAA electrically by making the polymer conductive with chemical treatment by the solution of platinic chloride and sodium borohydride. Thus, they could actuate it by electricity instead of the pH variation of the surrounding liquid environment. Although the resulting ionic polymer was slow in response relative to its pH-activated counterpart, it nevertheless proved the possibility of electrical stimulation of these polymers, making it attractive for robotics controls and manipulation. Because PVA-PAA copolymer is a negatively charged gel (due to the $COOH^-$ side group in PAA), it swells or shrinks osmotically depending on total ionic concentration inside the polymer; this is determined by the degree of ionization of the weak carboxylic acid group ($COOH^-$).

When the external environment (electrolyte solution) is acidic, the dissociation degree is low, and the polymer shrinks in an alkaline solution. It expands similarly to the lengths of the pH muscles mentioned earlier. A conductor such as platinum can be included in the polymer and made to have lower voltage for the evolution of hydrogen and oxygen. This class of ionic polymers is called electrically activated ionic polymer gels (or merely electroactive muscles). In these researchers' experiments, PVA-PAA fibers were treated eight times in platinum solution and submerged in a 0.01-N solution of NaCl (1% concentration). A counter-electrode of platinum wire was used and placed in a container holding the muscle and salt solution. A square wave of ±5-V amplitude at 40 mA and a 20-min period ($f = 0.0008$ Hz) were then applied. When fiber is negative, hydrogen evolves, causing the solution to become alkaline, and the fiber expands. When the fiber is positive, the solution surrounding and within fibers becomes acidic, causing its contraction.

Fragala and colleagues (1972) of the GE-Direct Energy Conversion Program experimented with weak acidic contractile polymeric membranes in a setup that forced a change in pH of the solution surrounding the membrane by the electrodialysis process. They developed mathematical formulations that adequately described the response of the artificial muscle in relation to applied field current. They concluded that a dissociation constant of more than 10^{-3} g-eq/L for weak acidic groups within the material was needed to obtain large deformation in the muscle membrane. Their apparatus consisted of a three-compartment container with a weak acidic polymeric muscle membrane immersed in a mixture of weak salt and acid solution in the middle compartment. This arrangement was separated by two weak cationic and anionic ion-exchange membranes, respectively, to make the other compartments fill with the same concentration of weak salt and acids as the muscle membrane.

Next, they inserted two electrodes (Ag/AgCl) in the ion-exchange compartments. They applied a voltage gradient to change the pH of the solution surrounding the muscle membrane through an electrodialysis process. In effect, they were activating a pH-sensitive muscle by varying the pH of the surrounding solution electrically. They studied muscle membrane response as a function of pH value, solution concentration, compartment size, certain cations and membrane fabrication.

The next section details a formulation used by Fragala et al. (1972) to develop analytical relations among the applied field, hydrogen ion and salt concentration that affect the pH and, ultimately, the muscle contraction and expansion.

The dissociation constant K of the acid groups in the muscle membrane is given by

$$K = \frac{\left(\overline{H^+}\right)\left(\overline{A^-}\right)}{\left(\overline{HA}\right)} = \frac{\overline{h}\alpha}{1-\alpha} \tag{1.1}$$

where

HA represents the acid group
α is the degree of dissociation
h is the hydrogen-ion concentration

The bar over the variable here and in what follows indicates values within the membrane. The hydrogen-ion concentration inside the membrane can then be written as

$$\overline{h} = (1-\alpha)K/\alpha \tag{1.2}$$

Assuming equilibrium is established between the interior of the muscle membrane and the surrounding solution instantaneously, we have

$$\frac{\overline{h}}{\overline{s}} = \frac{h}{s} \tag{1.3}$$

where s is the concentration of the salt cation.

Concentrations within the membrane are referred to as water volume absorbed in the membrane structure. Assuming that the Donnan equilibrium principle governs the distribution of free electrolytes between the interior of the muscle membrane and the surrounding solution,

$$\frac{\overline{s}}{s} = \left(\frac{(\alpha X)^2}{4s^2} + 1\right)^{1/2} + \frac{\alpha X}{2s} \equiv D(\alpha, s) \tag{1.4}$$

in which X is the total concentration (dissociated and undissociated) of weak acid groups in the membrane. This expression reduces to unity for $\alpha = 0$ and X/s for $\alpha = 1$. Combining Equations (1.3) and (1.4) gives the external solution hydrogen-ion concentration as

$$h = (1-\alpha)K/\alpha D(\alpha, s) \tag{1.5}$$

The total amount of hydrogen ion H (including undissociated) in the muscle compartment is

$$H = \bar{h}\bar{V} + hV + 1(1-\alpha)X\bar{V} \tag{1.6}$$

where V represents water volume. The first and third terms are the dissociated and undissociated hydrogen ions in the muscle membrane, respectively. The second term gives the acid added to the external solution. The total amount of salt cation S is then

$$S = \bar{s}\bar{V} + sV \tag{1.7}$$

Assuming water absorption by the membrane depends linearly on the degree of dissociation, then

$$\bar{V} = W(w_0 + k\alpha) \tag{1.8}$$

where W is the dry weight of the muscle membrane, w_0 its water content in milliliters per gram dry membrane and k is a proportionality constant. In terms of the total volume V_T, we also have:

$$V = V_T - W(w_0 + k\alpha) \tag{1.9}$$

Substituting in Equation (1.7) and rearranging yields

$$S = s\left[D(\alpha,s)W(w_0 + k\alpha) + V_T - W(w_0 + k\alpha) \right] \tag{1.10}$$

Substituting for $\bar{h}$, $\bar{V}$ and V in Equation (1.6) from Equations (1.5), (1.8) and (1.9) gives

$$H = \frac{1-\alpha}{\alpha}(Z + \alpha WC) \tag{1.11}$$

where

$$Z \equiv K\left[W(w_0 + k\alpha) + V_T - \frac{W(w_0 + k\alpha)}{D(\alpha,s)} \right] \tag{1.12}$$

and the quantity $X\bar{V}$ has been replaced by the equivalent quantity WC, where C is the total ion-exchange capacity in milliliters per gram dry membrane.

Equation (1.11) can be partially solved to give

$$\alpha = \frac{1}{2}\left(-R + \sqrt{R^2 + \frac{4Z}{WC}} \right) \tag{1.13}$$

where

$$R \equiv \frac{H}{WC} + \frac{Z}{WC} - 1 \tag{1.14}$$

Values of s and α can be computed numerically using Equations (1.10) and (1.13). All other parameters are known through experiments or membrane properties.

The differential equations describing the time rate of change of hydrogen and salt cations in the muscle membrane compartment with applied constant electric current I are given by

$$\frac{dH}{dt} = \frac{-h\mu_h}{h\mu_h + s\mu_s} I \tag{1.15}$$

and

$$\frac{d(H+S)}{dt} = -I \tag{1.16}$$

The multiplication factor for I in Equation (1.15) is the fraction of the total current carried by hydrogen ions through the cation membrane. μ_h and μ_s are the hydrogen-ion and salt-cation mobility, respectively. The positive direction for I is taken to be from the anion-exchange to cation-exchange membranes. Substituting the total charge removed, $Q = It$ and $\mu_s/\mu_h \equiv p$, then:

$$\frac{dH}{dQ} = \frac{-h}{h+sp} \quad \text{and} \quad \frac{d(H+S)}{dQ} = -1 \tag{1.17}$$

These equations can be solved numerically for H and S as a function of Q.

MIT's Yannas and Grodzinsky (1973) were among the first to study the deformation of collagen fiber from a rat-tail tendon in an electric field when constrained at two ends and submerged in an aqueous solution. They viewed their results as an electrophoresis or electro-osmosis phenomenon. They reported that, except at the isoelectric point, individual collagen molecules and macroscopic specimens such as fibers and membranes constituted from such molecules carry a net electrostatic charge. This charge can give rise to a very intense electric field of the order 10^8 V/cm at the molecular level, which means a very large potential drop over a distance defined by the Debye length of roughly 10–1,000 Å. The charged polyelectrolyte is, therefore, surrounded by the mobile counterions. Their experimental observation supported evidence of the model of an electric double layer formed by the primary charge fixed on the polyelectrolyte and the diffuse layer of mobile counterions in solution.

Yannas and Grodzinsky estimated the Debye length or effective thickness $1/k$ of the ionic layer surrounding individual fibrils' charged sites. Using the Poisson–Boltzmann equation with the linear Debye–Hückel approximation, the Debye length becomes:

$$\frac{1}{k} = \left(\frac{\varepsilon k T}{\sum_i c_i z_i^2 e^2} \right)^{1/2} \tag{1.18}$$

where
  c_i is the concentration of the ith species
  e is the electronic charge
  ε is the dielectric constant
  k is Boltzmann constant
  T is the absolute temperature
  z is the valance

According to this equation, an increase in ionic strength results in a decrease in Debye length. On the other hand, the observed drop in isometric force with ionic strength should, in terms of the preceding model, result from a weakening of interfibrillar repulsion and a corresponding decrease in lateral swelling. Grodzinsky and Melcher (1976) modeled the electromechanical transduction of collagen and other aqueous polyelectrolytes in the membrane form. In their study, they coupled membrane to mechanical load and observed the conversion of electrical to mechanical response and vice versa. Their model represented membrane at interfibrillar level with cylindrical pores relating externally measured potentials, membrane deformations, current flow and mass fluxes to pore radius, fibril diameter and the polyelectrolyte charge.

Molecules of collagen, like other protein polyelectrolytes, possess many ionizable groups capable of dissociating and attaining net charge in a variety of solvent media. On the electrical side, this primary charge can give rise to intense local electric fields as high as 10^8 V/cm. Therefore, collagen and other polyelectrolytes, whether in the form of isolated molecules, fibers or membranes, are susceptible to interactions with externally applied fields. Besides, polyelectrolyte structures may interact with each other through their internal fields. On the mechanical side, the fibril–electrolyte matrix composing macroscopic polyelectrolyte fibers and membranes is often flexible and elastic in the hydrated state. Therefore, forces due to electric fields can result in motion and deformation of the polyelectrolyte.

Collagen is the primary structural protein of the body found in the extracellular space (usually in fibril form) as a major constituent of vertebrate connective tissue. It also is found in a blood vessel and organ walls, the cornea and vitreous humor, basement membranes and other epithelial and endothelial linings. Fundamental to all electrochemical interactions with collagen and to our continuum models are the physiochemical origin and location of charge sites on the molecules. In rat-tail tendon collagen, there are about 350 acidic and 265 basic groups among the ~3,000 amino acid residues per tropocollagen molecules (predominantly carboxyl [–COOH] and amino [–NH$_2$] groups, respectively). These groups are capable of ionizing as a function of electrolyte pH and ionic strength and thereby leaving a net positive or negative charge fixed to the molecule. Titration methods could be used to determine the number of these fixed charges in the specimen.

Tanaka and colleagues (1980) of MIT used the Flory and Huggins (1953b) theory of shrinking–swelling to formulate further the mean free energy needed for phase transition of electrolytic gels. They used partially hydrolyzed acrylamide gel for their experimental work. The solvent used was a mixture of acetone and water. The polymer was highly sensitive to temperature, solvent composition, pH, added salt concentration and applied electricity. The degree of volume change of the gel was found to be 500%. This change is the result of phase transition of the system of the charged polymer network, counter-ions and fluid composition.

Three major competing forces on the gel, in turn, cause the phase transition:

1. positive osmotic pressure of counter-ions
2. negative pressure due to polymer-polymer affinity
3. rubber elasticity of the polymer network

Electric forces on the charged sites of the network produce a stress gradient along the electric field lines in the gel. At the critical stress, the gel shrinks or swells depending on whether stress developed is above or below critical stress.

In their experiment, they prepared polyacrylamide (PAM) gels by free-radical polymerization of acrylamide (monomer), using N, N'-methylene bisacrylamide (as cross-linker), ammonium persulfate and N, N, N', N'-tetramethylenediamine (TEMED) (as initiator) – all of which were dissolved in water. Gel formation was initiated after 5 min and, to remove excess monomer, the gel was immersed in water after 1 h. The gel was then hydrolyzed in a solution of TEMED (1.2% solution, pH 12) for one month. Approximately 20% of acrylamide groups were converted to acrylic acid groups, which in turn were ionized in water:

$$-CONH_2 \rightarrow -COOH \rightarrow -COO^- + H^+ \tag{1.19}$$

After complete hydrolysis, the gel was immersed in a 50% solution of acetone-water to reach equilibrium before cutting into desired dimensions. To energize the resulting gel, platinum electrodes were used with a D.C. voltage of 0–5 V; the gel was placed between electrodes. It is noted that PAM is an anionic type of polymer gel, which means it is negatively charged. The gel reached its equilibrium shape in one day. At 2.5 V D.C., the entire gel collapsed. By removal of the voltage, the gel swelled again. If the 50% acetone–water solution were to be replaced by 100% pure water, the gel would continuously swell beyond its original size in the presence of an electric field.

The mean-field theory formulated by Flory (1941, 1953a, 1953b, 1969) and Huggins (1941) explains the shrink-swell phenomenon according to:

$$F_g = vkT \left\{ \begin{array}{l} N_0 \dfrac{1-\phi}{\phi}\left[\ln\left(1-\phi\right) + \dfrac{\Delta F}{kT}\phi \right] \\[4mm] + \dfrac{1}{2}\left[2\alpha^2 + \beta^2 - 3 - (2f+1)\ln(\alpha^2\beta) \right] \end{array} \right\} \dfrac{\Delta Z_0}{L_0} \tag{1.20}$$

where the total free energy to be minimized is given by

$$F = F_g + F_e \tag{1.21}$$

where

F_g = free energy due to gel deformation
F_e = free energy due to work done against the electric field
v = number of polymers cross-linked in the network
N_0 = number of freely jointed segments of the polymer
f = number of ionized segments out of N_0 total
L_0 = length of the cylindrical polymer sample
D_0 = diameter of the cylindrical polymer sample
ϕ_0 = concentration of polymer (no interaction between segments or reference states)
ΔZ_0 = elemental disk's thickness before applied electricity
Z = distance of the element from the free end of the gel
β = factor increasing ΔZ_0 after application of electricity
α = factor increasing D_0 after application of electricity
T = absolute temperature
k = Boltzmann constant
ϕ = volume fraction of polymer network
ΔF = free energy decrease as a result of contact between two polymer networks

Free energy needed to expand the gel network against electric potential (work done against electric potential) is given by:

$$F_e = vfeE\left(\frac{Z}{L_0} \right)(\beta-1)\Delta Z_0 \equiv BvkT\left(\frac{\Delta Z_0}{L_0} \right)(\beta-1) \tag{1.22}$$

where e is the electron charge and $B = feEZ/kT$ is reduced electric potential. Minimizing the total free energy equation F, we get:

$$\alpha^2 = \beta^2 + B\beta \tag{1.23}$$

$$\frac{N_0}{\phi_0}\left[\ln\left(1-\phi\right) + \phi + (1-\tau)\phi^2/2 \right] - (f+1/2)(\alpha^2\beta) + 1/\beta = 0 \tag{1.24}$$

where $\tau = 1 - 2 - \Delta F/kT$ is the reduced temperature.

Therefore, the anisotropy of deformation of the disk is uniquely determined by B. Namely, for positive B, the disk is compressed more in its length than in diameter ($\alpha > \beta$), and, for negative B, the compression in the radial direction is more than axial ($\alpha < \beta$). By plotting α (radial expansion) versus β (axial expansion) of a rod-shaped gel, various values for B and τ can be obtained.

De Rossi and coworkers (1985) of the University of Pisa in Italy explained the dynamic behavior of these gels from a thermodynamic point of view according to the general expression:

$$dU = TdS - PdV + Fdl + \mu_i dn_i + \varepsilon dq \tag{1.25}$$

where mechanical, chemical and electrical energy terms are present, and Fdl is the mechanical work performed by a polymer fiber muscle during uniaxial elongation dl, and keeping S, Vn_i and q constant. εdq is the electrical work term, and $\mu_i dn_i$ is the chemical energy term.

They further proposed an analytical model describing mechanical parameters governing the kinetics of thermally cross-linked PVA-PAA polyelectrolyte gel. For a thin film of the specimen, they found a relation between swelling rate and linear dimension, and diffusion coefficient of the material.

Several rate processes are taken into account to analyze and describe the transient mechanical behavior of a polyelectrolyte gel in response to proton and salt concentration gradients generated by electrode reactions and delivered within the gel by electrochemical potential differences. These can include ion diffusion, diffusion-limited chemical reaction, ion migration and consequent electrical and mechanical realignment of the polymer network.

Typically, proton diffusion reaction and mechanical realignment of the polymer network are slow. A system of coupled differential equations is formulated to determine the kinetics of gel de-swelling under electromechanical stimuli. A comparison among proton diffusion and gel-swelling characteristic time constants might prove very useful in decoupling the chemical and mechanical problems, particularly in the case of specific gel systems in which these limiting time constants are considerably different.

De Rossi and colleagues analyzed the kinetics of free swelling of a partially dehydrated gel in the case of spherical and thin-film samples. The only problem with their formulations is that they consider constitutive equations used in linear infinitesimal elasticity formulations that are not appropriate for a large swelling and de-swelling deformation of ionic polymeric gels.

Osada and Hasebe (1985) of Ibaraki University in Japan were pioneers in synthesizing various polyelectrolytes and describing their dynamic behavior by including the frictional effects of the ion transport phenomenon. They mostly concentrated on the synthesis, modeling and characterization of these ionic gels in biomimetic locomotion and biomedical applications such as drug delivery systems (DDSs). They were among the first to label them as artificial muscles. Osada (1991) described electrochemomechanical systems based on hydrogels that contract and dilate reversibly under electric stimuli. These were essentially thermodynamic systems capable of transforming chemical energy directly into mechanical work or, conversely, transforming mechanical into chemical potential energy.

The isothermal conversion of chemical energy into mechanical work underlies the motility of all living organisms and can easily be seen, for instance, in muscle, flagella and ciliary movement. All these biological systems are characterized by extremely high efficiency of energy conversion. The high conversion efficiency of the biological systems is largely due to the direct conversion of chemical energy without unnecessary intermediate heat-producing components. Chemomechanical systems are the only artificial systems at present that can achieve this. The chemomechanical reaction may be used to generate mechanical energy on a macroscopic level and transform information as a signal or receptor, whereby microscopic deformation plays an essential role in switches and sensors. Osada further classified synthetic polymer gels according to principle and type of reaction involved:

1. hydrogen–ion transfer (pH muscle)
2. ion-exchange or chelation
3. REDOX reaction (REDOX muscle)
4. steric isomerization
5. phase transition or order-disorder transition
6. polymer-polymer association or aggregation
7. electrokinetic processes

Several chemomechanical systems such as polyelectrolyte fibers and membranes have been investigated in the past. These materials expand and contract upon changing their solubility or degree of ionization. There are also thermosensitive and optically sensitive polymers. However, from the standpoint of robotic controls, electroactive polyelectrolytes have been the focus of most researchers. What follows is an explanation of materials' behavior and their application as artificial muscles.

1.3.1 CONTRACTION BEHAVIOR

For a polyelectrolyte undergoing shape changes by applying D.C. current, the velocity of shape change is proportional to the charge density in the gel. This shape change may be the first model of an electroactive artificial muscle working in an aerobic and aqueous medium system. The gel contracts and dilates reversibly when stimulated electrically under isothermal conditions.

The electrical control makes use of cross-linked polyelectrolyte gels. The system is quite simple. For instance, a water-swollen polymer gel is inserted between a pair of electrodes connected to a D.C. source. When the electricity is turned on, the polymer gel starts to shrink. This procedure has been verified and quantified experimentally in the case of several ionic gels such as poly(2-acrylamido-2-methyl-1-propane sulfonic) acid (PAMPS). The degree of swelling changes as other media, such as a mixture of water and ethanol or glycerin or other solvents, is used compared with using pure water only.

1.3.2 MECHANISMS

A quantitative interpretation of the phenomena observed is that the application of an electric field causes the pumping of largely mobile ions, partly macro-network ions and surrounding hydrated water to the opposite directions until they reach the electrode. In general, the velocity of migration and the velocity of the gel contraction are governed by the quantity of mobile ions q_n and the gradient of electric field ΔE, which is the driving force pumping these ions. The force produced in the gel is proportional to the product of q_n and ΔE:

$$F = \Delta E \times q_n \tag{1.26}$$

If we ignore the effect of the elastic forces due to the covalently cross-linked network and if the gel is free to move as an aggregate of particles in the electric field, a slightly cross-linked and highly swollen polymer eventually attains a steady-state velocity U.

This value is proportional to the velocity of contraction of the gel v over some time:

$$v \times U = \Delta E \times \frac{q_n}{r} = \Delta E \times \frac{I}{\mu} \tag{1.27}$$

where r is a hydrodynamic frictional coefficient, which depends on the geometry of the gel, thickness of a double layer, the viscosity of the medium and others. If the average mobility of the medium in the gel is μ, the contraction rate may be proportional to the product of the gradient of the electric field and current I.

1.4 ROLE OF MICROPARTICLES IN CONTRACTION OF GELS

To establish the system with minimized response time, microparticles of the sodium salt of PAA gel were synthesized. The size change of the particle with time under various electric fields was measured. Experiments showed that there is a threshold voltage below which no contraction occurs. As the electric voltage increased, the contraction increased as well. This also affected the volumetric reduction in the size of the sample. Volume reduction by a factor of 30 or 3,000% was easily achieved within less than 1 min under 6 V D.C. current. After removal of the electrical current, the gel returned to its original size within 5 min.

Rate of volume change is a function of the particle's size. The relative volume change decreased with the increasing size of the particle. The experimental results verified that this rate is inversely proportional to the square of particle size. This finding is in line with Tanaka's proportionality with the square of the characteristic length of the sample.

Shahinpoor and Osada (1995a) formulated a theoretical model for a cylindrical gel sample describing the dynamics of contraction of ionic polymeric gels in the presence of an electric field. Their model considered the dynamic balance between the internal forces during the contraction. These forces are assumed to be due to the viscous effects caused by the motion of the liquid solvent medium within the polymer network, the internal forces due to the motion of the liquid in and out of the network and the electrophoretic forces due to the motion of the charged ions in the solvent as it exudes from the ionic polymeric gel network. The effects of the rubber elasticity of the network and ion-ion interactions were assumed negligible compared to inertial, viscous and electrophoretic effects.

The governing equations were then solved for the velocity of the liquid exudation (water, ethanol, acetone or mixture of other solvents) from the network as a function of the time and radial distance in the cylindrical sample. The relative weight of the gel was then related to the velocity by an integral equation. This problem can be solved numerically to obtain a relationship among the amount of the contraction as a function of time, electric field, strength and other pertinent materials and geometrical parameters. The results of the numerical simulation in the case of PAMPS polymer indicated close agreement with experimental data, as described later in Chapter 6.

Caldwell and Taylor (1989) of the University of Hull in the United Kingdom attempted to describe the force-velocity relationship of the contraction–expansion mechanism of the PVA-PAA copolymer gel. They simulated flexor and extensor muscle pairs of humans to power a robot gripper. They used water and acetone for expansion-contraction of the copolymer gel. In their experiment, they measured up to 30 N/cm^2 of swelling and contractile forces of the gel. They proposed a two-part theory for the contraction rate of the copolymer gel:

1. The diffusion coefficient controls the rate of movement of chemicals within the polymer strips, and its effect is as important as chemical concentrations on the dynamic rate. The diffusion coefficient depends on the solvent used, concentration, degree of cross-linking and temperature.
2. Film thickness has been shown experimentally to be proportional to dynamic rate by an inverse square relationship:

$$\Phi = \frac{K}{\Delta^2} + C \tag{1.28}$$

where
Φ = contraction rate
Δ = thickness
K, C = fixed constants

Caldwell and Taylor measured a maximum contraction rate of 11%/s for a 0.1-mm thick PVA-PAA strip as compared with an animal muscle of 24–1,800%/s, depending on the muscle. They approximated the force-velocity curve of their artificial muscle by a rectangular hyperbolic function, which resembles that of natural muscles as:

$$(F+a)(V+b) = (F_0 + a)b \tag{1.29}$$

where

F = force during contraction
F_0 = force during isometric contraction
V = velocity of contraction
a, b = constants

The power output, which is the product of the force and velocity of contraction, was the maximum at medium contraction velocities and low contractile forces. Maximum power-to-weight ratio measured was 5.8 mW/g as compared to 40–200 mW/g for natural muscles.

Caldwell and Taylor (1990) also found that the solvent content had a significant effect on the elastic modulus of the material. Increasing the liquid content caused a decrease in Young's modulus according to:

$$G = \frac{K_1}{\left(S_w{}^3 - C_1\right)} \tag{1.30}$$

where

S_w = linear swelling ratio (ratio of swollen to dry polymer length)
K_1, C_1 = constants

Therefore, by controlling the percentage of the solvent content and the external solvent concentration, these researchers could vary the gripper's compliance. Their gripper consisted of muscle cell chambers, wire tendons and a scissor mechanical gripper. The polymer muscles were bundles of 0.1-mm thick strips of PVA-PAA that contracted with acetone and expanded with water utilizing computer-controlled hydraulic solenoid valves. They used a PD controller to control the gripper position. The cycle time for opening and closing the gripper was under 15 s, with a measured positional accuracy of ±2°.

In 1993, Oguro et al. (1993) of Osaka National Research Institute in Japan (now called AIST) were the first to report deformation of ion-exchange membrane polyelectrolytes when they were plated by metals and placed in an electric field. They observed the bending of strips of these polyelectrolyte membranes toward anode electrode when placed in a weak electric field. Although the initial attempt was to achieve better efficiencies for fuel cell membrane applications, the accidental movement of this polymer membrane proved to be a breakthrough for the future of polymeric biomechanical sensors and actuators.

Shahinpoor and Mojarrad (1996) investigated the characterization, modeling and application of chemoactive (pH-driven) and electroactive ionic polymeric gels, using nonlinear theories and numerical simulation. Brock (1991a, 1991b, 1991c) of MIT also has been involved with microminiature packaging of pH-activated artificial muscles. Shiga and coworkers (1993) investigated the deformation of small-diameter PVA-PAA hydrogel rods under sinusoidal varying electric field in the electrolyte solution. They measured response time of the order of less than several hundred milliseconds.

Since much of the latest research in polyelectrolyte application involving feedback controls require electrically CP actuators, these materials will be briefly discussed in the following chapters.

2 Fundamentals of Ionic Polymer-Metal Composites (IPMCs) and Nanocomposites (IPMNCs)

2.1 INTRODUCTION

This chapter describes ionic polymer-metal composites and associated nanocomposites. Ionic polymers can be chemically transformed to stimuli-responsive multifunctional nanocomposites with a metal (*ionic polymer-metal composite* (IPMCs). IPMCs can also be manufactured with a conductor such as graphite, carbon nanotube, conductive polymers (synthetic metals), or graphene rather than a metal. These different variations of IPMCs are called *ionic polymer conductor composites* (IPCCs), ionic polymer conductor *nanocomposites* (IPCNCs) or *ionic polymer-metal nanocomposites* (IPMNCs). These stimuli-responsive multifunctional artificial muscles can exhibit large dynamic deformation if placed in a time-varying electric field (see Figure 2.1) (Shahinpoor, 1991; Oguro et al., 1992, 1993; Shahinpoor and Kim, 2001c). Typical experimental deflection curves are depicted in Figure 2.2(a)–(d) with a dopant called polyvinyl pyrrolidone (PVP). Typical frequency-dependent dynamic deformation characteristics of IPMCs are depicted in Figure 2.2(c).

Once an electric field is imposed on an IPMNC cantilever, in their polymeric network, the conjugated and hydrated cations rearrange to accommodate the local electric field, and thus the network deforms. In the simplest of cases, such as in thin membrane sheets, spectacular bending is observed (Figure 2.2(d)) under small electric fields such as tens of volts per millimeter. Figure 2.2(d) depicts typical force and deflection characteristics of cantilever samples of IPMNC artificial muscles.

Conversely, dynamic deformation of such polyelectrolytes can produce dynamic electric fields across their electrodes, as shown in Figure 2.3 (Shahinpoor, 1996c). A recently presented model by de Gennes et al. (2000) describes the underlying principle of electro-thermodynamics in such ionic polymeric material based on internal transport phenomena and electrophoresis. IPMNCs show great potential as soft robotic actuators, artificial muscles, and dynamic sensors in the micro-to-macro size range. In this section, the IPMC manufacturing techniques have been described. Furthermore, linear irreversible thermodynamic modeling, as well as the phenomenological laws, is presented. Later, we will present the electronic and electromechanical performance characteristics of IPMCs.

Figure 2.3 shows the dynamic sensing, energy harvesting and transduction responses of an IPMNC strip in a cantilever configuration subject to a dynamic impact loading. A vibrationally damped output electric response is observed and is highly repeatable with a high bandwidth of up to tens of kilohertz. Such direct mechanoelectric behaviors are related to endo-ionic mobility due to imposed stresses.

Manufacturing an IPMNC begins with the selection of an appropriate ionic polymeric material. Often, these materials are manufactured from polymers that consist of fixed covalent, ionic

DOI: 10.1201/9781003015239-2

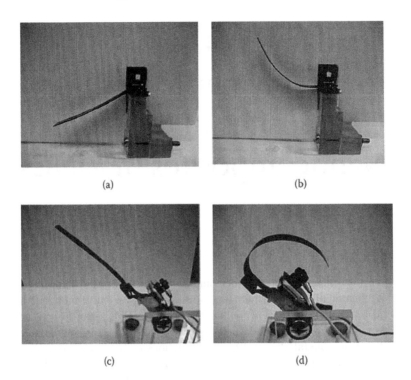

(a)   (b)

(c)   (d)

FIGURE 2.1 Successive photographs of an IPMC strip showing very large deformation. Samples in (a) and (b) are 1 cm × 4 cm × 0.2 mm, with 2 V; samples in (c) and (d) are 1 cm × 8 cm × 0.34 mm with 4 V. Note that $\Delta t = 0.5$ s between (a), (b) and (c), (d).

groups. The currently available ionic polymeric materials that are convenient to be used as IPMCs are as follows:

1. Perfluorinated alkenes with short side chains terminated by ionic groups (typically sulfo-nated or carboxylated (SO_3^- or COO^-) for cation exchange or ammonium cations for anion exchange (see Figure 2.4). The extensive polymer backbones determine their mechanical strength. Short side chains provide ionic groups that interact with water and the passage of appropriate ions.
2. Styrene/divinylbenzene-based polymers in which the ionic groups have been substituted from the phenyl rings where the nitrogen atom is fixed to an ionic group. These polymers are highly cross-linked and are rigid.

In perfluorinated sulfonic acid polymers, there are relatively few fixed ionic groups. They are located at the end of side chains to position them in their preferred orientation. Therefore, they can create hydrophilic nanochannels, so-called *cluster networks* (Gierke et al., 1982). Such configura-tions are drastically different in other polymers such as styrene/divinylbenzene families that limit, primarily by cross-linking, the ability of the ionic polymers to expand (due to their hydrophilic nature).

The preparation of ionic polymer-metal composites requires extensive laboratory work, includ-ing metal compositing employing chemical reduction. State-of-the-art IPMNC manufacturing tech-niques (Shahinpoor and Mojarrad, 2000) incorporate two distinct preparation processes: first, the *initial composting process* and the *surface electrode placement or process*. Different preparation

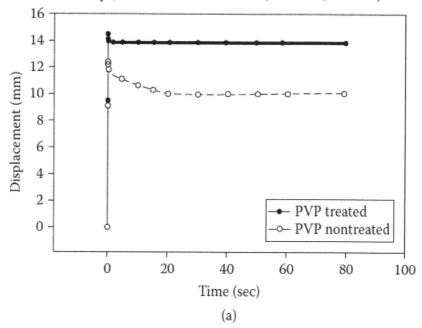

(a)

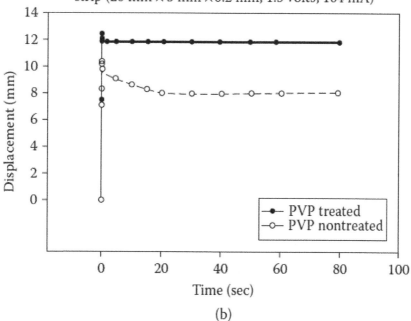

(b)

FIGURE 2.2 (a) and (b) Step response displacement characteristics of IPMC samples. (a) hydrated; (b) semidry. Note the use of a force dopant polyvinyl pyrrolidone (PVP). *(Continued)*

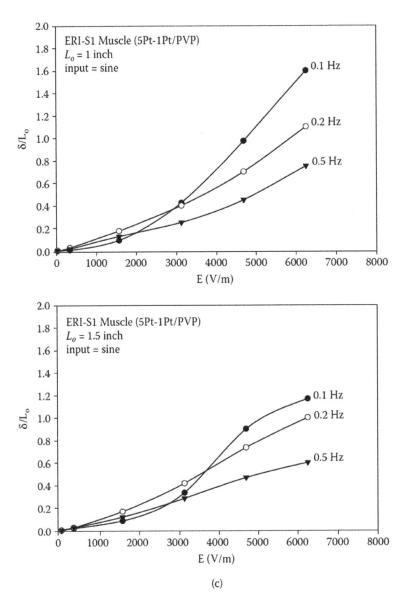

FIGURE 2.2   (c) Displacement characteristics of an IPMC, ERI-S1. δ: arc length; l_o: effective beam length; $l_o = 1.0$ in. (top) and $l_o = 1.5$ in. (bottom). *(Continued)*

processes result in morphologies of precipitated platinum that are significantly different. Figure 2.5 shows illustrative schematics of two different preparation processes (top left and bottom left) and two top-view scanning electron micrographs (SEMs) for the platinum surface electrode (top right and bottom right).

Note in Figure 2.5 that the top left-hand part is a schematic showing the initial process of making the ionic polymer metal nanocomposite. The top right-hand section shows its top-view SEM micrograph, while the bottom left-hand side shows a schematic of depositing surface electrodes on the ionic polymer. The bottom right-hand side shows its top-view SEM micrograph where platinum deposited is predominantly on top of the initial Pt layer.

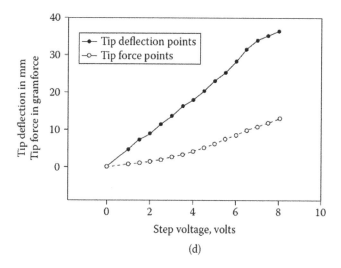

(d)

FIGURE 2.2 (d) Variation of tip blocking force and the associated deflection if allowed to move versus the applied step voltage for a 1 cm × 5 cm × 0.3 mm IPMC Pt–Pd sample in a cantilever configuration. *(Continued)*

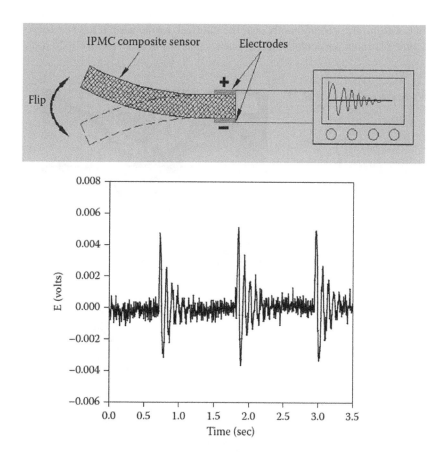

FIGURE 2.3 A typical sensing response of an IPMC strip of 0.5 cm × 2 cm × 0.2 mm manually flipped off by about 1 cm tip deflection in a cantilever form and then released.

$$- (CF_2CF_2)_n - CFO(CF_2 - CFO)_m CF_2CF_2SO_3^- \cdots Na^+$$
$$| \qquad \quad |$$
$$CF_2 \qquad CF_3$$
$$|$$

or

$$- (CF_2CF_2)_x - CFO(CF_2 - CFO)_m (CF_2)_n SO_3^- \cdots Na^+$$
$$| \qquad \quad |$$
$$CF_2 \qquad CF_3$$
$$|$$

FIGURE 2.4 Perfluorinated sulfonic acid polymers.

The initial composting process requires a suitable platinum salt such as $Pt(NH_3)_4HCl$ or $Pd(NH_3)_4HCl$ in the context of chemical reduction processes similar to those evaluated by several investigators, including Takenaka et al. (1982) and Millet (1989). The essential step in IPMC manufacturing is to metalize the boundaries or inner surfaces of ionic material. In this phase, specific chemical-oxidation-reduction means such as $LiBH_4$ or $NaBH_4$ are used.

The ionic polymeric material is soaked in a salt solution to allow platinum-containing cations to diffuse through the ion-exchange process. Later, a reducing agent such as $LiBH_4$ or $NaBH_4$ is introduced to platinize the material.

As shown in Figure 2.6, the metallic platinum particles are not homogeneously formed across the membrane but concentrate predominantly near the interface boundaries. It has been experimentally observed that the platinum particulate layer is buried microns deep (typically 1–20 μm) within the

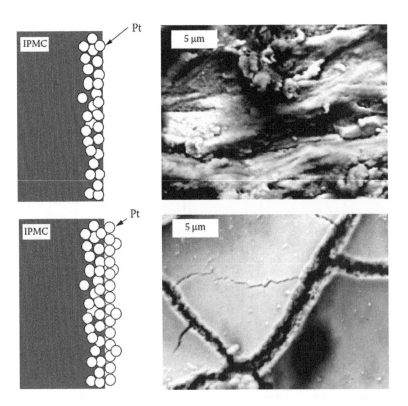

FIGURE 2.5 Two schematic diagrams showing different electrode material (gold, platinum, silver) penetration and manufacturing processes.

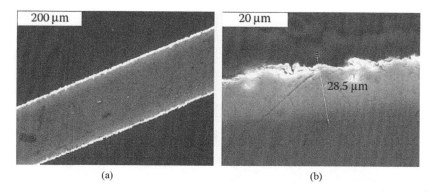

(a)   (b)

FIGURE 2.6 Two SEM micrographs showing the (a) cross-section and (b) close-up of a typical IPMC plated with platinum.

IPMNC surface and is highly dispersed. The fabricated IPMNCs can be optimized to produce a maximum force density by changing multiple process parameters. These parameters include time-dependent concentrations of the salt and the reducing agents (applying the Taguchi technique to identify the optimum process parameters that seem quite attractive; see Peace, 1993). The primary reaction is

$$\text{LiBH}_4 + 4\left[\text{Pt}\left(\text{NH}_3\right)_4\right]^{2+} + 8OH^- \Rightarrow 4\text{Pt}^{\circ} + 16\text{NH}_3 + \text{LiBO}_2 + 6\text{H}_2\text{O} \tag{2.1}$$

In the subsequent surface electrode placement process, multiple reducing agents are introduced (under optimized concentrations) to carry out the reducing reaction similar to Equation (2.1), in addition to the platinum layer formed by the initial compositing process. This reduction is shown in Figure 2.4 (bottom right), where the roughened surface disappears. In general, the majority of platinum salts stay in the solution and precede the reducing reactions and production of platinum metal. Other metals (or conductive media) also successfully used include palladium, silver, gold, carbon, graphite and nanotubes.

The surface morphology of the IPMCs can be described by the atomic force microscopy (AFM). Its capability of imaging the surface of the IPMC directly can provide detailed information with a resolution of a few nanometers. In Figure 2.7, several representative AFM images (surface analysis) reveal the surface morphology of the IPMCs. As can be seen, depending on the initial surface roughening, the surface is characterized by the granular appearance of platinum metal with a peak/valley depth of approximately 50 nm. This granular nanoroughness is responsible for producing a high level of electric resistance yet provides a porous layer that allows water movement in and out of the membrane.

During the AFM study, it was also found that platinum particles are dense and, to some extent, possess coagulated shapes. Therefore, the study was extended to utilize TEM (transmission electron microscopy) to determine the size of the deposited platinum particles. Figure 2.8 shows a TEM image on the penetrating edge of the IPMNC. The sample was carefully prepared in the form of a small size and was ion beam treated. The average particle size was found to be around 47 nm.

A recent study by de Gennes et al. (2000) has presented the standard Onsager formulation on the fundamental principle of IPMNC actuation/sensing phenomena using linear irreversible thermodynamics. When static conditions are imposed, a simple description of the *mechanoelectric effect* is possible based on the two forms of transport: *ion transport* (with a current density, J, normal to the material) and *electrophoretic solvent transport*. (With a flux, Q, we can assume that this term is water flux.)

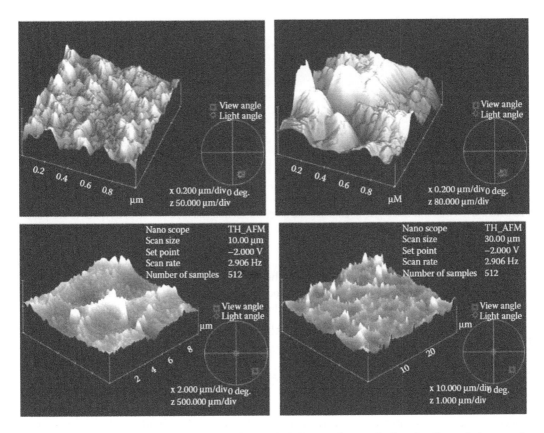

FIGURE 2.7 An atomic force microscopy surface analysis image taken on the functionally graded composite surface electrodes of some typical IPMCs, as shown in Figure 2.1. Note that the scanned area is 1 μm^2. The brighter/darker area corresponds to a peak/valley depth of 50 nm. The surface analysis image has a view angle set at 22°.

The conjugate forces include the electric field, $\vec{E}$, and the pressure gradient, $-\nabla p$. The resulting equation has a concise form of

$$J = \sigma \vec{E} - L_{12}\nabla p \qquad (2.2)$$

$$Q = L_{12}\vec{E} - K\nabla p \qquad (2.3)$$

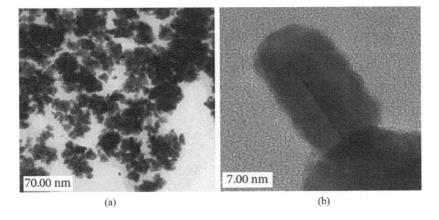

FIGURE 2.8 TEM micrographs of IPMC (a) Pt particles; (b) a Pt particle.

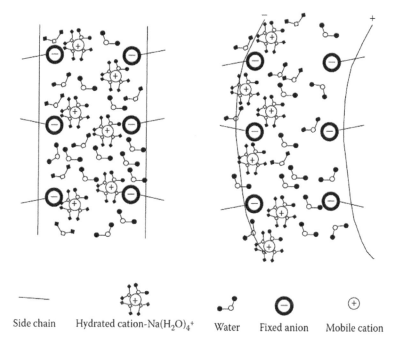

Side chain   Hydrated cation-Na(H$_2$O)$_4^+$ Water   Fixed anion   Mobile cation

FIGURE 2.9 A schematic of the typical IPMC and its actuation principle.

where σ and K are the material conductance and the Darcy permeability, respectively. A cross-coefficient is usually $L = L_{12} = L_{21}$, experimentally measured to be of the order of 10^{-8} {(m/s)/(V/m)}.

The simplicity of the preceding equations provides a close view of fundamental principles of actuation and sensing of IPMNC. We can illustrate it in Figure 2.9.

The IPMNC is composed of a perfluorinated ionic polymer, which is chemically surface composited with a conductive medium such as platinum. A platinum layer is formed a few microns deep within the perfluorinated ionic polymer.

Typically, the strip of perfluorinated ionic polymer membrane bends toward the anode (in the case of cation-exchange membranes) under the influence of an electric potential. Also, the appearance of water + cations on the surface of the expansion side and the disappearance of water on the surface of the contraction side are typical. This electrophoresis-like internal ion-water movement is responsible for creating effective strains for actuation. Water leakage through the porous Pt electrode reduces the electromechanical conversion efficiency.

2.2 PERFORMANCE CHARACTERISTICS

2.2.1 MECHANICAL PERFORMANCE

Figure 2.10 shows tensile testing results regarding normal stress versus normal strain on a typical IPMNC (H$^+$ form) relative to Nafion™-117 (H$^+$ form). Recognizing that Nafion-117 is the adopted starting material for this IPMNC, this comparison is useful. There is a little increase in the mechanical strength of IPMNC (both stiffness and the modulus of elasticity), but it still follows the intrinsic nature of Nafion. This increase means that, in the tensile (positive) strain, the stress/strain behavior is predominated by the polymer material rather than metallic powders (composite electrode materials).

Note in Figure 2.10 that the data show normal stress, σ_N, versus normal strain, ε_N, for IPMNC and Nafion-117. Note that both samples were fully hydrated when they were tested. Although the tensile testing results show the intrinsic nature of the IPMNC, a problem arises when the IPMNC

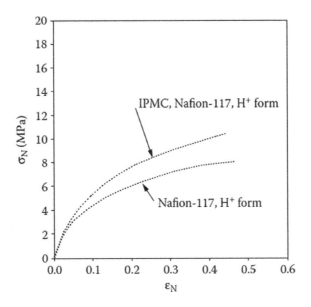

FIGURE 2.10 Tensile testing results.

operates in a bending mode. Dissimilar mechanical properties of the metal particles (the electrode) and polymer network seem to affect each other. Therefore, to construct the useful stress-strain curves for IPMCs, strips of IPMCs are suitably cut and tested in a cantilever configuration (see Figure 2.11(a)).

In a cantilever configuration, the end deflection, δ, due to a distributed load, $w(s, t)$, where s is the arc length of a beam of length L, and t is the time, can be related approximately to the radius of curvature, ρ, of the bent cantilever beam:

$$\rho \cong \frac{L^2 + \delta^2}{2\delta} \qquad (2.4)$$

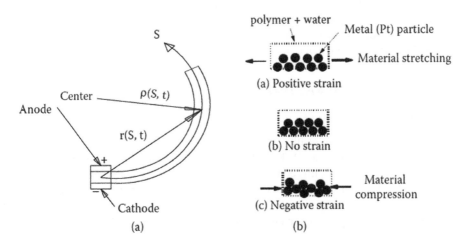

FIGURE 2.11 A cantilever configuration (left) of the IPMC (a) and an illustration of positive/negative strains experienced in the operation mode of the IPMC (b) in the cathode/anode sides of the electrodes, respectively.

Note that the radius of curvature ρ is, in turn, related to the maximum tensile (positive) or compressive (negative) strains in the beam as

$$\varepsilon \cong \frac{h}{2\rho} \tag{2.5}$$

where h is the thickness of the beam at the built-in end.

Note that in the actuation mode of the IPMNC, the tensile strain can be simply realized, but difficult to isolate. In the negative strain (material compression) illustrated in Figure 2.11(b), the metal particles become predominant to experience much higher stiffness and modulus of elasticity than those in the positive strain regime.

Thus, the mathematical description regarding the physics of the cantilever beam of the IPMC is somewhat challenging and should be addressed carefully. Experimental approaches are available and should be pursued.

Note in Figures 2.12 and 2.13 that swelling is an important parameter to affect the mechanical property; that is, swelling causes mechanical weakening, while electrical activation tends to stiffen the material due to redistribution of ions within the IPMC.

The stress, σ, can be related to the strain, ε, by merely using Hooke's law, assuming linear elasticity. (One can also consider other constitutive equations in which the stress σ can be related to the strain ε in a nonlinear fashion – i.e., rubber elasticity, which could be a future study.) It leads to

$$\sigma = \frac{Mh}{2I} \tag{2.6}$$

where σ is the stress tensor, M is the maximum moment at the built-in end and I is the moment of inertia of the cross-section of the beam.

Thus, the moment M can be calculated based on the distributed load on the beam or the applied electrical activation of the IPMNC beam. Having also calculated the moment of inertia I, which for

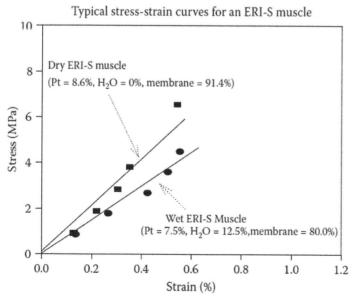

(w = 10.02 mm, t = 0.35 mm, L_{eff} = 49.09 mm, and Pt loading ~3 mg/cm²)
(The cantilever beam method was used)

FIGURE 2.12   Effect of swelling on the stress-strain characteristics of IPMCs.

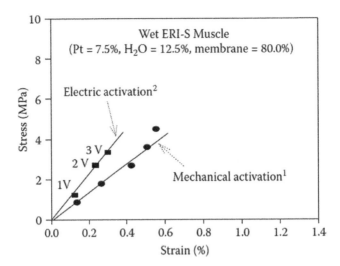

FIGURE 2.13 Stiffening of IPMCs due to the placement in an electric field or under electric activation.

a rectangular cross-section of width b will be $I = bh^3/12$, the stress σ can be related to the strain ε. The representative results are plotted in Figures 2.12 and 2.13. These figures include the effect of swelling and stiffening behavior under electric activation. Electric activation of the IPMCs in the electromechanical mode exhibits increased stiffness due to redistributed hydrated ions. Note the nonlinear characteristics of the electromechanical properties of the IPMCs. These features will be discussed in detail in this book.

2.2.2 ELECTRICAL PERFORMANCE

To assess the electrical properties of IPMNC, the standard A.C. impedance method that can reveal the equivalent electric circuit has been adopted. A typically measured impedance plot, provided in Figure 2.14, shows the frequency dependency of the impedance of the IPMCs. It is interesting to note that the IPMC is nearly resistive ($>50\ \Omega$) in the high-frequency range and fairly capacitive ($>100\ \mu F$) in the low-frequency range.

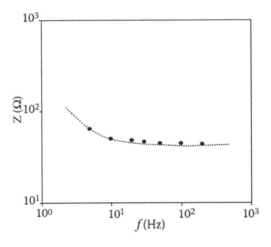

FIGURE 2.14 The measured A.C. impedance spectra (magnitude) of the IPMC sample. The moist IPMC sample has a dimension of 5-mm width, 20-mm length and 0.2-mm thickness.

Based on the previous findings, we consider a simplified equivalent electric circuit of the typical IPMC, such as the one shown in Figure 2.15 (Shahinpoor and Kim, 2000a). In this approach, every single-unit circuit (i) is assumed to be connected in a series of arbitrary surface resistance (R_{ss}, of the order of a few ohms per centimeter) on the surface. This approach is based on the experimental observation of the considerable surface electrode resistance.

We assume that there are four components of each single-unit circuit: the surface electrode resistance (R_s, of the order of a few tens of ohms per centimeter); the polymer resistance (R_p, of the order of a few hundreds of ohms per millimeter); the capacitance related to the ionic polymer and the double layer at the surface electrode/electrolyte interface (C_d, of the order of a few hundreds of microfarads); and an impedance (Z_w) due to a charge transfer resistance near the surface electrode. For the typical IPMNC, the importance of R_{ss} relative to R_s may be interpreted from $\sum R_{ss}/R_s \approx L/t \gg 1$,

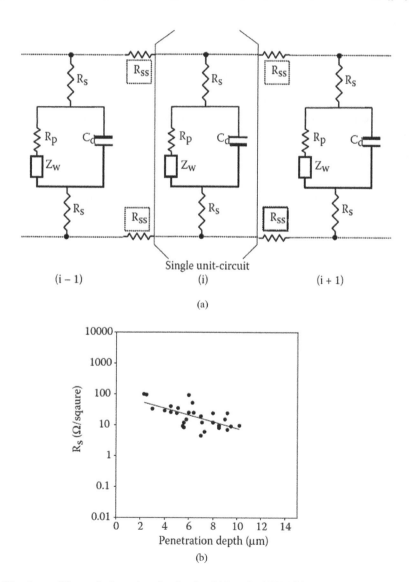

(a)

(b)

FIGURE 2.15 A possible equivalent electric circuit of (a) typical IPMCA and (b) measured surface resistance, R_s, as a function of platinum penetration depth (bottom). Note that SEM was used to estimate the penetration depth of platinum in the membrane. The four-probe method was used to measure the surface resistance, R_s, of the IPMCs. Clearly, the deeper the penetration, the lower the surface resistance is.

where notations L and t are the length and thickness of the electrode, respectively. It becomes a two-dimensional electrode. A thin layer of a highly conductive metal (such as gold) is deposited on top of the platinum surface electrodes (Shahinpoor and Kim, 2000a). To increase the surface conductivity of IPMCs.

Realizing that water in the perfluorinated IPMNC network is the sole solvent that can create useful strains in the actuation mode, another issue to deal with is the so-called *decomposition voltage*. As shown in Figure 2.16, the decomposition voltage is the minimum voltage at which significant electrolysis occurs.

Figure 2.16 contains the graph of steady-state current, I, versus applied D.C. voltage, E_{app}, showing that as the voltage increases, there is little change in current (obeying Faraday's law). However, a remarkable increase in D.C. is observed with a small change in voltage. Small overpotential (approximately 0.3–0.5 V) was observed even though the voltage causing water electrolysis is about 1.23 V. It should be noted that such water electrolysis leads to lower thermodynamic efficiency of the IPMC.

Figure 2.17 depicts the measured cyclic current/voltage responses of a typical IPMC (scan rate of 100 mV/s). As can be seen, a relatively simple behavior with a small hysteresis is obtained. The reactivity of the IPMC is mild such that it does not show any sharp reduction or reoxidation peaks within ±4 Volts. There is an exception for a decomposition behavior at ~±1.5 V, where the extra current consumption is due to electrolysis. The overall behavior of the IPMNC shows a simple trend of ionic motions caused by an imposed electric field.

Note that the scan rate is 100 mV/s. A simple behavior with a small hysteresis can be seen. It does not show any distinct reduction or reoxidation peaks within ±4 V, except for a decomposition behavior at ~±1.5 V, where the extra current consumption is apparently due to electrolysis.

Figure 2.18 depicts the frequency response of the IPMC in terms of the normal stress versus the normal strain. Its frequency dependency shows that as frequency increases, the beam displacement decreases. However, it must be realized that, at low frequencies (0.1–1 Hz), the effective elastic modulus of the IPMNC cantilever strip under an imposed voltage is also rather small.

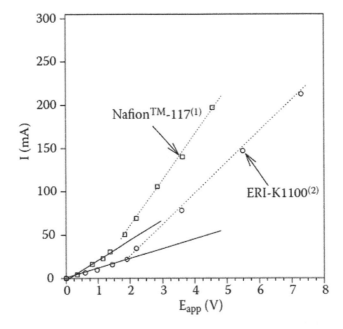

FIGURE 2.16 Steady-state current, I, versus applied voltages, E_{app}, on the typical IPMCs. ERI-K1100 stands for a proprietary IPMNC fabricated by Environmental Robots, Inc. It has a thickness of 2.9 mm and is suitably platinum/gold electrodes.

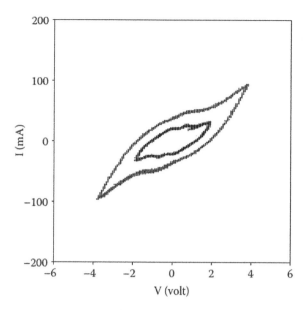

FIGURE 2.17 *I/V* curves for a typical IPMNC. Nafion™-117-based IPMC.

On the other hand, at high frequencies (5–20 Hz), such moduli are larger, and displacements are smaller. This observation is because, at low frequencies, water and hydrated ions have time to gush out of the surface electrodes. In contrast, at high frequencies, they are instead contained inside the polymer.

Therefore, the nature of water and hydrated ion transport within the IPMNC can affect the moduli at different frequencies and presents the potential application to smart materials with a circulatory system. This circulatory system is a biomimetic phenomenon in the sense that all living

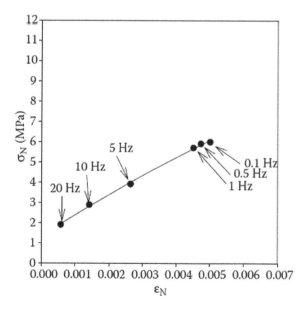

FIGURE 2.18 Frequency dependency of the IPMC in terms of the normal stress, σ_N, versus the normal strain, ε_N, under an imposed step voltage of 1 V. This Nafion-117 IPMC has a cation of Li$^+$ and a size of 5 × 20 mm.

systems have some kind of circulatory fluid to keep them smart and surviving. This is also of interest in a similar analogy to ionic hydraulic actuators (Shahinpoor and Kim, 2001d).

Encapsulation by highly elastic thin membranes such as Saran® F-310 (Dow Chemicals) or liquid latex has effectively maintained a reasonably constant polar medium for the cations' mobility and consistent performance.

The ionic polymer-metal composites or ionic polymer conductor composites are water-loving, living muscles. However, water can be replaced with ionic liquids or other polar liquids, as reported in several publications (see Bennet and Leo, 2003). A new family of encapsulated IPMNCs equipped with thin compliant electrodes showing higher efficiencies and power densities in actuation has been fabricated and tested.

2.2.3 Improved Force Performance of IPMCs

A fundamental engineering problem in achieving higher force density IPMCs is reducing or eliminating the leakage of the cations in the polar liquid within the macromolecular network out of the cathode surface electrode (made of finely dispersed platinum particles within or near the boundary region), so that water transport within the IPMC can be more effectively utilized for actuation. Figure 2.7 depicts that the nominal size of platinum particles in the IPMNC near the boundary is about 40–60 nm, much larger than that (~5 nm) of incipient particles associated with ion clusters. Thus, the embryonic particles coagulate during the chemical reduction process and eventually grow large, as schematically illustrated in Figure 2.19.

It is realized that there is a significant potential for controlling the reduction process in terms of platinum particle penetration, size and distribution. This process could be achieved by introducing active dispersing agents (additives) during the chemical reduction process. It is anticipated that the additives would enhance the dispersal of platinum particles within the ionic polymer and reduce coagulation. As a result, a better platinum particle penetration in the polymer with smaller average particle size and more uniform distribution could be obtained.

This uniform distribution makes it more difficult for water to pass through (granular damming effect) and emerge from the surfaces of the IPMNC samples. Thus, the water seepage or leakage out of the surface electrode could be significantly reduced. The use of active dispersing agents such as detergents, soaps, and, in particular, PVP during the platinum metallization process resulted in dramatically improved force density characteristics.

The results are shown in Figure 2.20(a) and (b), where the measured force of the improved IPMNC relative to the conventional one is reported. As seen, the additive-treated IPMC has shown

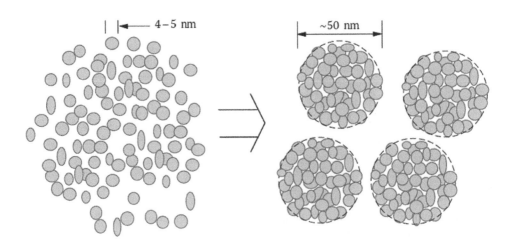

FIGURE 2.19 A schematic illustration of platinum coagulation during the chemical reduction process.

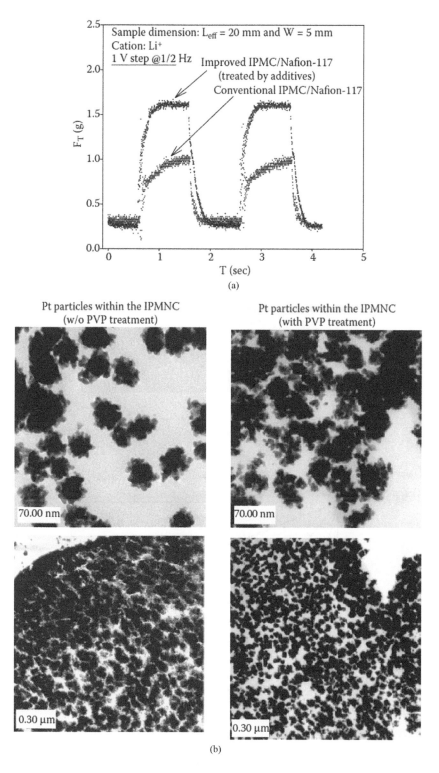

(a)

Pt particles within the IPMNC
(w/o PVP treatment)

Pt particles within the IPMNC
(with PVP treatment)

(b)

FIGURE 2.20 (a) Force response characteristics of the improved IPMC versus the conventional IPMC. Note that the improved IPMC is the one treated by an active dispersing agent. (b) TEM micrographs of two samples of IPMC with and without pyrrolidone (PVP) treatment. Note how the addition of PVP causes the nanoparticles of platinum not to coalesce and create a uniform and relatively homogeneous distribution of particles. This is believed to create more uniform internal electric fields and cause increased force capability of IPMCs and IPCCs.

a much sharper response to the input electric field, a dramatically increased force density generation by as much as 100%.

Figure 2.21(a) depicts an SEM, an IPMNC film treated with a dispersing agent (top), and its X-ray line scan (bottom). In contrast, Figure 2.21(b) depicts the profiles of platinum concentration versus penetration for different permeability coefficients.

Note that Figure 2.20(b) depicts TEM micrographs of two samples of IPMNC with and without PVP treatment, showing how the addition of PVP causes the nanoparticles of platinum not to coalesce and thus create a uniform and reasonably homogeneous distribution of particles. This is believed to create more uniform internal electric fields and cause increased force capability of IPMNCs and IPCNCs.

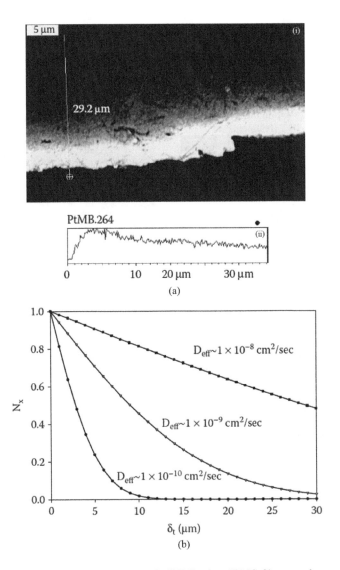

FIGURE 2.21 (a) A scanning electron micrograph (SEM) of an IPMC film sample treated with a (i) dispersing agent polyvinyl pyrrolidone or PVP and (ii) its X-ray line scan. As can be seen, the Pt penetration is deeper, more homogeneous and more consistent due to the use of the dispersant. (b) Profiles of platinum concentration versus depth penetration for different diffusivity coefficients D_{eff}. As can be seen, the Pt penetration is deeper, more homogeneous and more consistent due to the use of the dispersant PVP.

As can be seen, the Pt penetration is increasingly homogeneous and consistent using a dispersant during the reduction process. Note from Figure 2.21(a) and (b) that a good platinum penetration is achieved, meaning that an active additive enhances platinum dispersion and leads to better penetration in the polymer. A convenient way to handle this situation (free diffusion into finite porous slab or membrane) is to use an effective diffusivity, D_{eff}, and then consider it as a one-dimensional problem. Assuming fast kinetics for the metal precipitation reaction of $(Pt(NH3)4)^{2+} + 2e^- \geq Pt_0 + 4NH3$, the precipitated platinum concentration, N_x, can be expressed as

$$N_x = \frac{C_{Pt}(\delta_t)L}{C_{Pt,i}} = 1 - erf\left(\frac{\delta_t}{\sqrt{4D_{eff}t}}\right) \qquad (2.7)$$

where notations $C_{Pt}(\delta t)$, $C_{Pt,i}$ and δ_t are the platinum concentration, the platinum concentration at the interface and the particle penetration depth, respectively.

For a typical reduction time of $t = 15$ min (in Figure 2.21(a) and (b)), Equation (2.7) is plotted for values of $D_{eff} = 1 \times 10^{-10}$, 1×10^{-9} and 1×10^{-8} cm²/s, respectively. The effective diffusivity, D_{eff}, could be estimated to be of the order of 1×10^{-8} cm²/s for the improved IPMNC. This situation is somewhat complicated due to the simultaneous effect of a mass transfer and significant kinetics; nevertheless, the estimated value of $D_{eff} = 1 \times 10^{-8}$ cm²/s would be a convenient value for the engineering design of the platinum metallization process described here for the improved IPMNC.

In Figure 2.22, the results of the potentiostatic analysis are presented. The variation of current following applying an electric potential to the IPMCs (the PVP-treated IPMC and the conventional IPMC) is shown. It is useful to make a direct comparison between $Q_{t,PVP}$ (for the PVP-treated IPMNC), and Q_t (for the conventional IPMNC). The data shown in Figure 2.22 give $Q_{t,PVP}/Q_t \cong 1.1$. This means that the PVP-treated IPMNC consumes approximately 10% more charges. This raises a point that merely 10% increased consumption of charges is not the only reason to increase the force density by as much as 100%.

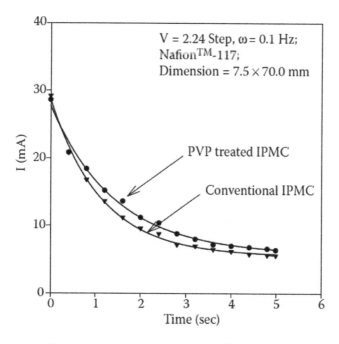

FIGURE 2.22 Potentiostatic coulometric analyses for the additive-treated IPMC and the conventional IPMNC. This graph shows that an increased current passage (Faraday's approach) can contribute to the observed improvement in the force characteristics of IPMC strips (see Figure 2.19).

An increase in force density of as much as 100% represents a very favorable gain for a 10% increase in consumed charge. Therefore, it can be concluded that it is the "granular damming effect" that minimizes the water leakage out of the porous surface electrode region when the IPMNC strip bends. Thus, such dispersing issues are rather crucial in increasing the force density of IPMCs.

2.2.4 A View from Linear Irreversible Thermodynamics

In connection with the phenomenological laws and irreversible thermodynamics considerations discussed in Section 2.2.1, when one considers the actuation with ideal impermeable electrodes, which results in $Q = 0$ from Equation (2.3), one has

$$\nabla p = \frac{L}{K} \vec{E} \tag{2.8}$$

Also, the pressure gradient can be estimated from

$$\nabla p \cong \frac{2\sigma_{max}}{h} \tag{2.9}$$

where σ_{max} and h are the maximum stress generated under an imposed electric field and the thickness of the IPMNC, respectively, the values of σ_{max} can be obtained when the maximum force (= blocking force) is measured at the tip of the IPMNC per a given electric potential.

In Figure 2.23, the maximum stresses generated, σ_{max}, under an imposed electric potential, E_o, for calculated values and experimental values of the conventional IPMNC and the improved IPMNC are presented. It should be noted that the improved IPMNC (by the method of using additives) is superior to the conventional IPMNC approaching the theoretically obtained values.

For theoretical calculation, the following experimentally measured values were used:

$L_{12} = L_{21} = 2 \times 10^{-8}$ {cross-coefficient, (m/s)/(V/m)},
$k = 1.8 \times 10^{-18}$ {hydraulic permeability, m^2} (Bernardi and Verbugge, 1992) and
$\vec{E} = E_0/h$ where $h = 200\ \mu m$ {membrane thickness}.

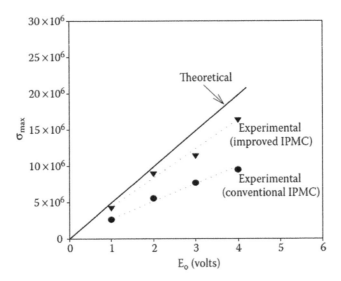

FIGURE 2.23   Maximum stresses generated by the IPMCs at given voltages.

TABLE 2.1

Current Capabilities of IPMNC Materials

Young's modulus, E	Up to 2 GPa
Shear modulus, G	Up to 1 GPa
Poisson's ratio, V	Typical: 0.3–0.4
Power density (W/volume))	Up to 200 mW/cc
Max force density (cantilever mode)	Up to 50 kgf/kg
Max displacement/strain	Up to 4% linear strain
Bandwidth (speed)	Up to 1 kHz in cantilever vibratory mode for actuation; up to 1 MHz for sensing
Resolution (force and displacement control)	Displacement accuracy down to 1 μm; force resolution down to 1 mg
Efficiency (electromechanical)	More than 25% (frequency-dependent) for actuation; more than 90% for sensing
Density	Down to 1.8 g/cm^3

Table 2.1 lists the current capabilities of IPMCs. These capabilities can be built into IPMNC samples by changing some 18 parameters. These parameters can be tweaked to manipulate the performance of IPMCs. This procedure will be discussed in the next chapter on manufacturing methodologies. Furthermore, the energy and power densities of IPMNCs and IPCNCs appear to be very compatible with mammalian muscles, as depicted in Figures 2.24 and 2.25.

The sample dimensions in creating these graphs were 20- × 5- × 1.6-mm thickness. A maximum square-wave voltage input at 16 V was applied, and the samples contained Li+ as a cation.

Note the locus of specific energy versus frequency of a broad spectrum of biological muscles, as reported by Robert Full of UC Berkeley (Full and coworkers, 2001–2002). IPMCs have a broad frequency spectrum, as shown in Figures 2.24 and 2.25. Figure 2.25 depicts the variation of power output as a function of the frequency of excitation for the same sample described in Figure 2.24.

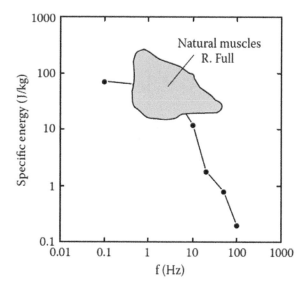

FIGURE 2.24 Specific energy as a function of frequency for typical IPMC samples.

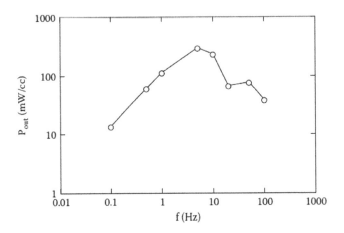

FIGURE 2.25 Power density output as a function of frequencies.

2.2.5 THERMODYNAMIC EFFICIENCY

The bending force of the IPMC is generated by the significant redistribution of hydrated ions and water. Typically, such a bending force is an electric field dependently distributed along the IPMNC strip's length, as noted in Figure 2.23. Figure 2.21 notes that a surface voltage drop occurs, which can be minimized (Shahinpoor and Kim, 2000a). The IPMC strip bends due to this ion migration-induced hydraulic actuation and redistribution.

The effectively strained IPMC exerts the bending force of the IPMC due to hydrated ions' transport. Typically, such force is field dependently distributed along the length of the IPMNC strip. The IPMNC strip bends due to this force. The total bending force, F_t, can be approximated as

$$F_t = \int_0^L f ds \qquad (2.10)$$

where f is the force density per unit arc length S and L is the effective beam length of the IPMNC strip. Assuming a uniformly distributed load over the length of the IPMNC, the mechanical power produced by the IPMNC strip can be obtained from

$$P_{out} = \frac{1}{2} \int_0^L f v dS \qquad (2.11)$$

where v is the local velocity of the IPMC in motion. Note that v is a function of S and can be assumed to linearly vary such that $v = (v_{tip}/L)S$, $0 \leq S \leq L$. Finally, the thermodynamic efficiency, $E_{ff, em}$, can be obtained as

$$E_{ff, em}(\%) = \frac{P_{out}}{P_{in}} \times 100 \qquad (2.12)$$

where P_{in} is the electrical power input to the IPMC – that is, $P_{in} = V(t)I(t)$, where V and I are the applied voltage and current, respectively.

Based on Equation (2.12), one can construct a graph (see Figure 2.26) that depicts the thermodynamic efficiency of the IPMC as a function of frequency. This graph presents the experimental results for the conventional IPMC and the improved additive (PVP)-treated IPMNC. It is of note

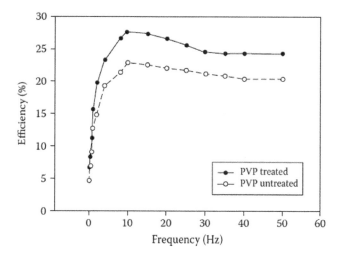

FIGURE 2.26 Thermodynamic efficiency of actuation of the IPMC as a function of frequency. Nafion-117 was used as a starting material.

that the optimum efficiencies occur at near 8–10 Hz for these new, improved samples of IPMCs. The optimum values of these IPMCs are approximately 25–30%. At low frequencies, the water leakage out of the surface electrode seems to cost the efficiency significantly. However, the additive-treated IPMNC shows a dramatic improvement in efficiency due to reduced water leakage out of the electrode surface. The essential sources of energy consumption for the IPMNC actuation could be necessary mechanical energy needed to cause the positive/negative strains for the IPMNC strip:

I/V hysteresis due to the diffusional water transport within the IPMNC,
Thermal losses – joule heating (see Figure 2.26),
Decomposition due to water electrolysis and
Water leakage out of the electrode surfaces.

The recent improvement in the performance of the IPMNC by blocking water leakage out of the porous surface electrode has resulted in higher overall thermodynamic efficiencies of all IPMNC samples tested in a frequency range of 0.1–50 Hz. It should be noted that the obtained values are favorable compared to other types of bending actuators – that is, conducting polymers and piezoelectric materials at similar conditions – that exhibit considerably lower efficiencies (Wang et al., 1999).

The samples used in Figure 2.26 have a dimension of 20-mm length, 5-mm width and 0.2-mm thickness. The applied potential is a 1-V step, and lines are least-square fits. It appears that at higher frequencies, the thermodynamic efficiency stabilizes and almost remains the same. This phenomenon, as well as the resonance state efficiencies, is currently under investigation.

Figure 2.27 displays IR thermographs taken for an IPMNC in action (the sample size of 1.2 × 7.0 cm). They show spectacular multispecies mass/heat transfer in a sample of IPMNC under an oscillatory step voltage of 3 and a frequency of 0.1 Hz. The temperature difference is more than 10°C. In general, the hot spot starts from the electrode and propagates toward the tip of the IPMNC strip (left to right). The thermal propagation is simultaneously conjugated with the mass transfer along with the possible electrochemical reactions. It clearly shows the significance of water transport within the IPMNC. These coupled transport phenomena are currently under investigation. Note in Figure 2.27 that the electrode is positioned on the left side of the IPMNC. The temperature difference was more than 10°C when a D.C. voltage of 3 was applied for the IPMNC sample size of 1.2 × 7.0 cm.

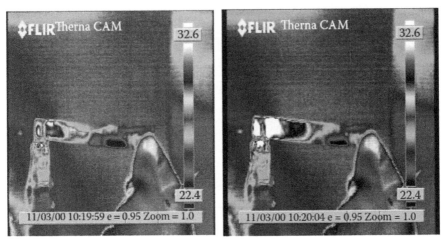

t = 0 sec; t = 5 sec

FIGURE 2.27 IR thermographs of an IPMC in action. The hot spot starts from the electrode and propagates toward the tip of the IPMNC strip.

2.2.6 Cryogenic Properties of IPMNC

To determine the cryogenic characteristics of IPMC sensors and actuators for the harsh space conditions, various samples of IPMNCs were tested in a cryo-chamber under very low pressures down to 2 torr and temperatures down to −150°C. This was done to simulate the harsh, cold, low-pressure environments in space. The results are depicted in Figures 2.28–2.33.

2.2.7 Internal and External Circulatory Properties of IPMCs

The families of IPMCs can induce electrically controllable autonomous changes in material properties by an intrinsic distributed circulatory system. Thus, they have the potential of creating a new class of structural nanocomposites of ionic polymers and conductors such as metals or carbon

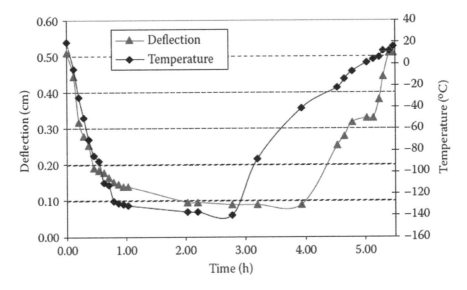

FIGURE 2.28 Deflection characteristics of IPMC as a function of time and temperature.

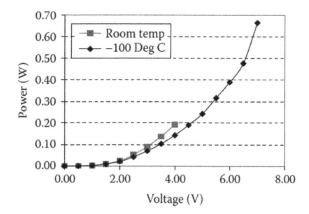

FIGURE 2.29 Power consumption of the IPMC strip bending actuator as a function of activation voltage.

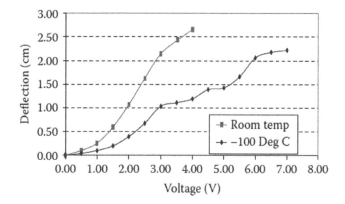

FIGURE 2.30 Deflection of the bending IPMC strip as a function of voltage.

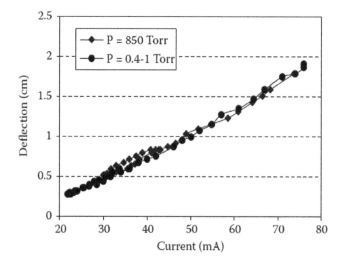

FIGURE 2.31 Deflection versus current drawn (top) and power input (bottom) at a high pressure of 850 torr and low pressure of 0.4–1 torr.

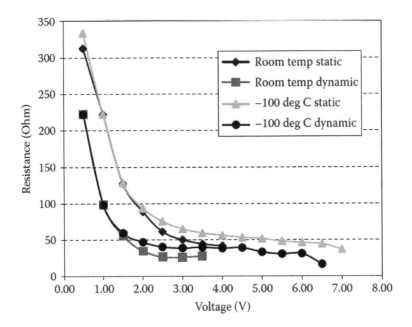

FIGURE 2.32 IPMC strip static (V/I) and dynamic (V/I) resistance at various temperatures.

nanotubes. These contain an embedded circulatory system capable of producing electrically controlled localized internal pressure changes. This procedure is done to hydraulically pump liquids containing ions and chemicals to various parts of the material to cause sensing, actuation and substantial changes in stiffness and conductivity and perform self-repair or healing.

Ionic polymers equipped with a distributed network of electrodes created by a chemical plating procedure, such as IPMCs, can create a central distributed circulatory system of ions, chemicals, water and polar and ionic liquids. The fluid motion is generated by electrically induced migration and redistribution of conjugated ions within its polymeric network of nanoclusters. Every part of the material can be reached by electric field-induced migration and redistribution of conjugated

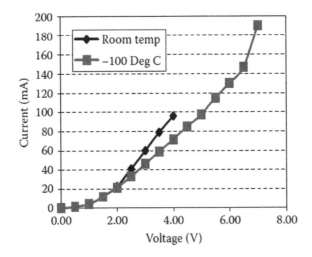

FIGURE 2.33 The relation between voltage and current for an IPMC strip exposed to room temperature = 20 and −100°C.

ions on a nanoscale for robotic motion action and feedback and embedded distributed sensing and transduction.

One of the essential characteristics of these nano-IPMNCs as smart multifunctional polymeric nanocomposites is allowing ionic migration on a molecular and nanoscale utilizing an imposed local intrinsic electric field within the material. This process then causes hydration or cations to move and create a local pressure or fluid motion while carrying additional water as hydrated water.

Such fluid circulatory migration in ionic polymers had been observed in our laboratories as early as 1994 in the form of water + cation appearance and disappearance on the cathode side of the IPMC strips under an imposed sinusoidal- or square-wave electric field for actuation. A pair of electrodes in the middle and around a strip of IPMC can make a soft parallel jaw robotic gripper. It enables it to bend and grab objects. This capability inspired us to think about the rapid nastic sensing and actuation of higher plants such as the carnivorous or insectivorous plants. Some higher plants, such as the Venus flytrap (*Dionaea muscipula*), may use the same ionic migration and water circulation for sensing and rapid actuation. The possible mechanism for nastic movement in higher plants, such as in the Venus flytrap (Figure 2.34), has been reported by Shahinpoor and Thompson (1994) and Shahinpoor (2013). They have shown that it is due to cationic migration. These plants use almost digital sensing because the trapped insect has to disturb the ionoelastic trigger hairs more than two or three times before the lobes close rapidly to trap the insect. And rapid actuation and deployment occur with ionic migration using the plant's sensing and ionic circulatory system. Mechanical movement of the trigger hairs (Figure 2.34(d)) puts into motion the ATP-driven changes in water pressure within these cells. These trigger hairs are located on the trap leaves. If these are stimulated twice in rapid succession, an electrical signal is generated that causes changes in the water pressure in different parts of the leaves (Shahinpoor and Thompson, 1994). The cells are driven to expand by the increasing water pressure, and the trap closes as the plant tissue relaxes. It is remarkable how these changes are similar to what actually happens in ionic polymer nanocomposites. One can even have cations in IPMCs to cause sensing and actuation. It is clear that ionic polymer nanocomposites have opened the door to the mysterious ion engineering world of nastic plant movements and rapid deployments, and this new ionic world now needs to be further explored.

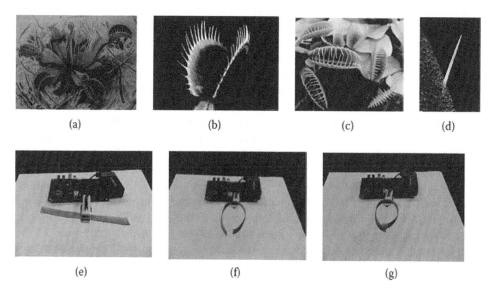

(a)   (b)   (c)   (d)

(e)   (f)   (g)

FIGURE 2.34 Venus flytrap (*Dionaea muscipula*). (a)–(c) A plant capable of rapid nastic deployment and movement based on its trigger hair; (d) digital sensing and an IPMNC gold strip (1 cm × 6 cm × 0.3 mm) performing similar rapid closure (e)–(g) under a dynamic voltage of 4 V.

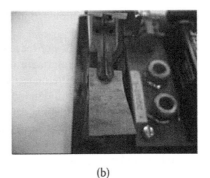

(a)   (b)

FIGURE 2.35 Migration of lithium cations to the surface on the cathode electrode side of a cantilever sample of IPMNC. (a) Sample bent downward, lithium ions appearing on the surface. (b) Sample bent upward with lithium ions disappearing by migration to the other side.

The first observation on the circulatory migration of chemicals to boundary surfaces of ionic polymer-metal nanocomposite samples occurred in 2000 in the Artificial Muscle Research Institute Laboratories. Shahinpoor and Kim (2002j) reported these observations in connection with such a circulatory system enabling sensing and actuation by creating internal pressure change and causing internal hydraulic actuation. One could consistently observe the color of the surface on the cathode side of a cantilever sample of IPMNC changed by applying a step electric field. Figure 2.35 depicts one such experiment.

If the imposed electric field were dynamic and oscillatory like sinusoidal, the color of the surfaces on the cathode side changed alternatively with the frequency of the applied dynamic electric field. When we changed the cations to other cations, such as sodium or calcium, the color of the migrated cations on the cathode side changed. The emergence of water on the cathode side was also always observable. We concluded that the mechanism of electrically induced bending was due to ionic polymer nanoscale energetics and ionic migration from one side of the cantilever film to the other side while carrying hydrated water or added mass to water in such hydraulic-type actuation. It was observed that the ionic migration and redistribution moved and redistributed water, chemicals, polar fluids, cations or ionic liquids contained within the macromolecular network. This action circulated water, chemicals, polar fluids, cations or ionic liquids within the materials. It helped transport ions and chemicals from one point to another to cause significant (>10) changes in the value of several properties, such as stiffness, conductivity, and material transport.

In fact, in the case of lithium cations, the color was greenish blue, which then indicated that it was Li^+ cations migrating under the influence of the imposed electric field and carried loose water and hydrated water along with it. These observations have been reported by Shahinpoor and Kim (2002j, 2004), Kim and Shahinpoor (2003b), as well as by Leo and Cuppoletti (2004) and Hawkins et al. (2004), and later became an everyday phenomenon to be observed with hydrated samples of IPMCs. Figure 2.36(a) depicts the essential mechanism in such electrically controllable ionic migration accompanied by water or ionic liquid movement. Figure 2.36(b) depicts both sensing and actuation characteristics of IPMCs, which will also be discussed in the next section and Chapter 7.

Furthermore, such ionic migration could also increase the local stiffness. We have observed stiffness changes of more than one order of magnitude in IPMCs, as depicted in Figure 2.1.

The migration of loaded or hydrated cations utilizing an imposed local electric field has been observed to cause deformation, stiffening, and substantial changes in local elastic modulus, substantial material transport within the material, ability to transport, and healing and repairing materials and chemicals to any location in the body or the body surfaces and skins.

A first-order analytical model for such nanocomposites can be presented by linear irreversible thermodynamics with fluxes such as a current density, J, and the flux of water + ions and chemicals,

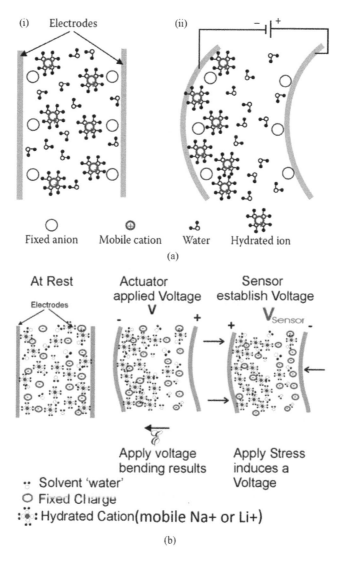

FIGURE 2.36 (a) The cation-transport-induced actuation principle of IPMCs. (i) Before a voltage is applied; (ii) after a voltage is applied. (b) The cation-transport-induced actuation and sensing mechanisms of IPMCs.

Q. The conjugated forces include the electric field, *E*, and the pressure gradient, $-\nabla p$. The resulting equations are Equations (2.2) and (2.3).

2.2.8 Near-D.C. Mechanical Sensing, Transduction and Energy-Harvesting Capabilities of IPMCs in Flexing Bending and Compression Modes

As discussed before, IPMCs have excellent sensing capability in flexing and compression. Furthermore, as will be discussed in detail in Chapter 7 of this book, IPMCs' active elements are capable of sensing relatively high frequencies and are capable of near-D.C. dynamic sensing and acceleration measurement, as shown in Figure 2.37.

In this sense, they are far superior to piezoelectric materials, which are only suitable for high-frequency sensing, while for low-frequency or near-D.C. sensing, piezoresistors are generally used.

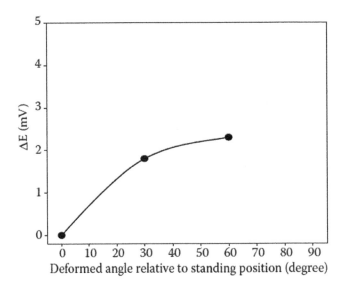

FIGURE 2.37 Near-D.C. sensing data in terms of produced voltages, ΔE, versus displacement. Note that the displacement is shown in terms of the deformed angle relative to the standing position in degree in a cantilever configuration. The dimension of the sample sensor is $5 \times 25 \times 0.12$ mm.

Thus, they span the whole range of frequencies for dynamic sensing and have a wide bandwidth. These issues will be detailed in Chapter 7.

Power harvesting capabilities of IPMCs are also related to near-D.C. or even high-frequency sensing and transduction capabilities of IPMCs. Figure 2.38 depicts a typical near-D.C. voltage and current production of IPMNC cantilevers.

The experimental result shows that almost a linear relationship exists between the voltage output and the imposed displacement of the IMPC sensors' tip.

IPMCs or IPMNCs sheets can also generate power under normal pressure. Thin sheets of IPMNC were stacked and subjected to normal pressure and typical impacts and observed to generate a large output voltage.

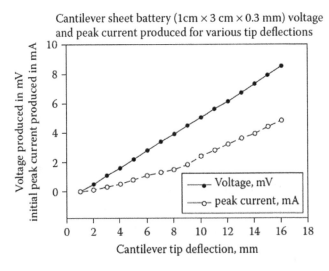

FIGURE 2.38 Typical voltage/current output of IPMC samples under flexing/bending. The IPMNC sample has a dimension of 10-mm width, 30-mm length and 0.3-mm thickness.

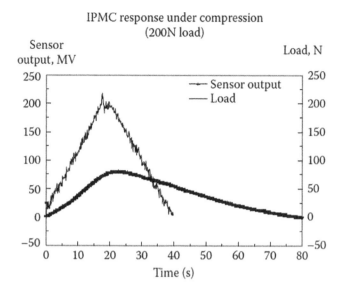

FIGURE 2.39   Voltage out due to the impact of 200-N load on a 2-cm × 2-cm × 0.2-mm IPMNC sample.

Endo-ionic motion within IPMC thin sheet batteries produced an induced voltage across these sheets' thickness when a normal or a shear load was applied. A material testing system (MTS) was used to apply consistent, pure compressive loads of 200 and 350 N across the surface of an IPMNC 2- × 2-cm sheet. The output pressure response for the 200-N load (73 psi) was 80 mV in amplitude, and for the 350-N load (127 psi), it was 108 mV.

This type of power generation may be useful in the heels of boots and shoes or places with a lot of foot or car traffic. Figure 2.39 depicts the output voltage of the thin sheet IPMNC batteries under a 200-N normal load. The output voltage is generally about 2-mV/cm length of the IPMNC sheet.

2.3   ADDITIONAL REPORTS ON FORCE OPTIMIZATION

The fabricated IPMCs can be optimized for producing a maximum force density by changing multiple process parameters, including bath temperature, (T_R), time-dependent concentrations of the metal-containing salt, $C_s(t)$ and the reducing agents, $C_R(t)$.

The Taguchi design of experiment technique was conducted to identify the optimum process parameters (Rashid and Shahinpoor, 1999). The analysis techniques for "larger-the-better" quality characteristics incorporate noise factors into an experiment involving such characteristics for the maximum force generated by the manufactured IPMCs in this case. Such an analysis allows us to determine the key factors and the possible best settings for consistently good performance. The force measurement configuration is depicted in Figure 2.40. The blocking force is measured at zero displacements.

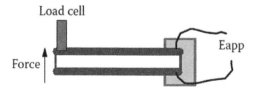

FIGURE 2.40   A blocking force measurement configuration.

Furthermore, the initial stretching of the ionic polymer samples before manufacturing also increases the force capability of IPMCs. One approach is to stretch the base material prior to the platinum composition process. Another approach is to stretch the base material uniaxially and carry out the conventional optimal IPMC manufacturing techniques.

Using such a stretching technique allows the particle penetration within the material to be much more useful to form a much denser platinum particle phase and distribution. The basic morphology of particle formation appears different, as can be seen in Figure 2.41.

In general, such a stretching method of manufacturing IPMCs seems to benefit the IPMC performance in terms of the blocking force in a cantilever configuration. The results are presented in Figure 2.42. Significantly improved generative forces are produced.

As far as making magnetic IPMCs is concerned, see Park et al. (2006). They describe how ferromagnetic metal salts can be embedded within the IPMCs molecular network and give rise to magnetically active IPMCs.

In connection with self-oscillatory phenomena in IPMCs, see D. Kim and K. J. Kim (2005).

2.3.1 RECENT REPORTS AND DISCOVERIES ON IPMCs

IPMCs enjoy tremendous attention from the scientific community. There have been many publications in recent years, which we will incorporate in this new edition of the book.

Chunga, Funga, Honga, Ju, Lin, and Wu published a paper entitled "A novel fabrication of ionic polymer-metal composites (IPMC) actuator with silver nano-powders", which was published in Sensors and Actuators, Elsevier, B 117: 367–375 (2006). They employed silver nanoparticles of 35-nm diameter in the fabrication of IPMCs. This IPMC actuator exhibits a large bending deformation curvature angle of more than 90° at a lower driving voltage of 3V.

Abdelnour, Mancia, Peterson, and Porfiri published a paper on "Hydrodynamics of underwater propulsors based on ionic polymer metal composites: a numerical study". This paper analyzed the hydrodynamics of a vibrating IPMC actuator in water to describe the thrust generation mechanisms of IPMC actuators. They computed the actuator's thrust as a function of its frequency of oscillation and maximum tip displacement. They showed that thrust generation of vibrating IPMC actuators is positively correlated to the vortex shedding.

Sang-Mun Kim, Il- Seok Park and Kwang J. Kim published a paper on "A Microscopic Model Developed for Ionic Polymer-Metal Composites with Porous Layered Electrodes", Journal of Applied Physics (2006).

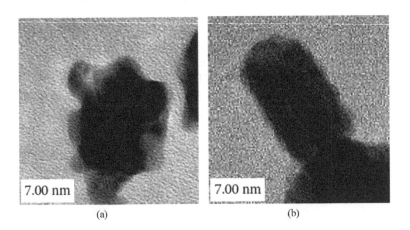

(a)   (b)

FIGURE 2.41 Two TEM micrographs show the intrinsic platinum particles for an IPMC mechanically stretched (a) before making the metal-ion polymer composite and (b) with no stretching. The 17% uniaxial stretching was performed.

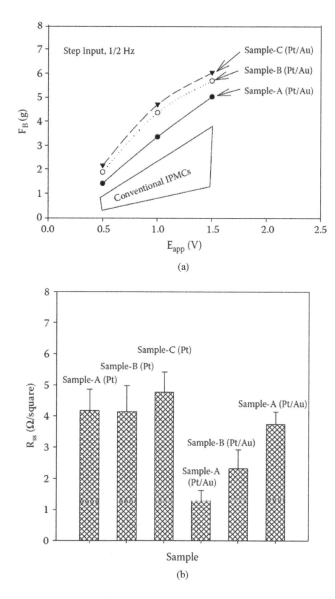

FIGURE 2.42 (a) Measured blocking forces and (b) surface resistance of IPMNC samples prepared. The standard sample size is 5 mm × 10 mm × 0.2 mm. The process information is given in terms of particles used, platinum, and the final material with gold electrodes.

Y. J. Kim (2006) has described a new type of perfluorinated carboxylate membrane/platinum composite and its electrochemical behavior. They introduced a new type of IPMC based on various compositions of perfluoroalkyl acrylate-acryl acid copolymer. They also employed different counter cation types by radical copolymerization of fluoroalkyl acrylate (F.A.) and acrylic acid (A.A.). It was published by Wiley Periodicals, Journal of Applied Polymer Science 99: 2687–2693 (2006).

X. Tan, D. Kim, N. Usher, D. Laboy, J. Jackson, A. Kapetanovic, J. Rapai, B. Sabadus, X. Zhou *reported* "An Autonomous Robotic Fish for Mobile Sensing", Proceedings of the 2006 IEEE/RSJ, International Conference on Intelligent Robots and Systems, October 9–15, 2006, Beijing, China.

G. Del Buffalo, L. Placidi, and M. Porfiri published a paper on "A mixture theory framework for modeling the mechanical actuation of ionic polymer-metal composites" in Smart Materials and Structures 17 045010 (14pp): (2008). The electrolytic solvent contains mobile ionic species, which interact with mobile ions and the solvent and between the solvent and the backbone polymer and are responsible for sensing and actuation. Their model consists of three coupled linear partial differential equations.

J. W. Paquette and K. J. Kim published their results in a paper entitled "Ionomeric Electroactive Polymer Artificial Muscle for Naval Applications", IEEE Journal of Ocean Engineering, 29(3): 729 (July 2004). A common ionomeric electroactive polymer operates in aqueous environments and is capable of transduction. Various materials, including IPMCs, are investigated.

T. Johnson, F. Amirouche, describe the "Multiphysics modeling of an IPMC microfluidic control device", in a publication in Microsystem Technologies 14:871–879 (2008) DOI 10.1007/s00542-008-0603-6, Published online: March 18, 2008, Springer-Verlag 2008.Their design has the potential to control temperature-sensitive IPMCs. The proposed fluidic control device's operation requires an understanding of the device's thermal properties under actuation conditions.

D. Bandopadhya, B. Bhattacharya, and A. Dutta report on "Characterization of IPMC as Passive and Active Damper as an Alternative Novel Smart Actuator". This publication first appeared on September 4, 2008. The acting passive and active damping characteristics of ionic polymer-metal composite (IPMC) are described. A simple proportional controller scheme has been proposed and employed to control the vibration of IPMCs with a sizeable, flexible link.

R. Tiwari, K.J. Kim, reported in their publication: "Energy Harvesting using Ionic Polymer Metal Composites". They considered vibration-based energy harvesters. These samples were treated with ionic liquids. Similar to bending, shear causes the temporary increase in charge concentration on the compressed side and decreases in concentration on the extended side leading to the motion of charge inside the membrane. In extension mode, due to the stretching of the polymer chain inside IPMC, a redistribution of ions inside the pendant causes electric field production.

D. Dogruer, R. Tiwari and K. Kim reported energy harvesting from IPMCs in a paper, "Ionic Polymer Metal Composite as Energy Harvester", published in 2008 IEEE International Conference on Robotics and Automation, Pasadena, CA, United States, May 19–23, 2008. Thus, ionic liquid (1-butyl-3-methylimidazolium)-treated IPMCs are used for energy harvesting experiments.

Soon-Gie Lee, Hoon-Cheol Park, Surya D. Pandita, and Youngtai Yoo described "Performance Improvement of IPMC (Ionic Polymer Metal Composites) for a Flapping Actuator" published in the International Journal of Control, Automation, and Systems 4(6): 748–755 (December 2006). Relations between length/thickness and tip force of IPMCs were identified for the actuator design. All IPMCs thicker than 200 μm were processed by casting Nafion™ solution. Evaporation due to electrically heated electrodes, gold was sputtered on both surfaces of the cast IPMCs by the physical vapor deposition (PVD) process. An insect wing was attached to the IPMC flapping mechanism for its flapping test. In this test, the wing-flapping device using the 800-μm-thick IPMC could create around 10°~85° flapping angles and 0.5~15 Hz flapping frequencies by applying 3–4 V.

Il-Seok Park and Kwang J. Kim described their results on "Multi-fields responsive ionic polymer-metal composite", Sensors and Actuators A 135: 220–228 (2007).

V. Panwar and A. Gopinathan discussed an ionic polymer-metal nanocomposite sensor using the direct attachment of an acidic ionic liquid in a polymer blend. Journal of Materials Chemistry C 7:9389–9397 (2019).

Y. Bahramzadeh Y and Shahinpoor M presented dynamic curvature sensing employing ionic–polymer-metal composite sensors. Their results were shown in the "Smart Materials and Structures Journal" in 2011.

M. Shahinpoor and K. Kim discussed the "Ionic polymer-metal composites: III. Modeling and simulation as biomimetic sensors, actuators, transducers, and artificial muscles", Smart Materials and Structures 13:1362–1388 (2004).

Y. Cha Y, Dolhamidi S, and M. Porfiri described the energy harvesting from an annular ionic polymer-metal composite underwater vibration. Meccanica 50(11):2675–2690 (2015).

F. Cellini F, J. Pounds, S. Peterson, and M. Porfiri discussed underwater energy harvesting from a turbine hosting ionic polymer-metal composites. Smart Materials and Structures 23:10 (2014).

G.H. Feng and R.H. Chen report on the fabrication and characterization of arbitrarily shaped IPMC transducers for accurately controlled biomedical applications in Sensors and Actuators B: Chemical 143 (1):34–40 (2008).

M. Shahinpoor and K.J. Kim reported the characteristics of ionic polymer-metal composites in Smart Materials and Structures 10:197–214 (2005).

J.H. Jeon, S.P. Kang, S. Lee, and I.K. Oh reported a novel biomimetic actuator based on SPEEK and PVDF in Sensors and Actuators B: Chemical 143(1):357–364 (2009).

Panwar, C. Lee, S.Y. Ko, J.O. Park, and S. Park discussed the dynamic mechanical, electrical and actuation properties of ionic polymer-metal composites using PVDF/PVP/PSSA blend membranes in Materials Chemistry and Physics 135(2–3):928–937 (2012).

V. Panwar, K. Cha, J.O. Park, and S. Park discussed the high actuation response of PVDF/PVP/PSSA-based ionic polymer-metal composites actuator. Sensors and Actuators B: Chemical 161(1):460–470 (2012).

V. Panwar V, B. S. Kang, J.O. Park, and S.H. Park reported new ionic polymer-metal composite actuators based on PVDF/PSSA/PVP polymer blend membrane. Polymer Engineering & Science 51:1730–1741 (2011) 14.

V. Panwar, J.H. Jeon, A. Gopinathan, H.J. Lee, Il-K Oh, and J.Y. Jo described a low-voltage actuator using ionic polymer-metal nanocomposites based on a miscible polymer blend in Journal of Materials Chemistry A 3:19718–19727 (2015).

K. Kukreti and V. Panwar discussed improvised energy harvesting from low-frequency vibrations of IPMCs in the International Journal of Management, Technology And Engineering 2:112–115 (2018).

J.H. Jeon, R.K. Cheedarala, C.D. Kee, and Il. K. Oh discussed dry-type artificial muscles based on pendent sulfonated chitosan functionalized graphene oxide for greatly enhanced ionic interactions and mechanical stiffness. Advanced Functional Materials 23(48):1–11(2013).

P. Khanduri, A. Joshi, L.S. Panwar, A. Goyal, and V. Panwar reported sensing performance of ionic polymer-metal nanocomposite sensors with pressure and metal electrolytes for energy harvesting applications. In: Hura G., Singh A., Siong Hoe L. (eds) Advances in Communication and Computational Technology. Lecture Notes in Electrical Engineering, Vol. 668. Springer, Singapore. https://doi.org/10.1007/978-981-15-5341-7_63(2021).

2.4 ELECTRIC DEFORMATION MEMORY EFFECTS, MAGNETIC IPMNCs, AND SELF-OSCILLATORY PHENOMENA IN IONIC POLYMERS

The preceding novel phenomena have been observed and established recently in our laboratories at the University of New Mexico and the University of Nevada-Reno. The novel electric deformation memory effect in connection with ionic polymer conductor composites (IPCNCs) and, in particular, IPMNCs, is first reported here.

An IPMNC sample is capable of storing geometric shape and deformation information for a given step voltage or imposed electric field, even when the field is turned off—provided the sample is completely isolated form any electric discharging systems. In this case, IPMNCs do not need to be trained for a given shape memory effect such as shape memory alloys (SMAs) or ferromagnetic shape memory materials. Rather, they have an infinite set of possibilities of deformation shapes versus voltage that can be memorized even when the electric field is removed. This creates, for the first time, potential for building "geometric computers" that store information in geometrical forms containing infinite amounts of information for a given voltage signal or electric field. The data presented here will establish that, from a neutral position and charge-free state, for any given voltage, a cantilever sample of IPMNC bends to a shape; if one removes the voltage, the shape will not change and the material remembers the shape permanently, provided the sample is completely isolated form any electric discharging systems.

The process is highly reversible. Any change in shape is due to environmental changes such as humidity or temperature and, in a controlled environment, we observe that after the voltage is removed, having allowed the sample to stabilize its shape, the shape stays almost permanently. Upon shorting out the electrodes on the two sides of the sample, the sample moves back to its initial configuration before the application of the step voltage.

The data enclosed here show that, from a neutral position and charge-free state (this is important because IPMNCs are such sensitive and large-capacity charge capacitors, and most anomalies are because we do not initially discharge the sample), for any given voltage, the cantilever sample bends to a shape. If the voltage is removed, the shape will not change, provided the sample is prevented from being electrically discharged by completely isolating it from any electric discharging systems. Typical observations are depicted in figure 2.43 (a, b, c, d, and e), in which the sample is shorted out

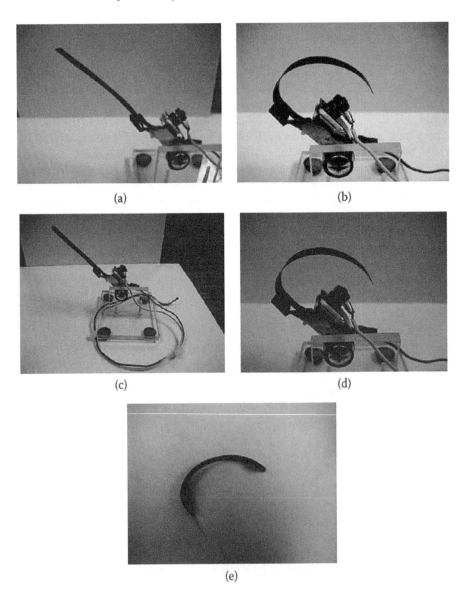

FIGURE 2.43 Sequence of events establishing electric shape memory effect. (a) Before the voltage of 4 V is applied to the sample of $10 \times 80 \times 0.34$ mm; (b) 4 sec after the voltage is applied; (c) 6 sec after the electrodes are shorted by connecting the lead wires together; (d) 4 sec later after the same voltage is applied again; (e) 10 min later after the muscle is completely detached from the electrodes with permanent deformation and laid on a table.

first and then reactivated and consequently completely detached from its fixtures and left on a table to show that the shape is memorized and is permanent.

In general, important observations are:

- The electric deformation memory effect is due to rearrangement of immobilized charges (double layer interactions) after the voltage is removed. Thus, in a way, the electric deformation shape memory effect is due to a kind of ionic solid phase transformation, which is capacitively induced.
- The direction of the IPMNC motion is always consistent with the direction of charges.
- The sample should be prevented from electrically discharging itself by completely isolating it from any electric discharging systems.

As far as making magnetic IPMNCs is concerned, the reader is referred to Park et al. (2006) to see how ferromagnetic metal salts can also be embedded within the IPMNCs molecular network and give rise to magnetically active IPMNCs. In connection with self-oscillatory phenomena in IPMNCs, see D. Kim and K. J. Kim (2005).

3 Ionic Polymer-Metal Nanocomposites (IPMCs and IPMNCs) Manufacturing Techniques

3.1 INTRODUCTION

This chapter describes the manufacturing procedures and technologies for ionic polymer-metal nanocomposites (IPMCs or IPMNCs). The performance of those IPMCs manufactured by different manufacturing techniques is described. Taguchi design of experiment methodologies (Peace, 1993) is used for optimization purposes. In particular, some issues, such as force optimization using the Taguchi design of experiment technique, effects of different cations on the electromechanical performance of IPMCs, electrode and particle size and distribution control, placement of near boundary surface electrodes, manufacturing cost minimization approaches by physical loading techniques, scaling and three-dimensional muscle production issues and heterogeneous composites, are also reviewed and compiled.

3.2 IPMNC BASE MATERIALS

3.2.1 In General

As we previously discussed in Chapter 2, the manufacturing of IPMC artificial muscles, soft actuators and sensors starts with *ion-exchange* (or "permeable" or "conducting") polymers (often called "ionomers"; Eisenberg and King, 1977; Eisenberg and Yeager, 1982). Ion-exchange materials are designed and synthesized to selectively pass the ions of a single or multiple charges – that is, cations or anions or both. Hence, depending upon the types of materials, particular passages of desirable cations or anions or both can be achieved; this means that ions are mobile within the polymeric materials. Ion-exchange materials are typically manufactured from organic polymers that contain covalently bonded fixed ionic groups (Eisenberg and Bailey, 1986).

Most popular ion-exchange materials used in industry are based upon copolymer of styrene and divinylbenzene, where the fixed ionic groups are formed after polymerization. An important parameter is the ratio of styrene to divinylbenzene, which controls the cross-linking process that limits ion-exchange capabilities and the water uptake. The known successful fabrication of the sulfonated ion-exchange membranes was based on grinding ion-exchange resins to microscale powders and mixing them with hydrophobic thermoplastic materials, including polyethylene polyvinylidene fluoride (PVDF), and sheeting them at elevated temperatures.

Recently, the inclusion of a hydrophobic polymer at the formulation stage is commonly used and is known as the paste method (Davis et al., 1997). A fine powder of hydrophobic polymer is mixed into paste with the liquid phase monomers in such a method. The initiator and a plasticizer typically control the final product. The introductory chemistry is shown in Figure 3.1, where the copolymerization is carried out with a predetermined ratio of styrene and divinylbenzene and, occasionally, with ethyl styrene. Also, the random intertwining of the polymer chains is illustrated

DOI: 10.1201/9781003015239-3

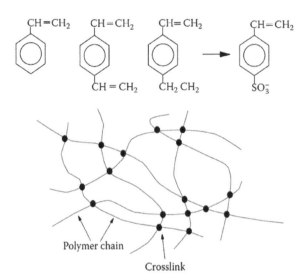

FIGURE 3.1   Styrene/divinylbenzene-based ion-exchange material (top) and its structural representation (bottom).

in a three-dimensional manner. The fixed anionic groups are typically sulfonated in nature. The effective sulfonation process can be carried out under highly concentrated sulfuric acid at elevated temperatures for a long time. As the reaction proceeds, swelling of materials and the evolution of heat will require precautions to prevent material weakening.

Although the chemical processes to make ion-exchange materials are well tailored, the remaining challenges are fabricating them into membrane format. Today, a popular method is to incorporate hydrophobic thermoplastic materials, such as PVDF or polyethylene, and press them out as a sheet form at elevated temperatures. Such a method can provide good mechanical strength and chemical stability. However, the drawback is to lower the electric conductivity of the membrane products. Heterogeneous fabrication techniques were also often incorporated to address such a problem.

Another exciting technique is to modify the membrane surfaces to improve ion selectivity by engineering them to distinguish monovalent and divalent cations for specific applications effectively. This application can be made by immobilizing the positively charged ionic groups at the surface to repel the doubly charged cations. As a result, only single charges can pass through the membrane. Large-scale manufacturers are Ionics Inc. in the United States and Tokuyama Corporation in Japan.

The bipolar membranes are also fabricated based upon the idea that a membrane has two layers side by side – cation conducting and anion conducting, respectively. A useful application of such a membrane is water splitting creating H^+ and OH^- under a small applied electric voltage. The fabrication of such a membrane is rather complicated.

Another popular ion-exchange material is perfluorinated alkenes with short side chains terminated by ionic groups (i.e., Nafion™ from DuPont, typically sulfonate or carboxylate [SO_3^- or COO^-] for cation exchange of ammonium ions for anion exchange). Such fluorocarbon polymers have linear backbones with no cross-linking and relatively few fixed ionic groups attached (Yeager, 1982). The large polymer backbones determine their mechanical strength. Short side chains provide ionic groups that interact with water and the passage of appropriate ions. When swollen by the water, Nafion undergoes phase separation ("clustering") (Gierke et al., 1982) on a supermolecular structure. When they are swollen, hydrophobic zones around the fluorocarbon backbones and hydrophilic zones around the fixed ionic groups coexist. Therefore, the ionic groups attract water and can move water under an electric voltage through nanoscale pores and channels where ions and water migrate within the polymer matrix.

One interesting feature is the cation dependency of water content in Nafion: in general, $H^+ > Li^+ > Na^+ > K^+ > Cs^+$ (Davis et al., 1997). This can also be interpreted that their moduli, in general, are H^+

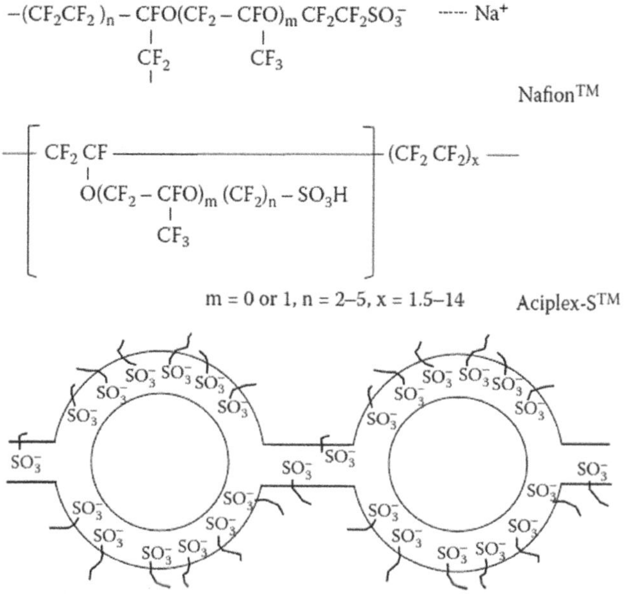

$$-(CF_2CF_2)_n - CFO(CF_2 - CFO)_m\,CF_2CF_2SO_3^- \quad \text{-----} \quad Na^+$$
$$\qquad\qquad\quad |\qquad\qquad |$$
$$\qquad\qquad\ CF_2\qquad\quad CF_3$$
$$\qquad\qquad\quad |$$

Nafion™

$$\left[\ \begin{array}{l} CF_2\,CF \text{————————} \\ \quad | \\ \quad O(CF_2 - CFO)_m\,(CF_2)_n - SO_3H \\ \qquad\qquad\ | \\ \qquad\qquad CF_3 \end{array}\ \right] (CF_2\,CF_2)_x \text{——}$$

m = 0 or 1, n = 2–5, x = 1.5–14 · · · · Aciplex-S™

FIGURE 3.2 A famous chemical structure of two perfluorinated ion-exchange materials (other cations can replace Na+) and their structural illustration (bottom). (From Davis, T. A. et al., 1997. *A First Course in Ion Permeable Membranes.* Electrochemical Consultancy; Gierke, T. D. et al., 1982. *ACS Symp. Ser.* 180:195–216; and Asahi Chemical Industry Co., Ltd.1995. Aciplex formula. Public release.)

< Li+ < Na+ < K+ < Cs+. Their popular chemical structure and properties are provided in Figure 3.2 and Table 3.1. A similar product, Aciplex-S™, from Asahi Chemical Industry Co., Ltd, of Japan, is also a perfluorinated cation-exchange membrane with a sulfuric acid functional group, and its chemical structure and properties are also included in Figure 3.2 and Table 3.1.

TABLE 3.1

Representative Properties of a Perfluorinated Ion-Exchange Material Nafion™-117 (H+ Form) and Aciplex-S™

	Nafion
Equivalent weight	1100 g/gmol SO_3^-
Thickness	200 μm
Conductivity	0.1–0.12 S/cm
Area resistance	5 Ω cm^2
Water uptake capacity	Up to 30% at room temperature
Volume expansion on hydration	12–15%
	Aciplex-S
Equivalent weight	950 g/gmol SO_3^-
Thickness	300 μm
Conductivity	0.1–0.12 S/cm
Area resistance	0.09–0.66 Ω cm^2
Water uptake capacity	Up to 20% at room temperature
Volume expansion on hydration	~15%

Sources: Kolde et al. (1995), Asahi Chemical Industry Co., Ltd. (1995). Aciplex formula. Public release.

Although there are several commercial ion-exchange material manufacturers (Davis et al., 1997), including Aqualitics, Asahi Chemicals, Asahi Glass, Du Pont, W. L. Gore, Ionics, Solvay, Sybron, and Tokuyama, popular products used as IPMC materials are Nafion from DuPont, Neosepta™ from Tokuyama, Aciplex from Asahi Chemical and Flemion™ or Selemion™ from Asahi Glass. At present, all these products perform reasonably well when the IPMNC chemical-plating technique properly treats them.

Since these materials exhibit the ion transport intrinsically, a vital process is electric charge transport through ions' materials. Therefore, it can be explained in terms of an ionic flux, Π_i,

$$\Pi_i = \frac{J_i}{z_i F} = \frac{E_D}{z_i F R_A} \tag{3.1}$$

Where

J_i is the current density (I/A)
z_i is the charge on the transported ion
F is the Faraday constant (96,485 C/mol)
E_D and R_A are the electric field across the material and the area resistance of the material (Ωm^2), respectively

The conductance of the material k_i ($\Omega^{-1}m^{-1}$) is also conveniently used as $R_A = L/k_i$ (where L = material thickness).

Another interesting property of Nafion can be recognized in terms of its functional groups: SO_3^- and COO^-. Table 3.2 compares the properties of the sulfonate and carboxylate Nafion typically used in the industry. Figure 3.3 depicts a general illustration of fixed ionic groups, counterions and co-ions for a typical cation-exchange polymer. In this cation-exchange polymer, the fixed ionic groups refer to the ion-exchange groups covalently bonded to the polymer molecular network. Typical cation-exchange polymers have carboxylate or sulfonate groups. Counterions are conjugated charges to the fixed ionic groups. For charge balance, counterions must be present within the polymer. They migrate within the polymer network under an imposed electric field. There also exist co-ions such as H^+ with the same charge as the fixed ionic groups, depending upon the polymers' pretreatment. Co-ions arise from the presence of salt (or salts) within the polymer.

Let us first emphasize the key role that water plays within the ionic polymer that determines the ion-exchange polymers' physical chemistry. When the polymer is dry, strong interactions between the counterions and the polymer's fixed ionic groups are dominant. As a result, the ion-exchange polymer exhibits low conductivity.

Once the ion-exchange polymer is swollen (or wet), it solvates the counterions and the fixed ionic groups such that it lowers the interactions between cations. Under these circumstances, the material conductivity dramatically increases. Note that routinely measured mechanical properties (i.e., thickness, tensile stress, burst strength and hydrostatic permeability) are important for inspecting ion-exchange polymers. Important parameters to affect these properties include the pretreatment of the polymers, the type of electrolyte solutions and the temperature.

TABLE 3.2

Properties of Sulfonate and Carboxylate Ion-Exchange Material Used in Industry

	pK_A	Water Content	Current Density	Electric Conductivity	Chemical Stability
SO_3^-	<1	High	High	High	Good
COO^-	3–4	Low	Low	Low	Good

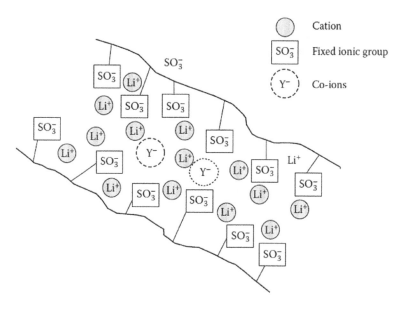

FIGURE 3.3 Illustration of the cation polymer showing fixed ionic groups (SO_3^-), counterions (Li^+) and co-ions (Y^-), respectively.

3.2.2 Water Structure within the IPMNC Base Materials

In this section, the critical role of the water content within the IPMC base materials is emphasized to determine their properties and their polymer structures. Also, the relationship between the electrolyte solutions in connection with the polymer properties is discussed. In Table 3.3, several important ion-exchange polymers that have been used as the base material for the IPMC artificial muscles are presented along with their characteristics (Oguro et al., 1993; Shahinpoor, 1993; Shahinpoor et al., 2002). In this table, actuation performance testing was performed with an IPMNC strip in a cantilever configuration while applying the electric field at the built-in end (Figure 3.4).

As depicted in Figure 3.4, an experiment was set to measure the blocking force per a given electric field. A PC-based test platform was used for actuation tests of all samples.

Note that the signal generation system utilized LabView's *Function Generator* in its virtual instrument. The output voltage from the power amplifier was supplied to the test sample for actuation. The test samples were IPMCs with platinum, and some test samples also had an additional surface coating with gold.

A digital oscilloscope was simultaneously used to monitor/store the input and output waveforms. The IPMC test sample was attached at one end to the load cell while the other end was placed at the contact platinum electrode, which typically formed the jaws of a forceps. The blocking force was

TABLE 3.3

Important Ionic Polymers Used as IPMNCs

	Water Content	H$_2$O/Fixed Charge	Cation Type	Polymer Type	Actuation Performance
Nafion-117	0.14	~12	Li$^+$	Perfluorinated sulfonate	>2 gf$_{max}$-cm/1 V
Neosepta C66	0.42	~9	Li$^+$	Polystyrene sulfonate	>1 gf$_{max}$-cm/1 V
Ionics CR-67	0.46	~12	Li$^+$	Polystyrene sulfonate/an acrylic fabric	>0.5 gf$_{max}$-cm/1 V
ERI-S1	0.17	~11	Li$^+$	Perfluorinated sulfonate	>2.5 gf$_{max}$-cm/1 V

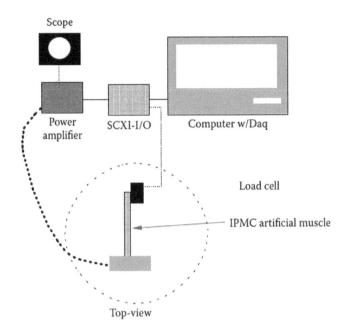

FIGURE 3.4 Experimental setup used.

measured at zero displacements. The IPMC test samples were positioned horizontally to eliminate the gravitational effects and influence.

The adsorption isotherm of Nafion (Figure 3.5) could be an effective means to understand the ionic polymer's water structure. During the absorption of the first four to five water molecules per an active site (i.e., SO_3^-), the absorption enthalpy seems constant. With a further increase in the water content, the absorption enthalpy decreases and reaches the saturated limit (Figure 3.6).

This fact could be interpreted as initial water interaction with the counterions (hydration only) and further rearrangement and expansion of the polymer structure during the expansion of microporous channels within the ionic polymer. The enthalpy reaches approximately −13 kJ/mol of water (H^+ form), much less than condensation (approximately −53 kJ/mol of water). This result is a clear

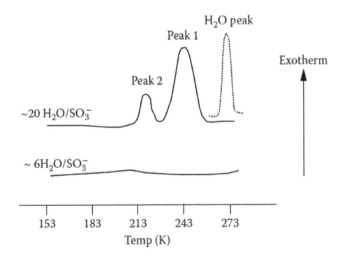

FIGURE 3.5 Absorption isotherms of Nafion™.

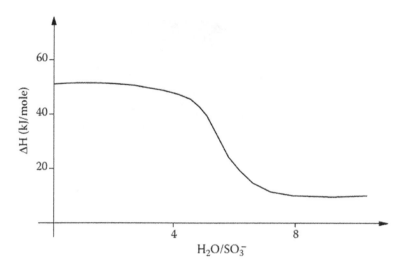

FIGURE 3.6 Enthalpy of H_2O absorption as a function of the water content (Nafion). (Escoubes, M., and M. Pineri. 1982. In *Perfluorinated Ionomer Membranes*, eds. A. Eisenberg and H. L. Yeager. Washington, DC: American Chemical Society.)

indication of an endothermic contribution arising from rearranging and expanding the polymer structure. It has been established that Nafion can have three different types of water (Escoubes and Pineri, 1982; Yoshida and Miura, 1992; Davis et al., 1997):

Type 1: water closely bound to the ions
Type 2: water weakly bound to the ions (or the polymer – possibly ether oxygens)
Type 3: free water

According to the differential scanning calorimetry (DSC) studies, the presence of several types of water structures within the ionic polymer is more plausible (Escoubes and Pineri, 1982; Yoshida and Miura, 1992; Davis et al., 1997). The heat absorbed (desorbed) by a sample as a function of temperature is monitored. Figure 3.7 illustrates the DSC responses for Nafion-117 (Li⁺ form) for

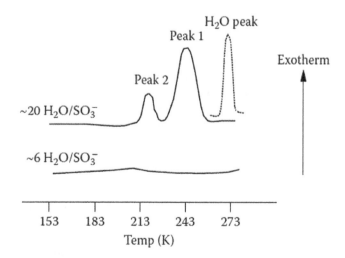

FIGURE 3.7 DSC data for cooling of Nafion (Li⁺ form). (Yoshida, H., and Y. Miura. 1992. *J. Membrane Sci.* 68:1–10.)

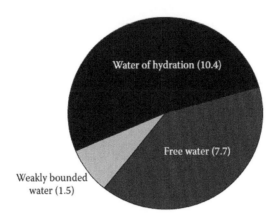

FIGURE 3.8 Number of water molecules per an ionic group in Nafion (Li+ form). The total number of water molecules is 19.6. (Davis, T. A. et al., 1997. *A First Course in Ion Permeable Membranes.* Electrochemical Consultancy.)

cooling. At the high water content, a sharp peak (peak 1) corresponds to the water freezing at around 240K. In other words, some water freezes at this temperature, 30–40°C lower than the normal water-freezing temperature (type 3). Also, it shows another peak that attributes to the freezing of more ordered (freezing bound) water, which appears to be weakly bonded to ions (type 2). Note that such temperatures are much lower than that of the normal water freezing. However, more importantly, this DSC experiment overlooks another type of water present in Nafion (type 1) that does not freeze and is water in the cations' hydration shell (or anions). The amount of these three different types of water for Nafion is given in Figure 3.8. One may note that the greater hydration expands the lattice to bring more water into the network.

For styrene/divinylbenzene type polymers, all the water is bound to ions, and therefore no phase changes associated with freezing can be realized. Thus, the styrene/divinylbenzene-type polymers have more ions and are considered continuous structures and closer ion spacing. In DSC data, usually, all the peaks are small and broad.

3.3   MANUFACTURING TECHNIQUES

3.3.1   In General

The current state-of-the-art IPMC manufacturing technique (Asaka et al., 1995; Shahinpoor and Mojarrad, 2000; Shahinpoor and Kim, 2001g; Kim and Shahinpoor, 2003a) incorporates two distinct preparation processes: *initial nanocomposite making* and *placement of surface electrodes*. Due to different preparation processes, the morphologies of precipitated platinum are significantly different. Figure 3.9 shows illustrative schematics of two different preparation processes (top left and bottom left) and two top-view scanning electron micrographs (SEM) micrographs for the platinum surface electrode (top right and bottom right).

The initial compositing process requires a suitable platinum salt such as $Pt(NH_3)_4HCl$ in the context of chemical reduction processes similar to the processes evaluated by several investigators, including Takenaka et al. (1982) and Millet et al. (1989). The IPMC nanocomposite manufacturing principle is to metalize the material's inner near boundary surface (usually, in a membrane shape, Pt nanoparticles) by a chemical-reduction means such as $LiBH_4$ or $NaBH_4$. The ion-exchange polymer is soaked in a salt solution to allow platinum-containing cations to diffuse via the ion-exchange process. Later, a proper reducing agent such as $LiBH_4$ or $NaBH_4$ is introduced to platinize the materials by molecular plating.

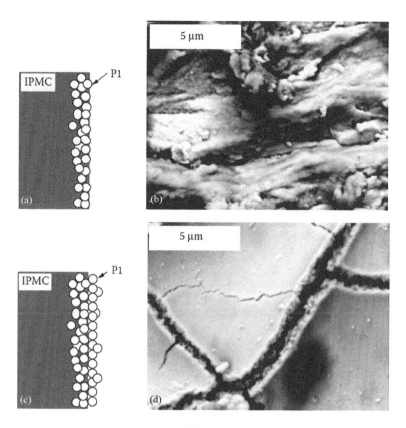

FIGURE 3.9 Two schematic diagrams showing different preparation processes. (a) A schematic representation shows the initial compositing process; (b) its top-view SEM micrograph. (c) A schematic representation shows surface electrode placement process; (d) the top-view SEM micrograph shows where platinum is deposited predominantly on top of the initial Pt layer.

As shown in Figures 3.10 and 3.11, the metallic platinum particles are not homogeneously formed across the material but concentrate predominantly near the interface boundaries. It has been experimentally observed that the platinum particulate layer is buried a few microns deep (typically 5–20 μm) within the IPMNC surface and is highly dispersed. A TEM image of the near-boundary region of an IPMNC strip on the penetrating edge of the IPMC shows a functional particle density gradient

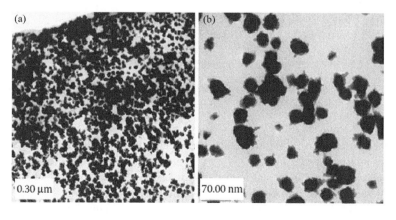

FIGURE 3.10 TEM micrographs show a cross-section of an (a) IPMNC and (b) its close-up.

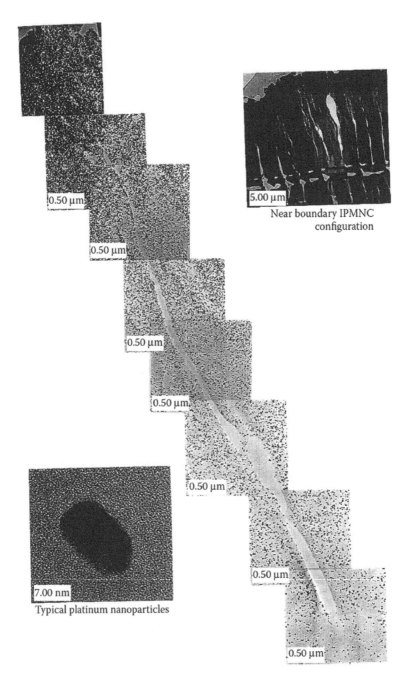

0.50 µm

5.00 µm

Near boundary IPMNC
configuration

0.50 µm

0.50 µm

0.50 µm

0.50 µm

7.00 nm

Typical platinum nanoparticles

0.50 µm

0.50 µm

FIGURE 3.11   TEM successive micrographs show the platinum particles reduced within the polymer matrix and typical nanoparticles of platinum.

where the higher particle density is toward the surface electrode. The range of average particle sizes was found to be around 40–60 nm.

Note in Figure 3.10 that these platinum particles reside within the polymer matrix. The top micrograph shows the higher particle density toward the surface electrode.

The fabricated IPMNCs can be optimized to produce a maximum force density by changing multiple process parameters. These parameters include time-dependent salt concentrations and the reducing agents (applying the Taguchi technique to identify the optimum process parameters that seem quite attractive) (Peace, 1993). The primary reaction is

$$LiBH_4 + 4\left[Pt(NH_3)_4\right]^{2+} + 8OH^- \Rightarrow HPt^o + 16NH_3 + LiBO_2 + 6H_2O \qquad (3.2)$$

In the subsequent process of chemically placing surface electrodes on and into the ionic polymeric substrate, multiple reducing agents are introduced (under optimized concentrations) to carry out the reducing reaction similar to Equation (3.2), in addition to the initial platinum layer formed by the initial compositing process. This nanocomposite-making process is clearly shown in Figure 3.9 (bottom right), where the roughened surface disappears. In general, most platinum salts stay in the solution and precede the reducing reactions and production of platinum metal. Other metals (or conducting media) that are also successfully used include palladium, silver, gold, carbon, graphite and nanotubes. Figure 3.12 includes several different IPMNC surfaces treated with different techniques.

3.3.2 IPMNC AND IPCNC MANUFACTURING RECIPE

An effective recipe to manufacture IPMNC materials (for a standard size of 5×5 cm^2 of Nafion-117) is described herein (Takenaka et al., 1982; Millet et al., 1989; Oguro et al., 1993, 1999; Asaka et al., 1995; Shahinpoor, 1995 (U.S. Patent); Onishi et al., 2000; Shahinpoor and Mojarrad, 2000, 2002). Essentially, the process is chemical plating of the ionomer by an oxidation-reduction (REDOX) procedure and utilizing a noble metal salt such as tetraamine platinum chloride hydrate [Pt(NH$_3$)$_4$] Cl$_2$ + H$_2$O, tetraamine palladium chloride hydrate [Pd(NH$_3$)$_4$]Cl$_2$ + H$_2$O or phenolic gold chloride [AuCl$_2$(Phen)]Cl.

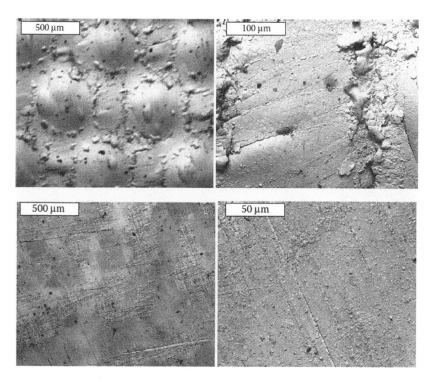

FIGURE 3.12 Various surface micrographs of IPMNCs use different chemical plating techniques and placing electrodes on the boundary surfaces.

The first step is to roughen the material surface, where it will serve as an effective electrode. The essential steps include sandblasting or sandpapering the polymer's surface to increase the surface area density where platinum salt penetration and reduction occurs, as well as ultrasonic cleaning and chemical cleaning by acid boiling (HCl or HNO_3-low concentrates).

The second step is to incorporate the ion-exchanging process using a metal complex solution such as tetraamine platinum chloride hydrate as an aqueous platinum complex ($[Pt(NH_3)_4]Cl_2$ or $[Pt(NH_3)_6]Cl_4$) solution. Although the equilibrium condition depends on the metal complex types, such complexes were found to provide good electrodes. The immersion and stirring time is typically more than 1 h.

The third step (initial platinum compositing process) is to reduce the platinum complex cations to a metallic state in the form of nanoparticles by using effective reducing agents such as an aqueous solution of sodium or lithium borohydride (5%) at favorable temperature (i.e., 60°C). Platinum black-like layers deposit near the surface of the material.

The final step (surface electrode placement process) is intended to effectively grow Pt (or other novel metals, a few microns of thickness) on top of the initial Pt surface to reduce the surface resistivity. Therefore, an additional amount of platinum is plated by the following process on the deposited Pt layer:

1. Prepare a 240-ml aqueous solution of the complex ($[Pt(NH_3)_4]Cl_2$) OR $[Pt(NH_3)_6]Cl_4$) containing 120 mg of Pt and add 5 ml of the 5% ammonium hydroxide solution (pH adjustment).
2. Plating amount is determined by the content of Pt in the solution. Prepare a 5% aqueous solution of hydroxylamine hydrochloride and a 20% solution of hydrazine monohydrate.
3. Place the polymer in the stirring Pt solution at 40°C. Add 6 ml of the hydroxylamine hydrochloride solution and 3 ml of the hydrazine solution every 30 min. In the sequence of addition, raise the temperature to 60°C gradually for 4 h. Note that gray metallic layers will form on the membrane surface. At the end of this process, sample a small amount of the solution and boil it with the strong reducing agent ($NaBH_4$) to check the endpoint.

Other metals (or conducting media) that are also successfully used include palladium, silver, gold, carbon, graphite and carbon nanotubes (see Figure 3.13). Also, it is noteworthy that electroplating was found to be very convenient (Kim and Shahinpoor, 2003).

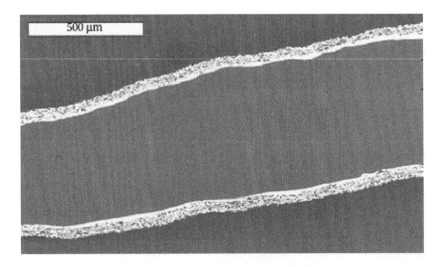

FIGURE 3.13 A cross-section of an IPMNC shows two electrodes (top and bottom) with porous expanded graphite and dense platinum. This IPMNC is manufactured by the solution casting and further treatment with porous graphite and chemical reduction of platinum as effective, compliant electrodes.

3.3.3 IPMNC Force Optimization

The fabricated IPMNCs can be optimized for producing a maximum force density by changing multiple process parameters, including bath temperature (T_R), time-dependent concentrations of the metal-containing salt, $C_s(t)$, and the reducing agents, $C_R(t)$. The Taguchi design of experiment technique was conducted to identify the optimum process parameters (Peace, 1993). The analysis techniques for "larger-the-better" quality characteristics incorporate noise factors into an experiment involving "larger-the-better" characteristics, for the maximum force generated by the manufactured IPMNCs in this case. Such an analysis allows us to determine the key factors and the possible best settings for consistently good performance. The beauty of this technique is just as applicable for attaining consistently high "larger-the-better" responses for process performance and end product functionality (Peace). In Table 3.4, experimental design and signal-to-noise (S/N) ratio data are provided.

TABLE 3.4
Experimental Design and Signal-to-Ratio Data

Run No.	A	B	C	D	E	F	G	H	I	J	K	L	M	S/N (dB)
1	1	1	1	1	1	1	1	1	1	1	1	1	1	46.75
2	1	1	1	1	2	2	2	2	2	2	2	2	2	33.85
3	1	1	1	1	3	3	3	3	3	3	3	3	3	52.90
4	1	2	2	2	1	1	1	2	2	2	3	3	3	46.08
5	1	2	2	2	2	2	2	3	3	3	1	1	1	43.58
6	1	2	2	2	3	3	3	1	1	1	2	2	2	46.12
7	1	3	3	3	1	1	1	3	3	3	2	2	2	45.35
8	1	3	3	3	2	2	2	1	1	1	3	3	3	39.38
9	1	3	3	3	3	3	3	2	2	2	1	1	1	37.01
10	2	1	2	3	1	2	3	1	2	3	1	2	3	41.42
11	2	1	2	3	2	3	1	2	3	1	2	3	1	43.60
12	2	1	2	3	3	1	2	3	1	2	3	1	2	33.57
13	2	2	3	1	1	2	3	2	3	1	3	1	2	45.05
14	2	2	3	1	2	3	1	3	1	2	1	2	3	46.12
15	2	2	3	1	3	1	2	1	2	3	2	3	1	46.93
16	2	3	1	2	1	2	3	3	1	2	2	3	1	45.02
17	2	3	1	2	2	3	1	1	2	3	3	1	2	43.78
18	2	3	1	2	3	1	2	2	3	1	1	2	3	44.26
19	3	1	3	2	1	3	2	1	3	2	1	3	2	42.77
20	3	1	3	2	2	1	3	2	1	3	2	1	3	44.78
21	3	1	3	2	3	2	1	3	2	1	3	2	1	37.69
22	3	2	1	3	1	3	2	2	1	3	3	2	1	37.94
23	3	2	1	3	2	1	3	3	2	1	1	3	2	50.14
24	3	2	1	3	3	2	1	1	3	2	2	1	3	35.00
25	3	3	2	1	1	3	2	3	2	1	2	1	3	39.75
26	3	3	2	1	2	1	3	1	3	2	3	2	1	35.93
27	3	3	2	1	3	2	1	2	1	3	1	3	2	42.27

Note: Samples 28 and 29 are prepared at levels of 2 and 3 for all factors listed.

TABLE 3.5
Signal-to-Noise (S/N) Response

	A	B	C	D	E	F	G	H	I	J	K	L	M
Level 1	43.45	42.59	43.29	43.28	44.01	43.75	42.96	42.67	42.44	43.64[a]	44.48[a]	41.03	41.61
Level 2	43.97[a]	44.11	42.04	43.79	42.35	41.03	40.23	41.65	42.51	39.48	42.26	41.63	42.54
Level 3	40.7	41.42	42.79	41.05	41.75	43.33	44.93[a]	43.79	43.16	43.16	41.37	45.54[a]	43.97
Delta	3.27[a]	2.69	1.25	2.74	2.37	2.72	4.7[a]	2.14	0.72	4.16[a]	3.11[a]	4.51[a]	2.36

[a] Maximum and minimum values at each level.

Tables 3.5 and 3.6 show the S/N response and factors engaged in our experiment. Based on the S/N computed for each factor and level, S/N response graphs are constructed and presented in Figures 3.14–3.16. From our analysis, the strong effects and elements to be considered are listed in Table 3.7. In Figures 3.17 and 29, manufactured samples were compared based on the Taguchi method (two added, sample 28 and sample 29, all level 2 and level 3, respectively) to establish that the force was compiled against the baseline condition (sample 28). The results allow one to optimize such a complicated bath process to produce much larger blocking forces that can be utilized for practical actuation applications.

TABLE 3.6
Factors Engaged in the Experiments

	Factors Engaged in our Experiment	Procedure
A	Surface roughening	Pretreatment
B	Boiling time in the water	Pretreatment
C	Platinum salt concentration	Ion exchange
D	Stirring time (Pt soaking)	Ion exchange
E	Reducing agent concentration (LiBH$_4$) – first	First metal reduction
F	Reducing bath temperature – first	First metal reduction
G	Reducing/stirring time – first	First metal reduction
H	Platinum salt concentration	Ion exchange
I	Stirring time (Pt soaking)	Ion exchange
J	Reducing agent concentration (LiBH$_4$) – second	Second metal reduction
K	Reducing bath temperature – second	Second metal reduction
L	Reducing/stirring time – second	Second metal reduction
M	LiOH concentration	Postprocess/ion exchange

TABLE 3.7
Strong Effects and Elements

First tier	(G) Reducing/stirring time in the first reduction process
	(J) Reducing agent concentration in the second reduction process
	(L) Reducing/stirring time in the second reduction process
Second tier	(A) Surface roughening
	(K) Reducing bath temperature in the second reduction process

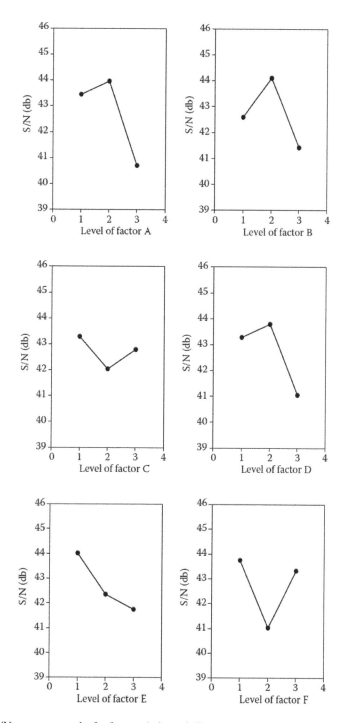

FIGURE 3.14 S/N response graphs for factors A through F.

Another interesting point is the relationship between the measured surface resistance and platinum particle penetration within the polymer matrix. In Figure 3.18, the measured surface resistances of 29 samples are plotted against platinum penetration. The platinum penetration is measured by SEM cross-sectioning, and Guardian Manufacturing's model SRM-232 measured four-point surface resistivities (sheet resistivity) of the samples. The surface resistivity of the samples appears to be linear despite scatters.

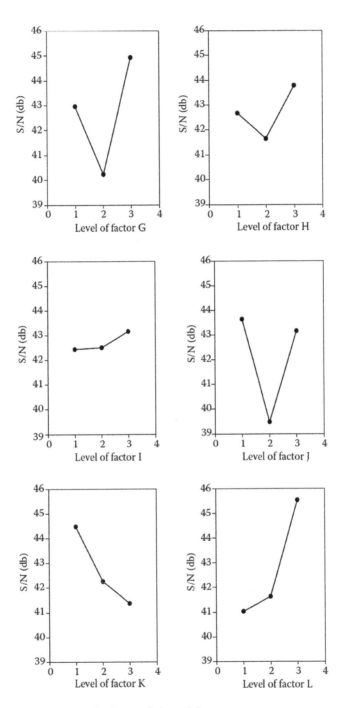

FIGURE 3.15 S/N response graphs for factors G through L.

Overall, it should be noted that samples with low surface resistivities tend to produce larger blocking forces.

An Instron 1011 table-top machine performed the tensile testing of the samples. The standard sample size was 9×55 mm^2. The strain rate was set at roughly 2.33 (s^{-1}). In Figure 3.19(a) and (b), tensile testing results are provided in terms of normal stresses, σ_N, and normal strain, ε_N, for several samples 1, 2, 3, 28 and 29, and Nafion-117 (dry and wet states).

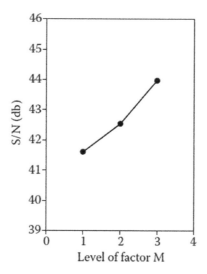

FIGURE 3.16 S/N response graph for factor M.

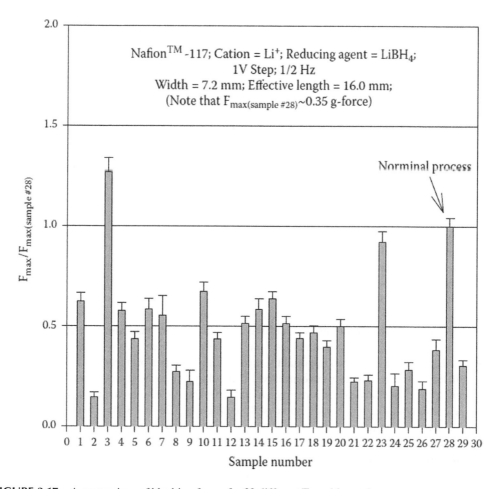

FIGURE 3.17 A comparison of blocking forces for 29 different Taguchi samples.

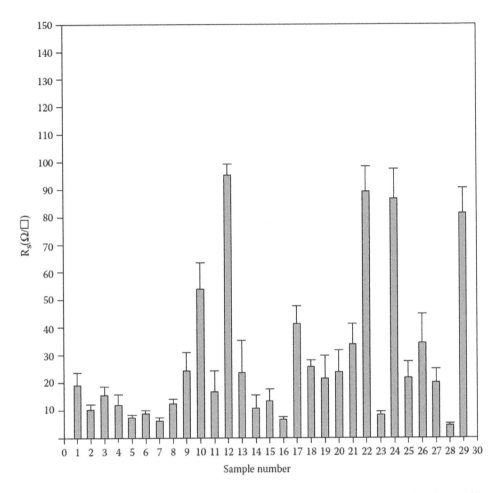

FIGURE 3.18 Four-point probe surface resistivity measurement on Taguchi samples. (Shahinpoor, M., and K. J. Kim.2001g. *Smart Mater. Struct. Int. J.* 10:819–833.)

All Taguchi IPMNC samples (these are wet samples) show Young's modulus of approximately 50–100 MPa at a normal strain of 0.02 (at 2%), similar to that of wet Nafion-117. The dry Nafion-117 (H⁺ form) is stiffer than that of all Taguchi samples, as expected. The approximate value of the modulus for dry Nafion-117 is 220–260 MPa at a normal strain of 0.02 (2%). The tensile strength of all Taguchi samples is around 10–13 MPa. The IPMNC appears to exhibit a little more stiffness and less yielding than Nafion-117 due primarily to the composited metal portion that interacts with the polymer matrix. The electric responses of the Taguchi samples were also investigated and are presented here in terms of the consumed amount of charges (current, I, vs. time) at applied step voltages of 1, 2, 3 and 4 V for two important samples: 3 (the best performance) and 28 (baseline).

Overall, the current responses were capacitive. This behavior may be attributed to the double layers adjacent to platinum particles or proximate metallic layers. As we increase the applied voltages, the charge consumption is significantly increased (see Figures 3.20 and 3.21). This increase means the IPMNC materials utilize charges but also store them within the materials. Further study is to investigate such an effect.

Note in Figure 3.19(a) and (b) that the actual normal stress, σ_N, and normal strain, ε_N, stress-strain curves for IPMNCs lie somewhere between the normal stress, σ_N, and normal strain ε_N, stress-strain curves for the wet and the dry Nafion ionomer samples.

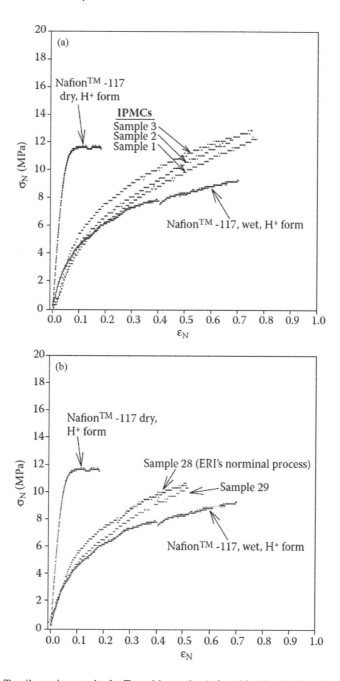

FIGURE 3.19 (a) Tensile testing results for Taguchi samples 1, 2 and 3. Also, Nafion-117, H+ form, both dry and wet, is shown. (b) Tensile testing results for Taguchi samples 28 and 29. Also, Nafion-117, H+ form, both dry and wet, is shown.

3.3.4 Effects of Different Cations

Realizing that IPMNC base material (polymer) properties differ depending upon different cations, IPMNC test samples were prepared by the processes described in the previous section with different cations. First, a batch of the IPMNC artificial muscle of 10×20 cm^2 dimensions was prepared with platinum. Then, nine samples were cut in a standard size of a 2×10-cm^2 strip for the ion-exchange

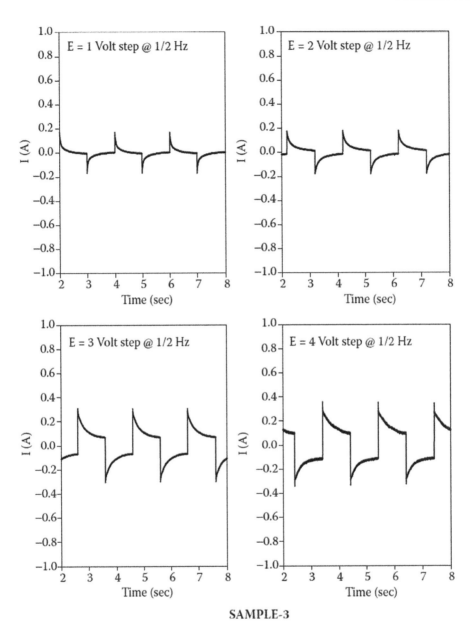

SAMPLE-3

FIGURE 3.20 Current responses to step voltages of 1, 2, 3 and 4 V for Taguchi sample 3.

process. Each IPMNC sample was treated to contain nine different counterions (Na+, Li+, K+, H+, Ca++, Mg++, Ba++, $R_nNH_4^+{}_{4-n}$ [tetrabuthylammonium (TBA)] and tetramethylammonium (TMA)], respectively) by soaking it in an appropriate salt solution (1.5 N of NaCl, LiCl, KCl, HCl, CaCl$_2$, MgCl$_2$, BaCl$_2$, CH$_3$(CH$_2$)$_3$NBr (TBA) and (CH$_3$)$_4$NBr (TMA), respectively) at moderate temperatures of 30°C for three days. All chemicals were obtained from Aldrich Chemical and used without further treatment. Test conditions were waveform = sinusoidal, E_{app} = 1.2 V$_{rms}$ and frequency of 0.5 Hz.

Figure 3.22 summarizes the test results in terms of the maximum force generated (= blocking force) by each IPMNC sample (at zero displacements) containing various cations, under a given voltage of 1.2 V$_{rms}$ relative to the Na+-containing IPMNC artificial muscle. A total of 13 measurements per each sample were taken. The error bars represent their corresponding standard deviations in Figure 3.22.

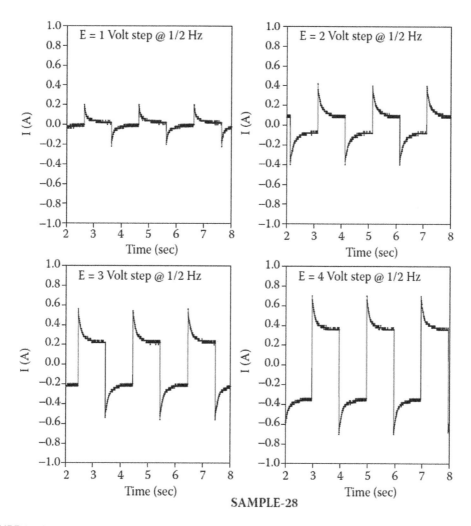

SAMPLE-28

FIGURE 3.21 Current responses to step voltages of 1, 2, 3 and 4 V for Taguchi sample 28.

As observed, the Li+-containing IPMNC is superior to the others, meaning that hydration processes with respect to mobile cations play a significant role in actuation behavior. Also, the samples with TBA and TMA show much smaller force generation capability. Knowing that such hydrophobic cations have a relatively large size and negligible hydration, the true transfer mechanism is close to electrophoretic movement without water. However, the maximum force generated at the tip of the cantilever under a given voltage of 1.2 V_{rms} is in the general order of the ones generated with counterions: Li+ ons:+ > (K+, Ca++, Mg++, and Ba++) > (H+, TBA, and TMA). This fact was observed for Nafion- and Flemion-based IPMNCs (Onishi et al., 2000).

In the set of Li+, Na+, and K+, one should realize that Li+ is undoubtedly the smallest bare ion in the set (its radius, $r[Li^+] < r[Na^+] < rK^+]$) but has the lowest mobility (the drag to its motion through the solution, $u[Li^+] < u[Na^+] < u[(K^+])$) (Moor, 1972; Atkins, 1982; Komoroski et al., 1982; Gebel et al., 1987; Moor et al., 1992).

All these observations could be due to the hydration phenomena (a tightly held sheath of water molecules bound under the electric field caused by cations/anions). Therefore, large organic hydrophobic ions such as TBA and TMA generate less force than Li+- and Na+-based IPMNCs. An important aspect regarding the use of such alkylammonium ions could be attributed to their large and bulky size relative to small cations investigated.

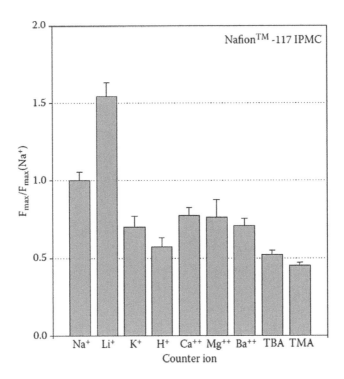

FIGURE 3.22 Effects of various cations on the actuation of the IPMNC muscle. Comparisons were made against Na^+ in terms of maximum force generated at zero displacements. A sinusoidal input voltage of $1.2\ V_{rms}$ and a frequency of 0.5 Hz were set for all experiments. The samples were cut in a standard size of $0.675 \times 2.54\ cm^2$. Na^+ was chosen as a reference since it is coordinated with four water molecules.

Overall, the hydration process within the membrane in connection with the electrophoretic effect is fairly complex in the sense that the mobile cations experience a large viscous drag and, at the same time, exert force due to the generated strain while they are moving through the water-containing polymer network. The situation can be interpreted that each cation with its connected clusters (Komoroski et al., 1982) is shearing or rubbing past other cations and networks. This process could increase the viscous drag and lower the conductivity.

Overall, considering the hydrated volume of each cation, $v[Li^+]$ erall,$^+] > (v[K^+], 0.5v[Ca^{++}]$, $0.5v[Mg^{++}]$ and $0.5v[Ba^{++}])$, and $v[H^+]$ (Moor, 1972; Atkins, 1982), one can see a general trend of force generation: Li^+, on$^+ > (K^+, Ca^{++}, Mg^{++},$ and $Ba^{++}) > H^+$ and (TBA and TMA). Another interesting note is that the relaxation behavior of IPMNCs is less observable in TBA and TMA.

Another issue is water uptake, depending upon the cation type (Komoroski and Mauritz, 1982). The type of ion-exchange materials used in this study, perfluorinated sulfonates, is presumably not cross-linked (Eisenberg and Yeager, 1982) (or partially cross-linked). Therefore, water uptake (swelling) is expected to be controlled by mobile cations and pretreatment. The ion-cluster phenomenon related to exchange sites, counterions and solvent phase as an independent phase (note that cross-linked polymers cannot form such a phase) could be another important factor in determining water uptake (Gierke and Hsu, 1982).

To clarify the importance of hydration processes, the sample's input power consumption containing different types of cations was measured. The test results are presented in Figure 3.23. It has the form of

$$\frac{F_{max}/P_{in}}{F_{max(Na^+)}/P_{in(Na^+)}} \tag{3.3}$$

This dimensionless form can be interpreted as the specific force generated per given applied input power.

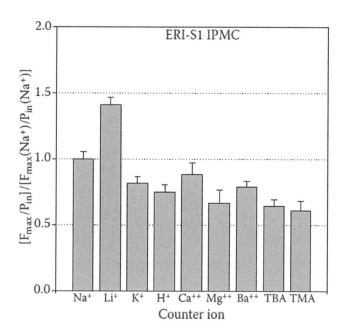

FIGURE 3.23 The specific force generated per an applied input power by the ERI-S1 IPMNC.

The observed magnitude of the specific force generated is still in the same range as $Li^+ \gg Na^+ >$ (K^+, Ca^{++}, Mg^{++}, Ba^{++}, H^+, TBA and TMA). In other words, the efficiency of the IPMNC with Li^+ is at least 40% higher than those of the other types of cations.

In Figures 3.24 and 3.25, relevant data are provided for force characteristics of the IPMNC and the dynamic force behavior under a constant voltage (Nafion-117-based IPMNC), respectively. Figure 3.24 reports the measured force under a step voltage of 1.2 V, clearly showing nearly no decay in the force generated for 2 min with proper electrodes. Figure 3.25 shows that the IPMNC artificial muscle responds very closely to the input sinusoidal, triangular and sawtooth waveforms, and those force responses are nearly equal in the traveling direction. However, in the case of square wave input, a very short delay of force is observed. Although it could be due to the capacitor charging process (Bar-Cohen et al., 1999; Kim and Shahinpoor, 1999), we believe that the water leakage through the porous electrode also contributes to such delayed responses (see Figure 3.24, bottom) (Moor, 1972).

3.3.5 ELECTRODE PARTICLE CONTROL

Based on our manufacturing experience, controlling the electrode consisting of platinum particles (concentration and size distribution) within the polymer matrix was also a nontrivial issue that can directly affect the performance of IPMCs. Here, we discuss the importance of a repetitive platinum reduction process the use of dispersing agents base material stretching (i.e., uniaxial mechanical stretching).

First, noting that platinum reduction is carried out in a wet batch system, repetitive platinum reduction steps (Asaka et al., 1995; Onishi et al., 2000) were found to be effective in producing IPMNCs exhibiting larger force characteristics. We believe that increased platinum concentration within the polymer leads to a higher double-layer charge capacity. Figure 3.26(a) and (b) presents the electric power consumption and current/voltage (I/V) behavior under a sinusoidal wave input to an IPMNC sample size of 12×51 mm^2 at different manufacturing stages. The following can be seen:

Bare Nafion has an insignificant current response under an applied voltage.
Repetitive platinum reduction improves the current responses.
Surface electroplating by gold further improves the current responses.

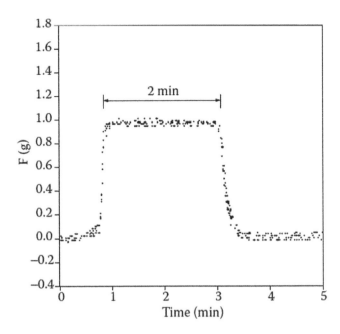

FIGURE 3.24 Force generated by an IPMNC with respect to time in minutes. The measured force under a step voltage of 1.2 V clearly shows nearly no decay in the force generated over 2 min. The dimension of the IPMNC sample is 0.25×0.75 in.^{2}. In this case, the surface electrode is effectively made to block water leakage (organic ions were incorporated as well).

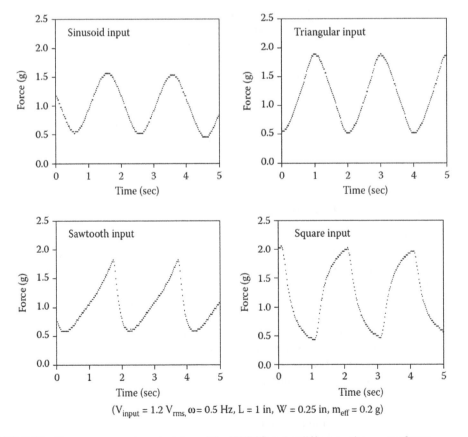

$(V_{input} = 1.2\ V_{rms,}\ \omega = 0.5\ Hz,\ L = 1\ in,\ W = 0.25\ in,\ m_{eff} = 0.2\ g)$

FIGURE 3.25 Force response characteristics of the IPMNC under different voltage waveforms.

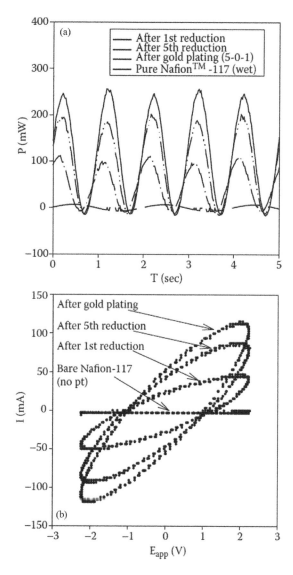

FIGURE 3.26 (a) Power consumption of IPMNC versus time. Note that the platinum reduction was completed five times, and final electroplating was done with gold. Also, phase shifts can be seen as the platings proceed. (b) Current/voltage characteristics of IPMNC. Note that the platinum reduction was completed five times, and final electroplating was done with gold. Also, phase shifts can be seen as the platings proceed.

Figure 3.27(a) contains an SEM cross-section (top) of the sample and its surface (bottom), respectively.

Second, the most significant aspect of having the porous electrode for the IPMNC is that it allows an effective water transport and internal circulation mechanism within the electrode region to create effective strains (porous nature of the compliant electrodes) (Shahinpoor and Kim, 2001b, 2002a, 2002b). However, it can also cause water leakage in the form of ion-driven water molecules gushing out of the porous electrode surface. Figure 3.27(b) depicts schematically such a phenomenon. This phenomenon is always visible and observable in any experiment involving wet IPMNC under the influence of an electric field across it. This process prevents the generated strains from effectively generating larger forces similar to a leaky hydraulic jack. Therefore, a key engineering problem is how to prevent such a water (solvent) leakage out of the porous electrode. Here, we describe a successfully developed process technique to address such a leakage problem to manufacture a high output force IPMNC.

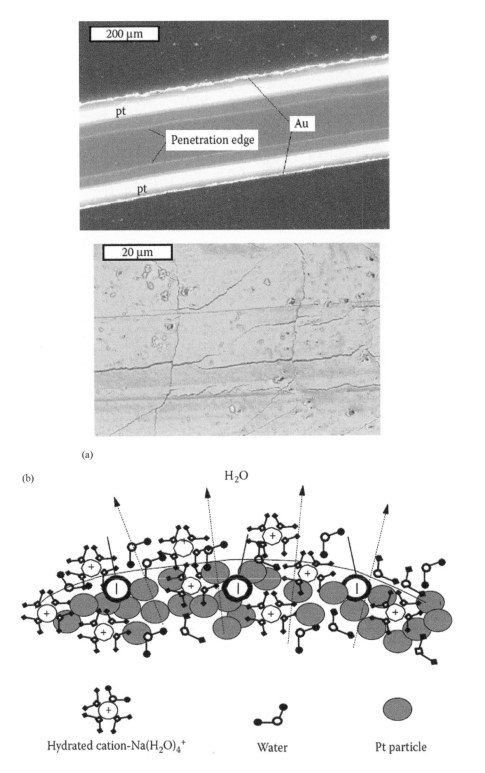

FIGURE 3.27 (a) An SEM cross section (top) of the sample and its surface (bottom), respectively. The cross-sectional view shows two platinum penetration leading edges and gold surface electrodes. The surface is fairly uniform but shows minor imperfect spots. (b) A schematic representation showing how the loose and ion-hydrated water is gushing out under an imposed voltage.

In these materials, a circulatory system of hydrated cations and water exists that may be employed to do various functions with IPMCs. Figure 3.27(b) depicts a schematic to show how water and hydrated cations can gush out of the surfaces of IPMNCs under an imposed voltage.

As discussed in a previous paper (Shahinpoor and Kim, 2001g), the nominal size of primary platinum particles of the IPMNC is found to be around 40–60 nm, which is much larger than that of incipient particles associated with ion clusters (~5 nm). Thus, this finding leads to a firm conclusion that incipient particles coagulate during the chemical reduction process and eventually grow large. If so, one can realize a significant potential to control this process (in terms of platinum particle penetration, size and distribution). To do so could be achieved by introducing effective dispersing agents (additives) during the chemical reduction process. One can anticipate that effective additives should enhance platinum particles' dispersion within the material and finally reduce coagulation. As a result, a better platinum particle penetration into the material can be realized. This process also creates a somewhat smaller particle size with fairly good distribution. Thus, the water leakage out of the surface electrode could be significantly reduced. A recent observation by Shahinpoor and Kim has been the identification of several effective dispersion agents or dispersants.

As a successful outcome, the use of the effective dispersing agent during the platinum metal-ization process has dramatically improved force characteristics showing a much sharper response (smaller time constants) to the input electric impetus and a dramatically increased force generation.

In Figure 3.28(a) and (b), the surface micrographs are presented for an IPMNC without dispers-ing agent treatment and for an IPMNC with dispersing agent treatment. The effect of the dispersing

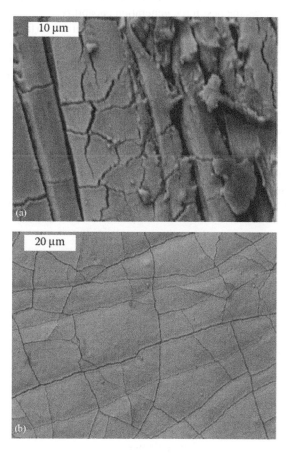

FIGURE 3.28 (a) SEM micrographs show surface morphologies of the IPMNC sample without using a dispersing agent. (b) SEM micrographs show surface morphologies of the IPMNC sample using a dispersing agent.

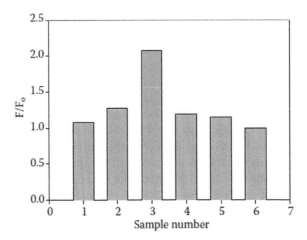

FIGURE 3.29 Comparisons of six different IPMNC samples treated with dispersing agents.

agent is to form uniformly electroded surfaces. Our effort has been extended to investigate various dispersing agents; however, results are somewhat scattered, as shown in Figure 3.29. Note in this figure that sample 6 is the one with no treatment. Other samples were treated by proprietary dispersing agents, including PVP, PVP/PMMA, PVA/PVP and a commercial detergent, CTAB. Sample 3, which was PVP treated, shows the best force generation capability.

The third issue is related to the strong diffusional resistance at the surface caused by the platinum compositing process that is a primary reason for limiting the platinum layer growth (Figure 3.30). One approach is to stretch the base material before the platinum composition process. By doing so, we anticipate that the base materials are plastically deformed and, as a result, the larger pores (higher permeability) could be created relative to the starting materials.

3.3.6 ADDITIONAL RESULTS ON STRETCHED IPMNCS TO ENHANCE FORCE GENERATION AND OTHER PHYSICAL PROPERTIES

In Chapter 2, the effect of increased permeability due to physical stretching of ionic polymer-metal composites (IPMNCs) on the physical properties of IPMNCs was briefly discussed as a phenomenon. Here, a detailed experimental discussion is presented on this observed phenomenon connected

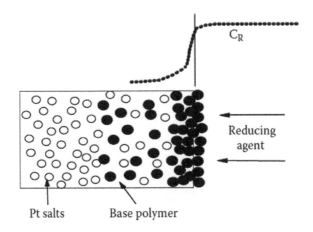

FIGURE 3.30 Platinum-reducing process. Dark circles represent platinum particles.

TABLE 3.8
IPMNC Samples Prepared

	Sample A (Pt and Pt/Au)	Sample B (Pt and Pt/Au)	Sample C (Pt and Pt/Au)
Base material	Nafion-117	Nafion-117	Nafion-117
Initial composition	Pt	Pt	Pt
Surface plating	Au	Au	Au
Cation	Lithium	Lithium	Lithium
Stretching (%)	5	6	19
Pt penetration	~20 sμm	~20 μm	~40 μm
Size	Width = 6.35 mm	Width = 6.35 mm	Width = 6.35 mm
	Length = 25.4 mm	Length = 25.4 mm	Length = 25.4 mm
	Thickness = 163 μm	Thickness = 159 μm	Thickness = 150 μm

with increasing the force density of IPMNC samples in cantilever mode and terms of the blocking force. In this new effort, as-received ionic polymers or, in particular, Nafion-117 membranes were mechanically stretched following the machining direction (uniaxially).

Initially, three samples, sample A (Pt), sample B (Pt) and sample C (Pt), were prepared for 5, 6 and 19% permanent stretching, respectively. The underlying principle of this effort is to attempt to increase the permeability of polymeric base materials. Platinum compositing was carried out based upon ERI-recipe with PVP.

Further, three more samples with surface plating were prepared by gold plating. Those samples were noted as sample A (Pt/Au), sample B (Pt/Au) and sample C (Pt/Au), respectively. A total of six samples were prepared in this effort. Note that Li$^+$ was chosen as an effective cation for all samples.

Table 3.8 shows the detailed information regarding the manufacturing processes involved. Figures 3.31 and 3.32 depict the difference in particle geometry and distributions before and after stretching.

Note that the mechanical stretching of raw ionomers using a testing machine (17% uniaxial stretching was performed) affects nanoparticles' morphology embedded within the macromolecular network.

The change in the blocking force is rather significant, as shown in Figure 3.33. Figures 3.33 and 3.34 show the measured blocking forces as a function of electric potential imposed across the

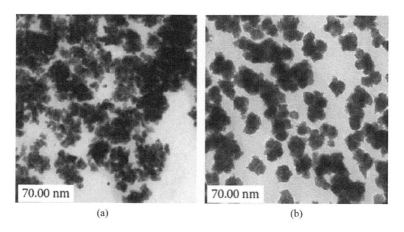

(a)   (b)

FIGURE 3.31 Platinum particle geometry and distributions (a) before and (b0 after stretching 17% uniaxially.

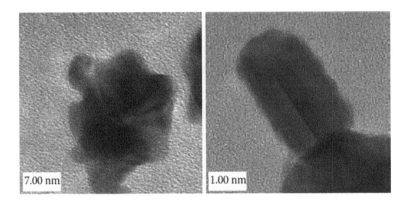

FIGURE 3.32 Change in blocking force for typical samples after stretching.

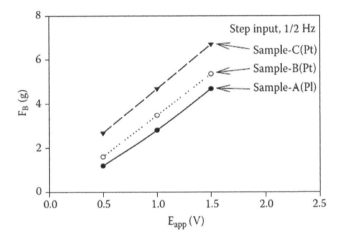

FIGURE 3.33 Change in blocking force for typical samples after stretching.

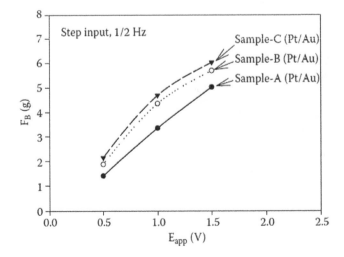

FIGURE 3.34 Change in blocking force for typical samples after stretching. A factor of almost two increases the force.

IPMNCs. Overall, the blocking forces are fairly large (up to 10-gram force [gf]) for sinusoidal and step inputs at 0.5 Hz. Note that the effective length was set at 12.7 mm.

Figures 3.35 and 3.36 depict additional results for blocking force increase upon IPMNC stretching. Figures 3.37–3.40 depict SEM of stretched IPMNC samples. Figure 3.41 depicts the sequential increase in blocking force upon adding dispersant polyvinyl pyrrolidone (PVP) and further stretching to 19%.

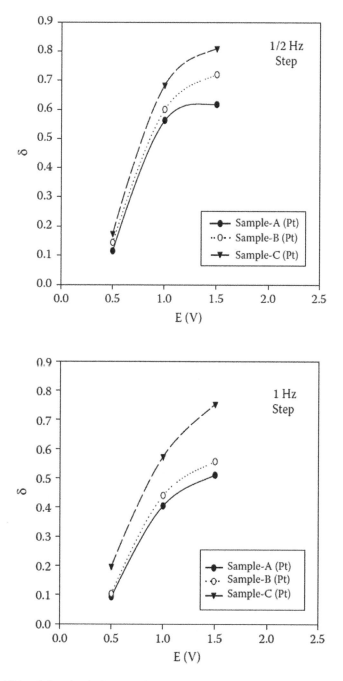

FIGURE 3.35 Additional data for the increase in blocking force for typical samples after stretching.

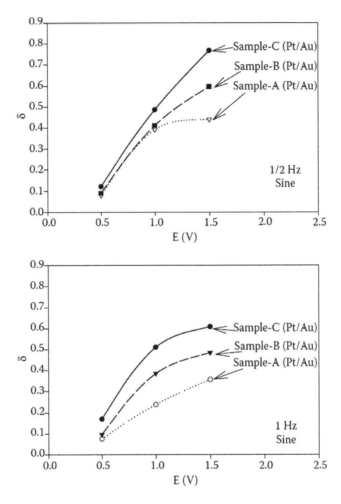

FIGURE 3.36 Additional data for the increase in blocking force for typical samples after stretching.

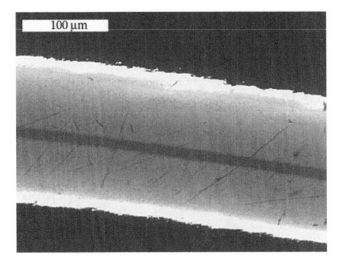

FIGURE 3.37 SEM of 4% stretched IPMNC samples.

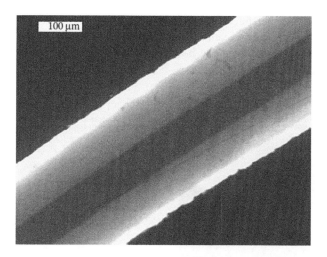

FIGURE 3.38   SEM of 5% stretched IPMNC samples.

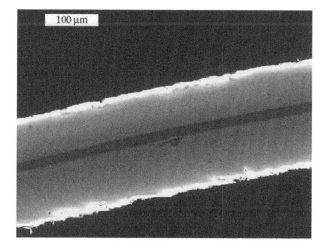

FIGURE 3.39   SEM of 17% stretched IPMNC samples.

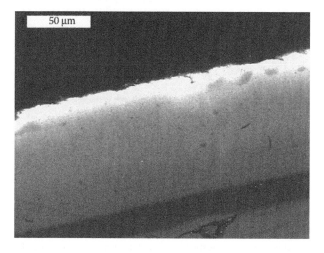

FIGURE 3.40   SEM of 17% stretched IPMNC samples (close-up).

It appears that much higher forces can be obtained. Note that stretching probably induced a larger void volume to increase permeability significantly. As a result, platinum particles are well distributed within the electrodes near the boundaries of IPMNCs. These new developments are being further investigated.

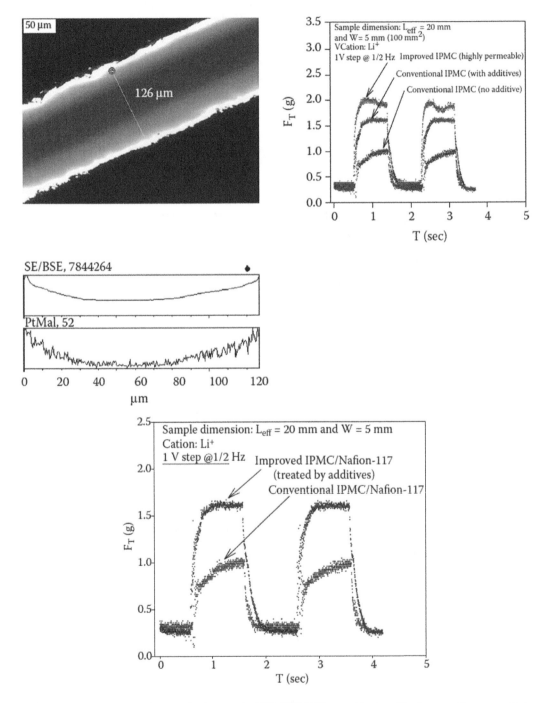

FIGURE 3.41 (a) SEM micrograph of a stretched IPMNC (19% stretching) and its force performance. Note that an additive, PVP, was added. (b) Force response characteristics of the improved IPMC versus the conventional IPMC. Note that the improved IPMC is one treated by an active dispersing agent.

3.3.6.1 Platinum-Palladium – New Phenomenon

Experimentation with a combination of platinum and palladium to explore possible improvements in the physical characteristics of IPMNCs has led to a peculiar phenomenon. This phenomenon is depicted in the SEM picture of these new types of IPMNCs in Figure 3.42. As the X-ray scan indicates, Pt and Pd's metallic particles tend to deposit at a certain depth of about 25–30 μm, as shown

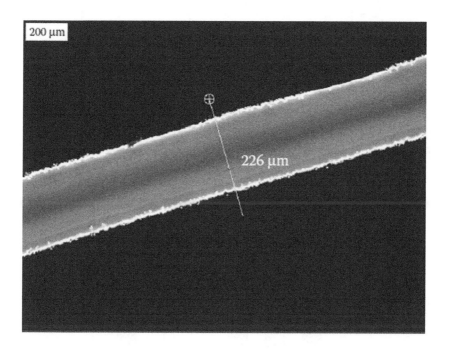

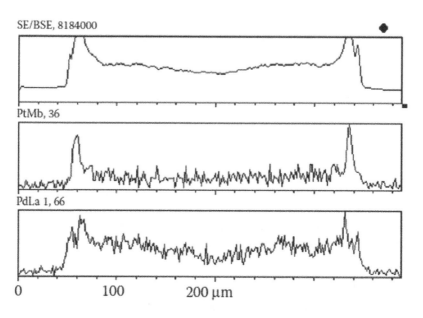

FIGURE 3.42 Concentration of Pt–Pd particles during chemical plating at a certain depth away from the surface of IPMNCs.

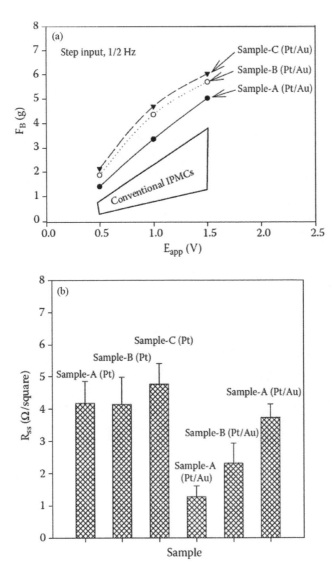

FIGURE 3.43 (a) Measured blocking forces and (b) surface resistance of IPMNC samples prepared. The standard sample size is $5 \times 10 \times 0.2$ mm^3. The process information is given in terms of particles used, platinum, and the final plated material, gold.

in the figure. We have repeated this experiment several times and have repeatedly observed this phenomenon. We are still exploring an explanation as to why this occurs. However, at this time, we do not have an explanation. The resulting IPMNCs appear to behave almost the same as the platinum or palladium ones by themselves. Additional experimental results on stretching and the effect of surface electrodes are given in Figure 3.43.

3.3.7 EFFECTIVE SURFACE ELECTRODES

One electrochemical method to study IPMNC artificial muscles is to use the A.C. impedance method that reveals the structure of an equivalent electric circuit. Figure 3.44 presents a simplified equivalent electric circuit of the typical IPMNC artificial muscle.

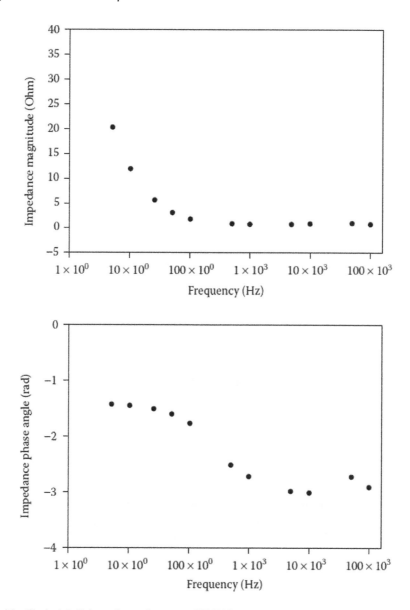

FIGURE 3.44　Typical A.C. impedance data on an IPMNC.

In this connection, each single unit circuit (i) is assumed to be linked in a series of arbitrary surface resistances (R_{ss}) on the surface. This approach is based on the experimental observation of the large surface resistance (typically, $\Sigma R_{ss}/L \sim 1\text{--}2\ \Omega/\text{cm}$; L is the surface electrode). In general, it can be assumed that there are four components to each single unit circuit:

surface-electrode resistance ($R_s \sim$ tens of ohms per centimeter)
polymer resistance ($R_p \sim$ hundreds of ohms per millimeter across the membrane)
capacitance related to the ionic polymer and the double layer at the surface–electrode/electrolyte interface ($C_d \sim$ hundreds of microfarads)
impedance (dynamic resistance, Z_w) due to a charge transfer resistance near the surface electrode

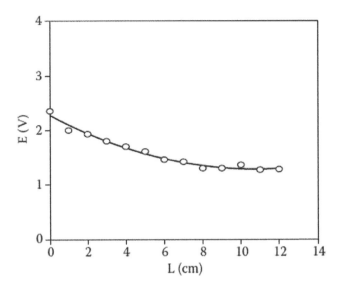

FIGURE 3.45 A typical set of data shows the voltage, E, drop along the surface-electrode direction, L.

Figure 3.44 shows typical A.C. impedance data of IPMNCs. For the typical IPMNC artificial muscles, the importance of R_{ss} relative to R_s may be interpreted as

$$\text{s}\frac{\sum R_{ss}}{R_s} \approx \frac{L}{t} > 1 \tag{3.4}$$

where t is the thickness of the platinum surface electrode, considering that the typical value of t is ~1–10 μm, Equation (3.4) is valid. The equation states that significant overpotential is required to maintain the effective voltage condition along the typical IPMNC muscle's surface. In other words, the voltage drop along the surface electrode direction is appreciable. A typical set of data is shown in Figure 3.45.

One way to solve this problem is to overlay a thin layer of a highly conducting metal (such as silver or copper) on top of the platinum surface electrode. Figure 3.46 shows a schematic diagram of the standard silver (or copper) deposited IPMNC artificial muscle fabricated in this manner. The typically measured impedance plot should show the imaginary part against the real part at different A.C. frequencies.

Note in Figure 3.46 that the silver layer was electrochemically deposited onto the Pt particle electrode that had been chemically composited. The silver layer is the bright metal electrode with a typical thickness of silver, approximately 1–2 μm. Therefore, the surface resistance is significantly reduced. Platinum particles are usually formed in dark black. Note that PIEM stands for the perfluorinated ion-exchange membrane.

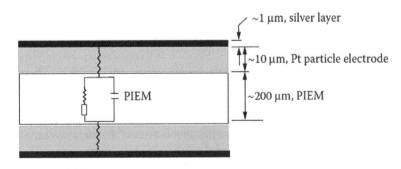

FIGURE 3.46 A schematic diagram of the typical silver- (or copper-) deposited IPMNC artificial muscle fabricated.

In Figure 3.47, the measured platinum X-ray counts across the membrane are provided in terms of N_x ($= C_o/C_i$) along with a typical X-ray spectrum. Notations C_i and C_o represent X-ray counts at the interface and ones within the membrane, respectively. Clearly, the diffusion-dominated platinum precipitation process is shown. JEOL 5800LV SEM measured the X-ray counts operated with Oxford Isis-Link imaging software.

Another good tool to characterize the IPMNC artificial muscle's surface morphology is atomic force microscopy (AFM). Its capability to image the IPMNC artificial muscle's surface directly can provide detailed information with a few nanometers' resolutions. Thus, we attempted to reveal the surface morphology of the IPMNC artificial muscles using AFM. Figure 3.48 depicts an AFM image of an IPMNC sample.

Digital Instruments' AFM NanoScope IIIa was used. A tip (ultralevers) from Park Scientific Instrument was utilized in an air-contact mode under low voltage. As can be seen, the surface is characterized by a granular appearance of platinum metal with a peak/valley depth of approximately 50 nm.

This granular nanoroughness seems responsible for producing a high level of electric resistance but still provides a porous nature that allows water movement in and out of the membrane. During

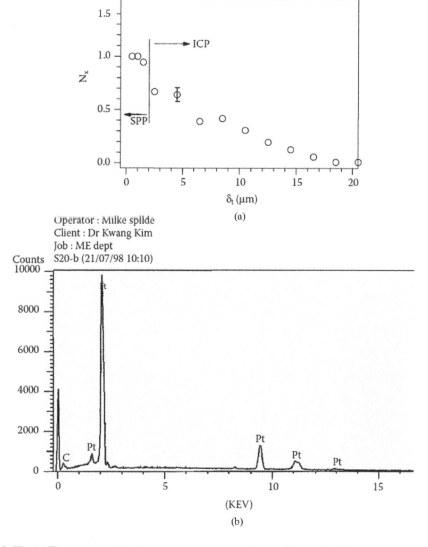

FIGURE 3.47 (a) The measured platinum concentration profile and (b) a typical X-ray spectrum.

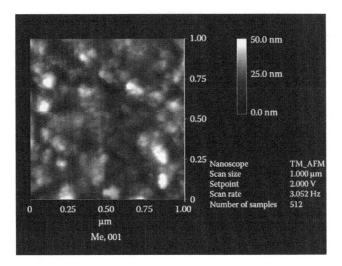

FIGURE 3.48 An atomic force microscopy image taken on a typical muscle's surface electrode is shown in Figure 3.48. The scanned area is 1 μm^2. The brighter/darker area corresponds to a peak/valley depth of 50 nm. The surface analysis image has a view angle set at 22 degrees.

the AFM study, it was also found that platinum particles are dense and, to some extent, possess coagulated shapes.

The electrochemical deposition of silver (or copper) on top of the IPMNC muscle was straightforward. It requires a rectifier and silver (or copper) solution. The rectifier (MIDAS, pen-type) controls the D.C. voltages and currents within appropriate ranges. Careful approaches were taken to obtain a thin and uniform silver (or copper) layer. Silver solution concentration, deposition time and solution temperatures were varied to obtain an optimized thickness of approximately 1–2 μm.

Figure 3.49 shows a schematic diagram illustrating the silver deposition process and a typical X-ray spectrum taken for the silver surface. As can be seen in the spectrum, pure silver peaks are apparent. The silver surface is much brighter and smoother than that of the platinum-based IPMNC

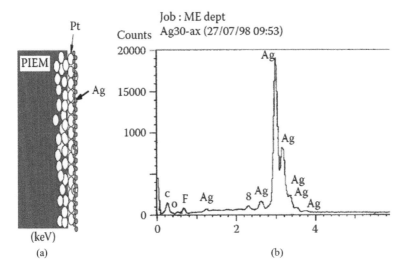

FIGURE 3.49 A schematic diagram illustrating (a) the silver deposition process and (b) a typical X-ray spectrum taken on the surface. Low voltage was applied to carry out the silver deposition.

TABLE 3.9

Test Samples

	Sample 1 (Baseline)	Sample 2	Sample 3
Cation type	Li⁺	Li⁺	Li⁺
Surface roughening	Yes	Yes	Yes
Pretreatment	1 N HCl boiling	1 N HCl boiling	1 N HCl boiling
Platinum deposition process	Initial compositing process; surface electroding process	Initial Compositing process; surface electroding process	Initial compositing process; surface electrode process
The top layer deposition process	None	Silver, electrochemical	Copper, electrochemical
Comments		Thin and bright color	Thin and copper-like color (initially); later turned into bluish color (corrosive)

artificial muscle and shows a typical silver-like color. As a result, the deposited silver reduces the IPMNC artificial muscle's surface resistance by a factor of approximately 10 (typically < 1 Ω/cm).

Three test samples were prepared by the processes described previously: sample 1 with platinum only (baseline), sample 2 with platinum/silver and sample 3 with platinum/copper. First, one batch of the IPMNC artificial muscle (100×200 mm^2) was prepared with platinum before silver (or copper) deposition. Then, sample 1 was cut in a standard size of a 5×20-mm^2 strip for baseline testing. Also, two strips were cut to 10×50 mm^2 for the electrochemical deposition process. Silver and copper solutions were prepared by dissolving appropriate concentrations of AgNO$_3$ and CuSO$_4$ in water, respectively. After the electrochemical deposition was completed, samples 2 and 3 were cut into 5×20-mm^2 strips. Table 3.9 gives detailed process information for each sample.

First, a baseline test was performed for a sinusoidal waveform with sample 1. The results showed that the IPMNC artificial muscle responds very closely to the input sinusoidal waveform, meaning that force responses are nearly equal in the traveling direction. Figure 3.50 shows a typically measured

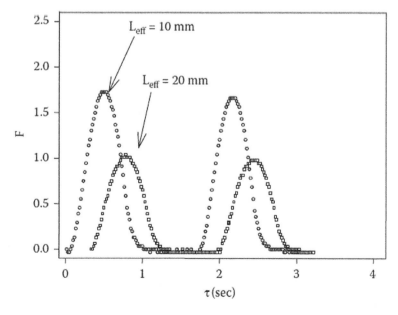

FIGURE 3.50 The IMPC artificial muscle response for sinusoidal wave input at 2 V$_{rms}$. Tests were performed in air and without preloading.

force response to an input sinusoidal waveform of 2 V_{rms}. It shows the results of the effective length for 10 and 20 mm. As expected, a larger force was observed for an effective length of 10 mm.

Figure 3.51(a) and (b) summarizes the test results that contain the measured surface resistance of test samples 1, 2 and 3 and their maximum forces relative to one produced by sample 1. Multiple measurements were performed, and error bars represent their corresponding standard deviations. As expected, overlayers of the silver and copper significantly reduce the surface resistance of the IPMNC muscle.

It should be pointed out that the electrochemical deposition of silver (or copper) is attractive; in fact, it can produce a thin homogenous metal phase on top of the platinum particle surface electrode. This process eliminates the problems associated with large voltage drops away from the contact electrode. The bonding between the platinum surface and silver (or copper) was favorable. However, the copper layer eventually became bluish due to copper oxidation (as expected). A simple comparison was made for the actuation performance of samples 2 and 3 relative to sample 1. Multiple measurements were performed.

In general, the IPMNC artificial muscle with a silver (or copper) deposited shows significantly improved actuation performance (roughly 10–20% more force relative to one without silver or

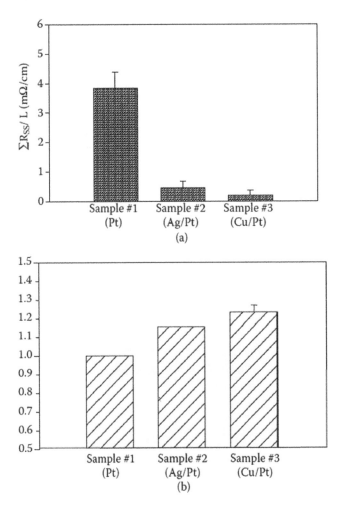

FIGURE 3.51 The measured surface resistance of the samples prepared in this study (a) and the ratio of the measured maximum force of samples 2 and 3 compared to sample 1 (b). The effective length, l_{eff}, was set at 20 mm for all samples.

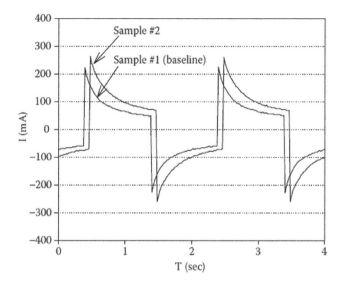

FIGURE 3.52 The current responses of IPMNC muscles (sample 1 vs. sample 2). A step voltage of 2.2 V was applied.

copper layers). For samples 2 and 3, it is interesting to note that more active water movement was found relative to sample 1.

The electromechanical dynamics associated with the IPMNC artificial muscles are complex. However, our approach to alleviating the inherent particle surface electrode resistance seems practical and effective. The fact that the IPMNC artificial muscle becomes more active with reduced surface electrode resistance is attributed to enhanced current passages, as seen in a current versus time curve (Figure 3.52).

3.3.8 AN ECONOMIC APPROACH – PHYSICAL METAL LOADING

A novel fabrication process of manufacturing IPMNCs equipped with physically loaded electrodes as biomimetic sensors, actuators and artificial muscles that can be manufactured at about one-tenth of the typical cost has been developed. The underlying principle of processing this novel IPMNC is to physically load a conductive primary powder layer into the ionic polymer network forming a dispersed particulate layer. This primary layer functions as a major conductive medium in the composite. Subsequently, this primary layer of dispersed conductive material particles is further secured within the polymer network with smaller secondary particles via chemical plating, which uses reducing agents to load another phase of conductive particles within the first layer. In turn, primary and secondary particles can be secured within the ionic polymer network and reduce the potential intrinsic contact resistances between large primary particles. Furthermore, electroplating can be applied to integrate the entire primary and secondary conductive phases and serve as another effective electrode (Shahinpoor and Kim, 2001e). The essence of such physically loaded and interlocked electrodes for IPMNC is depicted in Figure 3.53.

The principal idea of processing this new IPMNC so that primary and smaller secondary particles can be secured within the polymer network has two parts:

First, physically load a conductive primary powder (Ag in this case) into the polymer network, forming a dispersed layer, functioning as a major conductive medium near boundaries.

Subsequently, secure such primary particulate medium within the polymer network with smaller particles (Pd or Pt in this case) via a chemical plating process.

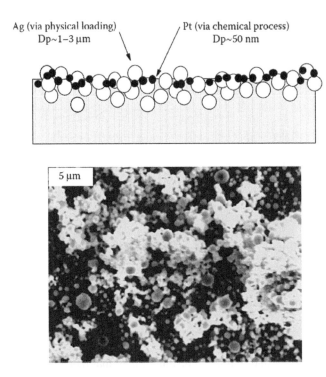

FIGURE 3.53 A schematic process illustration of the PLI-IPMNCs (left) and silver particle used (right). (Shahinpoor, M., and K. J. Kim. 2002. In *Proceedings of SPIE 9th Annual international Symposium on Smart Structures and Materials.* SPIE publication no. 4695, paper no. 36.)

Furthermore, an electroplating process can be applied to integrate the entire conductive phase intact, serving as an effective electrode. Figure 3.53 illustrates such a process.

The processes developed are:

1. A silver-based spherical powder (MOx-Doped Ag; Superior MicroPowders EM10500X-003A; $D_{10} < 0.8$ μm, $D_{50} < 1.5$ μm, $D_{90} < 2.5$ μm; $A_{sur} < 6$ m²/g) is dispersed in isopropyl alcohol (99%). Using a standard airbrush (VEGA), the powder is sprayed onto the backing material.
2. The isopropyl alcohol is then allowed to evaporate completely (it takes approximately an hour).
3. The ion-exchange polymer is the first surface treated with sandpaper. The standard size of the polymer sample is about 5×5 cm².
4. The ion-exchange polymer is placed between the backing materials facing the powder-coated side.
5. Pressing is carried out at 2 tons using a temperature-controlled hot press (RAMCO, 50-ton capacity) at 120–130°C for 15 min.
6. Steps 1–4 are repeated three times. Usually, the low electric surface sheet resistance is obtained ($R < 1$ Ω/square by the four-probe method).
7. The preferred process is to impregnate small noble metal particles (i.e., platinum or palladium, $D_p \sim 50$ nm) between the primary particles to fixate them within the ion-exchange polymer further. This process is to introduce metallic ions [Pd(NH$_3$)$^{2+}$] or [Pt(NH$_3$)$^{2+}$] into the ion-exchange polymer initially and, later, reduce them to a metallic state.
8. As a final step, a conductive metallic layer (i.e., gold or palladium) is further electroplated on the top of the interlocked electrode layer.

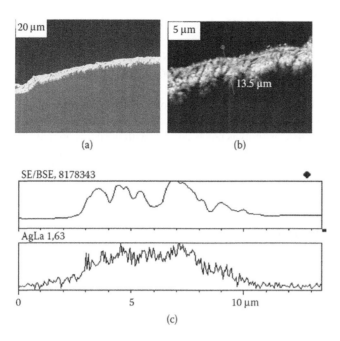

FIGURE 3.54 SEM micrograph of IPMNC and (a) its close-up and (b) X-ray line scan (c), respectively. The silver penetration is about 7–8 μm for this sample, and the majority of silver particulates are in tack within the ion-exchange polymer.

Figure 3.54 includes an SEM photograph (cross-section) of a sample IPMNC, its close-up and an associated X-ray line scan. Ag particles' penetration is approximately 7–8 μm, as can be measured by the X-ray line scan.

The surface SEM micrograph provided in Figure 3.55 shows fairly uniformly distributed Ag particles at the surface. Ag particles stay intact to function as the highly conductive surface electrode. Overall, the surface resistance is lowered as Ag particles' penetration increases and reaches slightly below 1 Ω/square. Note that surface resistance was measured by the four-probe method

FIGURE 3.55 SEM micrograph showing Ag surface properly electroded and composited onto an ion-exchange polymer.

(Guardian Manufacturing, #SRM-232). The performance of the IPMNC manufactured by this newly developed technique was gauged by measuring the blocking force, F_b, in a cantilever configuration under a certain voltage across the IPMNC strip.

In Figure 3.56, representative data are provided for the cases of step voltage of 1 and 1.5 V at 0.5 Hz, respectively.

In general, the performance of the physically loaded IPMNC shows slightly less force generation than the conventional IPMNC, and it still produces a fairly good output force. However, one can note that the response sensitivity is not as good as with the conventional IPMNC. This could be attributed to the Ag electrode with a larger water leakage than the conventional IPMNC (Shahinpoor and Kim, 2001e). In other words, the Ag electrode made by this new technique is fairly porous and permeable to water. Currently, this issue is being investigated. Table 3.10 compares the physically loaded IPMNC and the current state-of-the-art IPMNCs.

It has been estimated that the cost reduction is due primarily to lower platinum loading and secondarily to the significant reduction in labor.

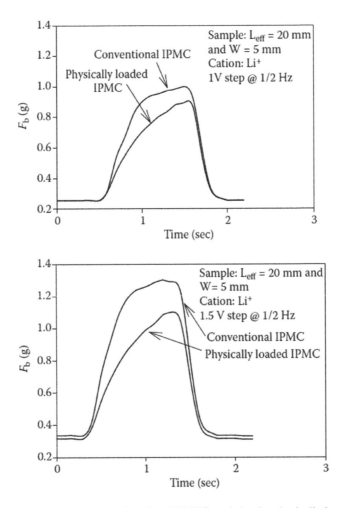

FIGURE 3.56 Force response characteristics of the IPMNC made by the physically loaded technique (top: 1-V responses and bottom: 1.5-V responses, respectively). The blocking force, F_b, was measured at the cantilever configuration's tip with slightly preloaded conditions of 0.27 and 0.32 g, respectively.

TABLE 3.10

Comparisons between Proposed Physical Loaded Manufacturing Technique and Current State-of-the-Art Manufacturing Technique for IPMNCs

	The Current State-of-the-Art Manufacturing Technique	The Proposed Physical Loaded Manufacturing Technique
Fundamental processes	Chemical metal-reducing processes	Physical metal loading processes
Choice of metal	Typically Pt, Au and Pd	Not limited (Ag or graphite)
Process parameters	Multiple parameters, including chemical concentrations, temperature, reaction time, and preliminary treatments	Only a few parameters, including particle loading pressure and temperature
Estimated material price	~$10/cm^2	<$0.1/cm
Nominal production time	48 h	2 h

3.3.9 Scaling

It is well understood that all commercially available (as-received) perfluorinated ion-exchange polymers are in the form of hydrolyzed polymers, are semicrystalline and may contain ionic clusters. These polymers' membrane form has a typical thickness in the range of approximately 100–300 μm. Such a thin thickness of commercially available membranes permits fast mass transfer for use in various chemical processes. Such as-received semicrystalline membranes are not melt-processable, so they are not suitable for fabricating three-dimensional electroactive materials or other composite forms.

In the previous work (Kim and Shahinpoor, 2001a), the authors reported a newly developed fabrication method that can scale up or down the IPMNC artificial muscles in a strip of micro- to centimeter size thickness. We have adopted a recently developed technique by Moor et al. (1992) for dissolving as-received ion-exchange membranes in appropriate solvents. By carefully evaporating solvents out of the solution, recast ion-exchange membranes were obtained (Gebel et al., 1987; Moor et al.). Several samples are shown in Figure 3.57.

The preparation of a solution recast Nafion film sample includes:

1. DuPont liquid Nafion solution was purchased.
2. According to the manufacturer's specification, this solution contains 10% Nafion and 90% solvent of an approximately one-to-one mixture of 2-butanol and water.
3. It was noted, initially, that during the solvent evaporation, the solidified Nafion developed surface cracks. Therefore, an essential trick was introducing an additive that makes the solvent mixture act like an azeotrope.
4. The use of DMF was successful. Subsequently, when given multiple layers of liquid Nafion dried, an approximately 2-mm thick sample was prepared successfully.
5. First, a known quantity of liquid Nafion with an additive is placed in a Teflon (polytetrafluoroethylene, PTFE)-coated Pyrex glass.
6. Second, appropriate annealing was performed at an elevated 70°C temperature to create crystallinity. Usually, the temperature is raised to 150°C for further curing. The annealing process can tailor the mechanical and chemical stability of the solution recast Nafion film.

The IPMNC sample was prepared by using 2-mm thick recast Nafion fabricated by following the process just described. Later, platinum plating was done on both sides of the sample with a particle penetration depth of ~20 mm. Figure 3.58 shows photographs of the IPMNC samples with

An eight-finger synthetic muscle.
It has a thickness of approximately 2 mm.

A rod shape synthetic muscle.
It has a rectangular cross-section of
approximately 8 mm × 8 mm.

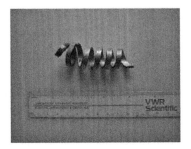

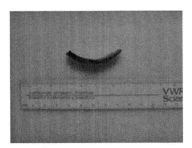

A coil type synthetic muscle.

A circular shape synthetic muscle.

FIGURE 3.57 Various IPMNCs with three-dimensional shapes.

the primary platinum electrode and secondary gold electrodes. Another fabrication method is also
described in Figure 3.59.

The mechanical tensile behavior of an as-received Nafion (fully hydrated H⁺ form, 16% H$_2$O)
was first measured and presented in Figure 3.60. As can be seen, the as-received Nafion clearly
shows the electrostatic cross-linking and crystallinity-induced tensile characteristics exhibiting an
unclear yielding and strain-induced further crystallinity, somewhat similarly to cross-linked elas-
tomeric behavior. The normal stress-strain curve for the recast membrane (fully hydrated H⁺ form,
9% H$_2$O) is superimposed. It clearly shows a distinct, plastically deformed behavior that indicates
weakened (or eliminated) electrostatic cross-linking relative to dominating elastic forces.

The force/displace measurement was done separately in a cantilever beam configuration: the tip
force (truly the maximum force) was measured by a load cell at the zero displacement condition
(blocking force), and the tip displacement (truly the maximum displacement) was measured without
a load applied. In Figure 3.61 (left), an IPMNC sample (2-mm thickness, 15-mm effective length
and 5 mm-width) is provided. Note that the applied step voltages across the sample (0.5 Hz) are 2, 4
and 6 V, respectively (1, 2 and 3 V/mm, respectively). Although these applied electric fields are very
small, the sample's responses are excellent in terms of useful forces being generated.

In Figure 3.61 (right), the measured displacement is presented against the force (displacement
vs. tip force = generative force). The useful meaning of this graph attributes the possible maximum
work output (= mechanical energy stored in the IPMNC beam,

$$U_m = \int_0^{\delta_{max}} F_T \delta \mathrm{d}s \cong \frac{1}{2}\delta_{max}|F_T|)$$

of the IPMNC, although more elaborate interpretations may be needed (i.e., by simultaneous measurement of the tip velocity and displacement or curvature). In this configuration, internal stresses

FIGURE 3.58 The IPMNCs made with solution recast Nafion. (a) The photos show the fabricated eight-finger IPMNC (*Octopus-IPMNC*). It (2-mm thickness) can easily sustain the eight U.S. quarters. (Note that a U.S. quarter has a mass of 5.3 g.) The diameter of this IPMNC is approximately 10.5 cm. The electrode is centered. (b and c) The photos show the IPMNC in action without applying load and with a load. As can be seen, a quarter is lifted. The time interval between the frames is approximately 1 s. A step voltage of 2.8 V was applied ($E = 1.4$ V/mm). Typical REDOX metal-reducing techniques manufactured these IPMNCs. Platinum was composited initially, and gold was plated later. The cation is Li$^+$.

are usually built when transverse generative strain is converted into bending motion, which lessens the mechanical output energy. (Note that flexural strength is generally lower than tensile or compressive strength since the thickness is small.)

Based upon these measurements, we can define the electromechanical coupling factor, k (or thermodynamic efficiency, E_{ff}), as:

$$E_{ff} = k^2 = \frac{U_m \left(= \text{stored mechanical energy}\right)}{U_e \left(= \text{electrical input}\right)} \tag{3.5}$$

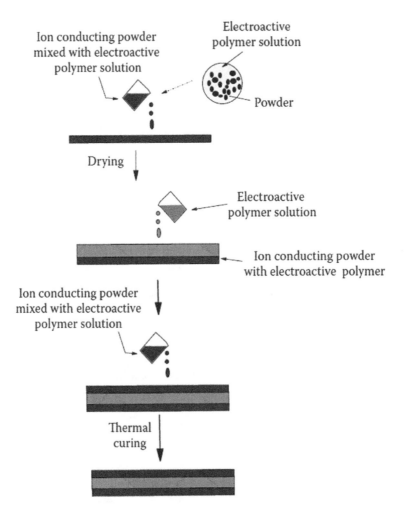

FIGURE 3.59 An illustrative process diagram for the ion-conducting, powder-coated electrode made by the solution recasting method. First, the ion-conducting powder (i.e., carbon, silver, platinum, palladium, gold, copper and any other conducting powders) is mixed with the electroactive polymer solution (e.g., liquid Nafion). The powder is fine and uniformly dispersed within the electroactive polymer solution. After forming a thin layer, the electroactive polymer solution undergoes the drying process of solvents, and therefore the residual consists of the ion-conducting powder dispersed within the polymer. Second, the electroactive polymer solution (without the powder) is added on top of the ion-conducting powder layer and dried. This is repeated until the desired thickness is obtained. Later, a layer of the ion-conducting powder is formed by the same method described previously. As a final step, the ion-conducting, powder-coated electrode is cured under the elevated temperature. If necessary, the surface conductivity can be enhanced by adding a thin layer of novel metal via electroplating or electroless plating.

Based upon Equation (3.5), a graph was constructed (see Figure 3.62) that shows the thermodynamic efficiency of the IPMNC as a function of frequency. Note that this graph presents the experimental results for the conventional IPMNC and the additive (PVP)-treated improved IPMNC (particle controlled). It is of note that the optimum efficiencies occur at near 8–10 Hz for these new IPMNCs. The optimum values of these IPMNCs are approximately 25–30%. At low frequencies, the water leakage out of the surface electrodes seems to cost the efficiency significantly. However, the additive (PVP)-treated IPMNC shows a dramatic improvement in efficiency since less water transports out of the surface electrodes. The important sources of energy consumption for the IPMNC actuation could be from necessary mechanical energy needed to cause the positive/negative strains for the IPMNC

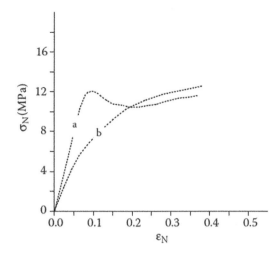

FIGURE 3.60 Tensile testing results (normal stress, σ_N, vs. normal strain, ε_N). Note that both samples were fully hydrated when they were tested. (a) solution recast membrane; (b) as-received membrane.

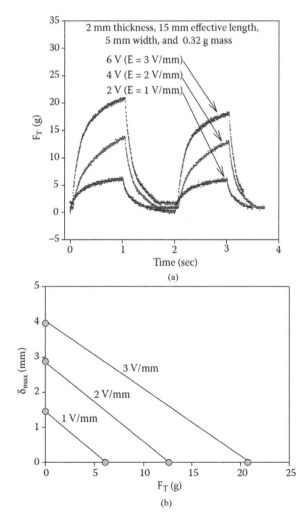

FIGURE 3.61 Force responses of (a) the solution recast IPMNC sample and (b) its conjugated graph showing tip displacement δ_T versus blocking force, F_T (bottom). Note that the frequency is 0.5 Hz, and step voltages of 2, 4 and 6 V were applied.

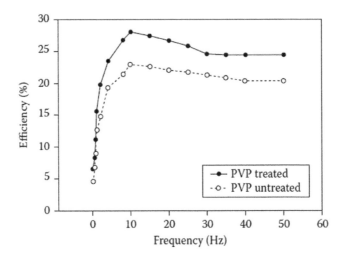

FIGURE 3.62 Thermodynamic efficiency of the IPMNC as a function of frequency.

strip I/V hysteresis due to the diffusional water transport within the IPMNC thermal losses – joule heating decomposition due to water electrolysis water leakage out of the porous electrodes.

Dramatic improvement has been made to fabricate a new generation of IPMNCs with a much-improved performance by blocking the water leakage out of the surface electrodes by compliant surface electrodes. The overall thermodynamic efficiencies of all IPMNCs tested in a frequency range of 0.1–50 Hz are comparable to those of biological ones. It should be noted that the obtained values for efficiency are favorable, realizing that other types of bending actuators – that is, conducting polymers and piezoelectric materials at similar conditions – exhibit considerably lower efficiencies (Wang et al., 1999).

The samples used in Figure 3.62 have a dimension of a 20-mm length, 5-mm width and 0.2-mm thickness. The applied potential is 1-V step. Lines are least-square fits. Resonant efficiencies are not included in this figure. It appears that, at higher frequencies, the thermodynamic efficiency stabilizes and almost remains the same. This phenomenon, as well as the resonance state efficiencies, is currently under investigation.

3.3.10 Technique of Making Heterogeneous IPMNC Composites

Platinum is not the only noble metal that can be produced by chemical reduction processes. Other noble metals such as palladium, gold and silver have been tried. Although there are not enough data regarding those noble metals as effective materials used to place electrodes on IPMNC samples, there are enough indications that they are also effective. Also, a heterogeneous technique of manufacturing the nanocomposites of IPMNCs, such as an alternative placing the Pt composite first, palladium next, then again platinum and, subsequently, palladium, etc. had been tried to control the penetration depth. Overall, those laminating procedures to make laminated nanocomposites of IPMNCs appeared to be very promising and effective.

Another interesting technique is to combine conducting polymers (polypyrrole) and metals to create a system of the base polymer/metal/conducting polymer. We have fabricated such an IPMNC system. One finding was that this procedure significantly reduced the input power consumption while maintaining the generative blocking forces' same level. Such an effect could be attributed to less I/V hysteresis acting on the metal particle interacting surfaces. The results are briefly presented in Figure 3.63.

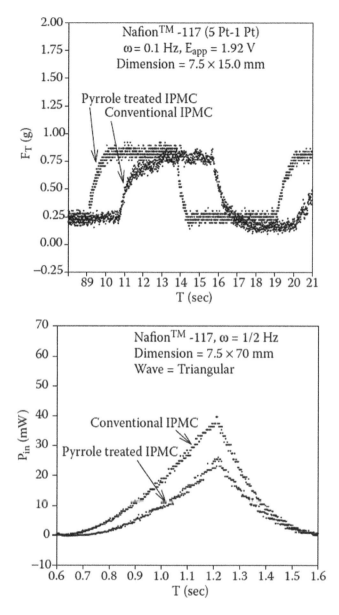

FIGURE 3.63 Force characteristics and input power consumption of IPMNCs (conventional IPMNC and pyrrole-treated IPMNC). Polymerization of pyrrole was carried with the presence of a catalyst within the base polymer Nafion.

4 Ionic Polyacrylonitrile (PAN) Fibrous Artificial Muscles/Nanomuscles

4.1 INTRODUCTION

This chapter covers the fabrication, characterization, testing and application of chemo (pH-activated) electromechanical artificial muscles made from ionic *poly-acrylonitrile* or simply polyacrylonitrile (PAN) material. PAN, in its nonionic form, can commercially be obtained under the trade name Orlon™. It is made by polymerizing acrylonitrile monomer:

$$
\begin{array}{c}
\text{---CH}_2\text{CH---CH}_2\text{CH---CH}_2\text{CH---CH}_2\text{CH---} \\
\mid \qquad \mid \qquad \mid \qquad \mid \\
\text{CN} \quad\;\; \text{CN} \quad\;\; \text{CN} \quad\;\; \text{CN}
\end{array}
\tag{4.1}
$$

As defined in the Merck Index, PAN is composed of white fibers that stick at 235°C. It becomes yellow at ironing temperatures above 160°C. It has flammability similar to that of rayon and cotton. More importantly, it has very good resistance to mineral acids and excellent resistance to common solvents, oils, greases, neutral salts and sunlight. It has fairly good resistance to weak alkalides but is degraded by strong alkalides. It also resists attacks by molds, mildew and insects.

PAN material is primarily used in the textile industry to manufacture clothing fabrics and artificial silk. Lately, these materials can be obtained in fibers made of thousands of smaller filaments and fibrils, each less than 10 μm in diameter. Companies like Mitsubishi Rayon Co. have constantly improved their techniques in manufacturing better fibers of PAN.

As the industry moves toward manufacturing these materials in finer and smaller filaments, the potential for improved artificial muscle fibers also increases, as will be explained later in this chapter. This chapter includes a method of encapsulation and activation of PAN fiber bundle artificial muscles by an electric field.

4.2 PAN FABRICATION

PAN fibers are initially nonactive and require a combination of physical and chemical processes to convert them into a gel-like contractile fiber material. First, to achieve active PAN muscles, fibers are cut to a desired length and number of fibers from a spool of Orlon materials. Additional length is added to allow for shrinkage due to annealing, end connections and bundling. The fibers are carefully placed in a Pyrex or similar nonsticking container to not contaminate the fibers with any types of oil or corrosive contaminant and placed in a laboratory oven to be annealed and oxidized for 6 h at about 200°C. It is best to use a convection oven to circulate the chamber air to maintain a uniform temperature throughout the fibers. Annealing time depends on the oven type and density of the fibers' and filaments' diameter; it can vary from 200 to 240°C for 6–2 h, respectively. The fiber color changes from dark brown to black, indicating oxidation of the polymer chain. The appearance of lighter color fibers indicates an incomplete annealing process.

The fibers then are bundled at both ends via a nylon line and epoxy adhesive to secure the ends and later attach to any structure to be moved or manipulated. Depending on the application and the

DOI: 10.1201/9781003015239-4

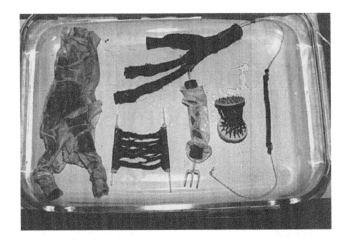

FIGURE 4.1 Assortment of pH muscles made from PAN fibers. Clockwise from left: encapsulated biceps muscle, triceps muscle, linear fiber bundle, linear platform muscle, encapsulated fiber bundle and parallel fiber muscles.

types of muscles needed, the end connection can be varied. For example, in the case of encapsulated fiber bundles, we used rubber stoppers with holes to inject pH solutions (weak acids and alkalides) and pure water into and out of the muscle (Figures 4.1 and 4.2). The rubber stoppers also provided a means of securing fibers by adhesives and nylon wrapping lines. Each was carefully constructed for a specific application. We then hydrolyzed by boiling the resulting bundle in a 1-N solution of NaOH for 30 min to complete the fibers' gelation and activation.

At this point, the bundle appears at its elongated or expanded state because of the effect of the OH ions on the active ionic polymer chain. After letting the muscle cool to room temperature, the fiber bundles are thoroughly rinsed of any residual alkali solution. Further expansion is observed because of the increased availability of water molecules to penetrate within gel-like fibers. (Note that PAN gel is a hydrophilic material like most ionic gels.) The PAN muscle is now ready to be stored in pure water (or slightly acidic solution since any alkalinity for a long duration will eventually degrade the PAN muscle) for any number of future experiments. In this way, various types of

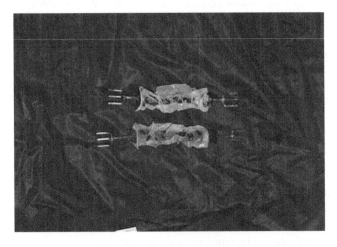

FIGURE 4.2 Two PAN muscles (50-fiber bundles) encapsulated with latex membrane incorporating three-way glass fittings at each end to allow for the transport of pH solutions within the fiber bundles (relaxed length is 3 in.).

PAN muscles can be manufactured, as shown in Figure 4.1, that can be attached to other structures for robotic manipulation. Thus, an optimum manufacturing procedure turned out to be:

1. Oxidation at 210°C for 75 min (annealed and cross-linked PAN)
2. Hydrolysis with 1-N NaOH at 95°C for 30 min

The addition of weak acids such as HCl 1 M causes contraction of the bundle. The muscle can be contracted and expanded using weak acids and bases with a rinse cycle in between. Regular contraction of 100–200% from the original length is commonly achieved. Depending on the fibers' packaging, one could improve the response time by further segmentation of the fibers. This fiber's total length, in effect, divides each fiber's total length into smaller pieces, allowing rapid diffusion of pH fluids along and within each segment of the fiber. Compared to a long fiber, where fluid will take a longer time to reach the entire fiber, this method proved useful in fabricating fast-response muscles similar to biological muscle fibers with each filament comprising smaller segments called sarcomeres. The additive effect of each segment improved the speed of response. In biological muscles, the maximum tension force due to contraction occurs at about 70–130% of the relaxed state of the sarcomere (refer to Appendix B of the first edition).

As seen in the preceding figures, latex sheets were commercially acquired, cut to size and patterned for specific muscle groups. After encapsulating the bundle, it was then sealed using a special silicon rubber adhesive to allow for frequent expansion and contraction of the fiber bundles in the latex enclosure. To completely seal the assembly, epoxy resin was used at each end of the bundle followed by a nylon string wrapping. This proved to be a most reliable method for a large loading condition to avoid damage to the individual fibers and muscle assembly. In more advanced versions, each end of the bundle was especially fitted with rubber stoppers with a concentric hole and with a specially designed three-way glass fitting to allow for transport of acids and bases into the fiber bundle (Figures 4.2 and 4.3).

4.3 PAN CHARACTERIZATION

Several test fixtures and structures were designed to evaluate the PAN muscles for loading and displacement characteristics, such as the isotonic and the isoionic test apparatus shown in Figures 4.4 and 4.5.

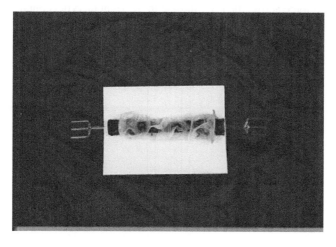

FIGURE 4.3 Single 100-fiber bundle encapsulated PAN muscle with three-way fittings designed especially for the transport of pH solution and water within the fiber bundle (relaxed length is 6 in.).

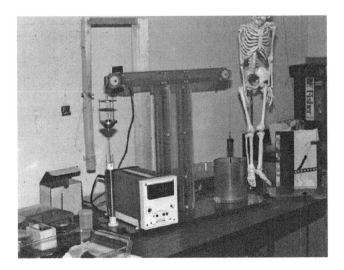

FIGURE 4.4   Isotonic (left) and isoionic (right) test fixtures for PAN fiber artificial muscles.

The isotonic apparatus consisted of a container made of the acrylic polymer where the PAN muscle was fastened on its bottom to a center eyebolt. This container was such that it could be removed and detached from the PAN muscle to empty the pH solutions after completing the test. The other end of the PAN fiber was connected to a balancing counterweight to neutralize the metal container's weight and transducer attachments on the other side of the apparatus. The counterweight was then attached to a nylon string, which was routed around two pulleys (to minimize the effect of friction) and then attached to a metal container on the other side of the apparatus. This container was secured from the bottom to a linear variable differential transducer with 0.001 mm accuracy. The transducer was directly connected to a digital indicator that converted the analog signal from the transducer (resulting from the displacement of the weight container) and at the same time conditioned the signal to correct for any temperature or pressure variation in the environment.

4.3.1   ISOTONIC CHARACTERIZATION

Consider Figure 4.4. To operate in isotonic mode, a known weight was added to the left (metal) container to put a tension force across the PAN muscle in the right (acrylic) container. The muscle was

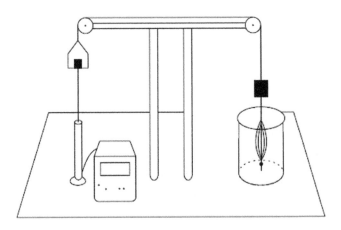

FIGURE 4.5   Schematic diagram of the isoionic test fixture.

FIGURE 4.6 Isoionic test fixture shown with a 25-fiber bundle PAN muscle.

then sprayed with a 1-M solution of HCl to initiate contraction. This caused the known weight on the other side to be lifted (pulled), indicating a positive reading on the digital displacement indicator. When full contraction of muscles was reached, the final readout was recorded for the known weight in gram force (gf) units. This can be repeated for any weights to get a relation between constant tension and muscle contraction. Similarly, the test can be repeated using a weak alkaline solution such as 1-M NaOH and recording the muscle relaxation versus a known tensile force. This is known as isotonic characterization.

4.3.2 ISOIONIC TEST

As seen in Figure 4.6, this fixture consists of an all-Lexan-made apparatus, especially resistant to corrosive materials. The apparatus is made up of a needle calibrated to show various displacements as the PAN muscle expands and contracts to show a correlation between displacement and a stable pH environment. A displacement can be recorded by applying a constant pH solution and letting the muscle reach the equilibrium position. Repeating this for several different pH conditions will give a relation to isoionic characterization of the muscle.

4.3.3 PAN SYNTHETIC MUSCLES' CAPABILITY MEASUREMENTS

Initial evaluation was performed for the linear strain capability of the PAN muscles under no applied load. Both regular PAN and nano-PAN exhibited large strain generation capabilities. Their linear strain capability was approximately 80 and 45% for the regular PAN muscles and the nano-PAN muscles (between 1-N HCl and 1-N NaOH). A total of ten cycles were carried out. The regular PAN showed nearly no degradation in performance. However, the nano-PAN fibers showed a significant reduction in their performance. At the end of ten cycles, only 20% strain generation capability was observed. Also, the structural deformation was observed for the nano-PANs.

4.3.3.1 SEM Studies

Figures 4.7–4.9 show annealed (cross-linked) PAN before activation. The process condition was heated at 210°C for 75 min. Note that the PAN fiber diameter is approximately 7 μm. Also, the surface morphology shows somewhat structured textures believed to be the consequence of polymeric cyclic macromolecular reconfiguration with attendant dehydrogenation. However, it can be expected that more close-chain systems may be achieved at higher temperatures.

Figures 4.10–4.13 show activated PANs at low-pH conditions (1-N HCl) after ten cycles. It can be seen that under such a condition, the activated PANs show a somewhat larger diameter, ~8 μm, than that of annealed and cross-linked PANs. This could be due to the polymer network's electrostatic force after the activation (ionic nature). Another interesting fact is the

FIGURE 4.7   PAN (before activation) heated at 220°C for 1 h, 15 min.

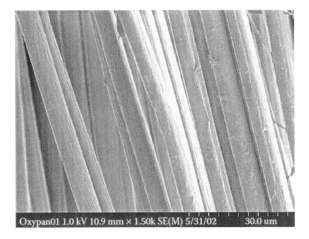

FIGURE 4.8   Annealed PAN (close-up).

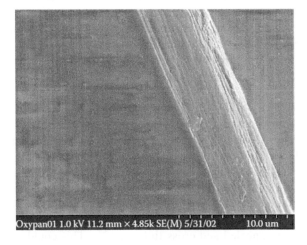

FIGURE 4.9   Annealed PAN, a single fiber (close-up). Its surface shows a texture that is believed to be an oxidized state of the fiber.

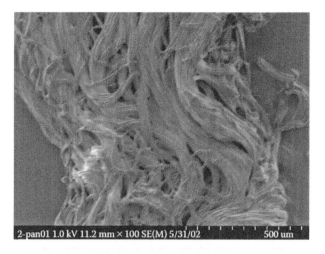

FIGURE 4.10 Activated PAN at a low-pH condition (1-N HCl).

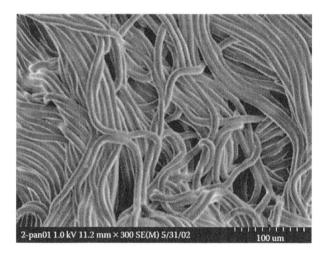

FIGURE 4.11 Activated PAN at a low-pH condition (1-N HCl).

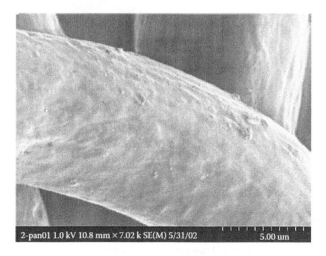

FIGURE 4.12 Activated PAN at a low-pH condition (1-N HCl).

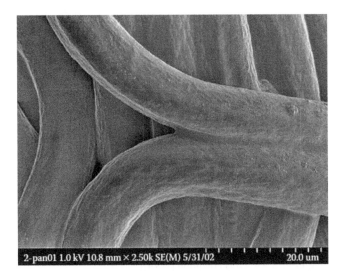

FIGURE 4.13   Activated PAN at a low-pH condition (1-N HCl).

changes in the surface textures (see Figures 4.9 and 4.12). Besides, some fibers stick together, as shown in Figure 4.13.

In Figures 4.14 and 4.15, the micrographs of activated PANs at a high pH condition (1-N NaOH) are shown. In these figures, the PAN fibers appear to be covered by salts (i.e., NaOH) but maintain their shape very well. Figures 4.16 and 4.17 clearly show that the PAN fibers' pilled-off close-up micrographs indicate that inner fiber (i.e., PAN) looks very clean with no visible damage. Figure 4.18 shows a single fiber having a diameter of approximately 9 μm.

The surface is very rough (Figure 4.19). This salt layer may act as a mass transfer resistance for actuation. Therefore, when the pH is reversed, such a salt layer may dominate the PAN muscles' response time.

Figures 4.20 and 4.21 show the raw nano-PAN fibers. Such fibers have a dimension of 250–300-nm diameter. These fibers were produced by an electrospinning technique based on

FIGURE 4.14   Activated PAN at a high-pH condition (1-N NaOH).

FIGURE 4.15 Activated PAN at a high-pH condition (1-N NaOH).

FIGURE 4.16 Activated PAN at a high-pH condition (1-N NaOH).

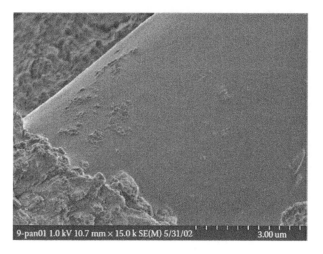

FIGURE 4.17 Activated PAN at a high-pH condition (1-N NaOH).

FIGURE 4.18 Activated PAN at a high-pH condition (1-N NaOH) (close-up).

FIGURE 4.19 Activated PAN at a high-pH condition (1-N NaOH).

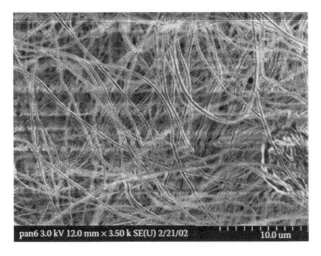

FIGURE 4.20 Raw nano-PAN fibers.

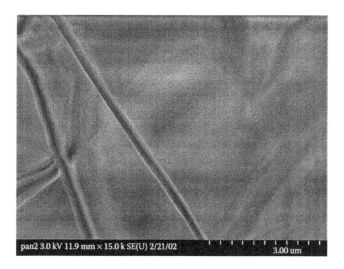

FIGURE 4.21   Raw nano-PAN fibers (close-up).

joint collaborative research and development between ERI and Santa Fe Science and Technology (SFST) Corporation.

In Figures 4.22 and 4.23, the activated nano-PAN fibers under a low pH condition (1-N HCl) are presented. As can be seen, the nano-PAN fibers were stuck together and deformed ("solvated" deformation) and lost their mechanical stability. This could be due to the lack of polymer cyclization with low dehydrogenation. A high-temperature heating process ($220°C < T < 250°C$) may enhance polymer cyclization and the close-chain system to increase the mechanical strength of the nano-PAN fibers.

We explored higher temperature processes for the nano-PAN fibers. In Figures 4.24 and 4.25, activated nano-PAN fibers at a high pH condition (1-N NaOH) were shown. The solvated and deformed nano-PAN fibers can be seen

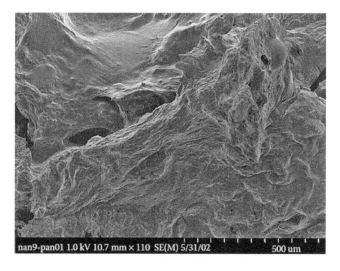

FIGURE 4.22   Activated nano-PAN at a low-pH condition (1-N HCl).

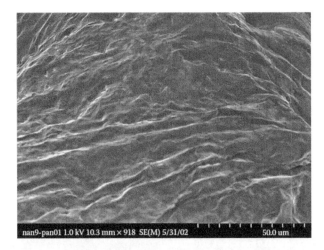

FIGURE 4.23   Activated nano-PAN at a low-pH condition (1-N HCl) (close-up).

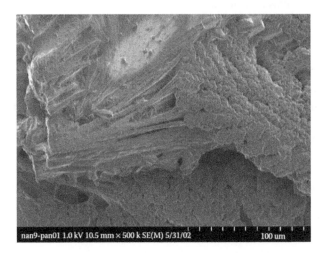

FIGURE 4.24   Activated nano-PAN at a high-pH condition (1-N NaOH).

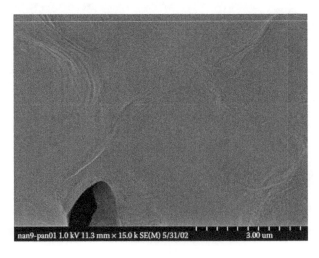

FIGURE 4.25   Activated nano-PAN at a high-pH condition (1-N NaOH) (close-up).

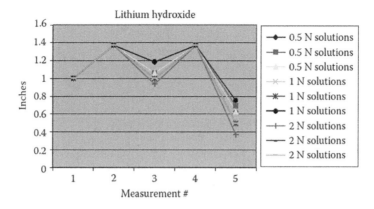

FIGURE 4.26   PAN fiber length change (lithium hydroxide and HCl).

4.3.4 EFFECTS OF DIFFERENT CATIONS

The effect of Li$^+$, Na$^+$ and K$^+$ on the contraction/elongation behavior of activated PAN fiber was also studied during this period. PAN fibers were tested in alkaline and acidic solutions of different normalities to determine their optimum contraction and elongation properties. Li$^+$-based PAN fibers exhibited the largest elongation/contraction performance. These results have been depicted in Figures 4.26–4.29.

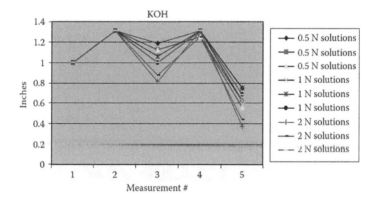

FIGURE 4.27   PAN fiber length change (potassium hydroxide and HCl).

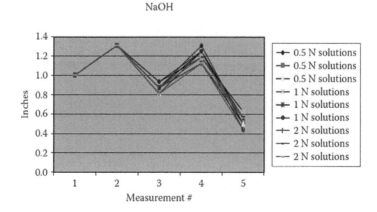

FIGURE 4.28   PAN fiber length change (sodium hydroxide and HCl).

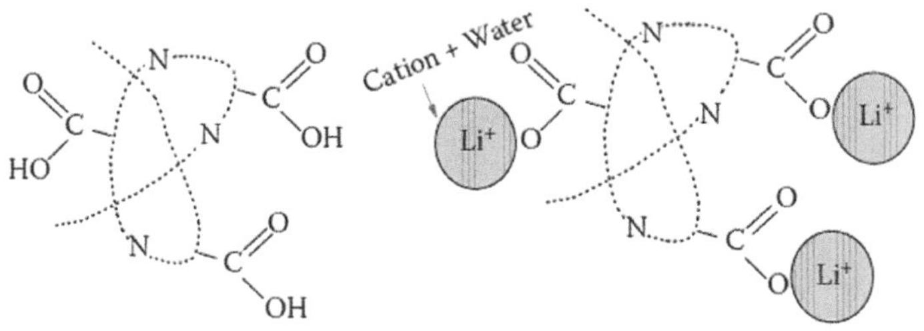

FIGURE 4.29 PAN elongation behavior is explained by osmotic pressure behavior.

4.3.5 ELECTRIC ACTIVATION OF PAN FIBERS

The effort was to design and fabricate a spring electrode configuration (Figure 4.31) using thin conductive wires (Figure 4.30) in a helical spring configuration with the PAN fiber encased in the middle. The whole assembly was encased inside a flexible membrane with some printed or embedded electrodes on the inside wall. Design issues were first to find the spring constants of both electrodes

![SEM image](nan9-pan01 1.0 kV 10.5 mm × 250 SE(M) 5/31/02   200 um)

FIGURE 4.30 Thin wires used as effective spring electrodes.

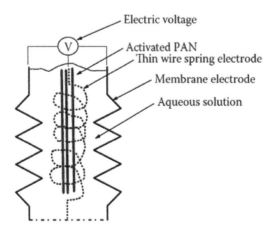

FIGURE 4.31 A PAN electrification configuration using a thin wire electrode and counter-membrane electrode.

and the selection of membrane materials. Conceptual development of hybridizing PAN and IPMNC for contractile fibers was recommended by ERI technical staff and was initiated at ERI laboratories.

4.3.6 ADDITIONAL RESULTS

PAN fibers were oxidized at 220°C for 90 min in the air. In the first experiment, these fibers were measured at 1 in. in length; each was activated in boiling 1-N KOH for 30 min. The fibers were then soaked in distilled water for 30 min to obtain a base length. Then, several fibers were placed in each of 0.5-, 1- and 2-N KOH for 30 min and measured. Next, the fibers were again put into distilled water for 30 min and then measured. Following this, the fibers were soaked in 0.5-, 1- and 2-N HCl solutions (corresponding fibers from the alkaline – for example, fibers from the 0.5-N KOH – were placed in 0.5-N HCl) and measured. Also, this process was repeated using NaOH and LiOH for the boiling and alkaline-soaking media.

The PAN fibers, regardless of whether activated in KOH, NaOH or LiOH, increased from their 1-in. initial length after being activated and soaked in distilled water. Lengths then decreased after the fibers were soaked in the bases. When the fibers were again soaked in distilled water, their lengths generally increased back to the originally distilled water lengths.

The fibers soaked in NaOH were the exception because they did not generally reach their previous distilled-water lengths. When the fibers were soaked in HCl, there was a much greater decrease in length than occurred with the alkaline media (see Figures 4.26–4.29 and Tables 4.1–4.3 for exact lengths and trends).

Fibers treated with LiOH (see Table 4.1 and Figure 4.26) had the largest increase in length following immersion in distilled water. Fibers soaked in all three media generally had the same decrease in length following immersion in the alkaline solutions as also occurred following immersion in HCl.

Especially noticeable with the fibers treated with LiOH was that greater displacement in the lengths occurred using the 2-N solutions. The lengths of fibers treated with NaOH (see Table 4.3 and Figure 4.28) were close to the same regardless of the normality of the solutions.

These findings are important to describe casual pH hysteric behaviors that have been reported previously. In Figure 4.29, the importance of osmotic pressure is illustrated. It should be noted that the maximum displacement could be determined when conditions are switched between pure water to acidic conditions.

TABLE 4.1
Experimental Results (Lithium Hydroxide)

	Fiber 1 (0.5 N)	Fiber 2 (0.5 N)	Fiber 3 (0.5 N)	Fiber 4 (1 N)	Fiber 5 (1 N)	Fiber 6 (1 N)	Fiber 7 (2 N)	Fiber 8 (2 N)	Fiber 9 (2 N)
Measurement 1 (in.)[a]	1	1	1	1	1	1	1	1	1
Measurement 2 (in.)[b]	1 3/8	1 3/8	1 3/8	1 3/8	1 3/8	1 3/8	1 3/8	1 3/8	1 3/8
Measurement 3 (in.)[c]	1 3/16	1 1/16	1 1/16	1	1	1 3/16	15/16	1	1
Measurement 4 (in.)[d]	1 3/8	1 3/8	1 3/8	1 3/8	1 3/8	1 3/8	1 3/8	1 3/8	1 3/8
Measurement 5 (in.)[e]	5/8	11/16	5/8	1/2	1/2	3/4	3/8	1/2	1/2

[a] Original fiber length.
[b] Length after 30 min of immersion in distilled water.
[c] Length after 30 min of immersion in varying normality of LiOH.
[d] Length after 30 min of immersion in distilled water.
[e] Length after 30 min of immersion in varying normality of HCl.

TABLE 4.2

Experimental Results (Potassium Hydroxide)

	Fiber 1 (0.5 N)	Fiber 2 (0.5 N)	Fiber 3 (0.5 N)	Fiber 4 (1 N)	Fiber 5 (1 N)	Fiber 6 (1 N)	Fiber 7 (2 N)	Fiber 8 (2 N)	Fiber 9 (2 N)
Measurement 1 (in.)[a]	1	1	1	1	1	1	1	1	1
Measurement 2 (in.)[b]	1 5/16	1 5/16	1 5/16	1 5/16	1 5/16	1 5/16	1 5/16	1 5/16	1 5/16
Measurement 3 (in.)[c]	1 3/16	1 1/8	1 1/8	1	1 1/16	1	13/16	7/8	1
Measurement 4 (in.)[d]	1 5/16	1 1/4	1 1/4	1 5/16	1 5/16	1 5/16	1 5/16	1 1/4	1 5/16
Measurement 5 (in.)[e]	3/4	5/8	9/16	5/8	3/4	11/16	3/8	7/16	11/16

[a] Original fiber length.
[b] Length after 30 min of immersion in distilled water.
[c] Length after 30 min of immersion in varying normality of KOH.
[d] Length after 30 min of immersion in distilled water.
[e] Length after 30 min of immersion in varying normality of HCl.

Note from Figure 4.29 that the left molecular structure is in a neutral state while the right molecular structure is under an alkaline solution. Therefore, if there is pure water in contact with the alkaline PAN, there will be an osmotic pressure-driven water influx.

Again note from Figure 4.31 that this configuration provides the membrane electrode functioning as a cathode and the thin conducting wire in a helical spring from being an anode shedding H[+]. It can cause the PAN fiber to contract within the helically configured wire and flexible membrane, and the wire will also contract like a helical spring with the PAN fiber. Once the polarity is changed, the PAN fiber tends to expand and the compressed helical spring will help it expand in a resilient manner.

4.3.7 ADDITIONAL EXPERIMENTAL RESULTS

PAN fibers were temperature treated at 230°C for 2 h. They were then boiled for 30 min in either 1-N NaOH or 1-N LiOH and stored in 1-N HCl. An apparatus was set up (see Figure 4.32) composed

TABLE 4.3

Experimental Results (Sodium Hydroxide)

	Fiber 1 (0.5 N)	Fiber 2 (0.5 N)	Fiber 3 (0.5 N)	Fiber 4 (1 N)	Fiber 5 (1 N)	Fiber 6 (1 N)	Fiber 7 (2 N)	Fiber 8 (2 N)	Fiber 9 (2 N)
Measurement 1 (in.)[a]	1	1	1	1	1	1	1	1	1
Measurement 2 (in.)[b]	1 5/16	1 5/16	1 5/16	1 5/16	1 5/16	1 5/16	1 5/16	1 5/16	1 5/16
Measurement 3 (in.)[c]	7/8	7/8	13/16	15/16	7/8	15/16	13/16	7/8	13/16
Measurement 4 (in.)[d]	1 5/16	1 1/8	1 1/4	1 3/16	1 1/4	1 1/4	1 1/8	1 3/16	1 1/8
Measurement 5 (in.)[e]	1/2	1/2	1/2	5/8	9/16	7/16	7/16	9/16	9/16

[a] Original fiber length.
[b] Length after 30 min of immersion in distilled water.
[c] Length after 30 min of immersion in varying normality of NaOH.
[d] Length after 30 min of immersion in distilled water.
[e] Length after 30 min of immersion in varying normality of HCl.

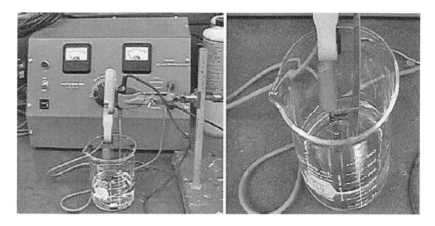

FIGURE 4.32 Photographs of the experimental setup.

of a stainless steel anode, a spring (serving as a counter-electrode) with one PAN fiber attached to it, a beaker filled with 150 mL of electrolyte solution and a 25-Amp Midas® Rectifier, which supplied the voltage. The electrolyte solution was either 0.1-N NaCl or 0.1-N LiOH, depending on the experiment.

The stainless steel anode was attached to the negative lead and held upright in the beaker of solution using a clamp. The spring with the fiber attached to the inside of it was attached to the positive lead. After the fiber and spring had been immersed in solution, the fiber was measured to see whether it had expanded at all. The spring and fiber were placed about 1 cm from the stainless steel anode. Note that the spring with the fiber was acting as the anode in this case.

The voltage was then turned on to 5 V for an approximate period of 1 min. The fiber was then measured. Following this, the polarity was switched and the spring electrode with the fiber became the cathode. Again, after 1 min at 5 V, the length of the fiber was measured.

4.3.7.1   Results

It was found that PAN fibers that had been boiled in lithium hydroxide, used with a 0.1-N sodium chloride electrolyte solution, had the best response to electrical activation (see Table 4.4 and Figure 4.33). These fibers, initially 1 in. long, expanded on average about 30%, to 1.3 in. when placed in the solution. After acting as an anode for 1 min at 5 V, the fibers, on average, were reduced in size to about 0.85 in. Then, as the fibers acted as cathodes, they increased in length to, on average, 1.05 in.

In contrast, fibers that had been boiled in lithium hydroxide and used with a 0.1-N LiOH solution did not do so well with electrical activation (see Table 4.5 and Figure 4.34). Upon being placed in the solution, the fibers expanded on average from 1 to 1.2 in. After acting as anodes, the fibers on

TABLE 4.4
LiOH PAN in NaCl Solution

	Initial Length (in.)	Initial Expansion (in.)	After Being Anode (in.)	After Being Cathode (in.)
Trial 1	1.000	1.200	0.938	1.188
Trial 2	1.000	1.380	0.938	1.188
Trial 3	1.000	1.100	0.875	1.063
Trial 4	1.000	1.200	0.750	0.875

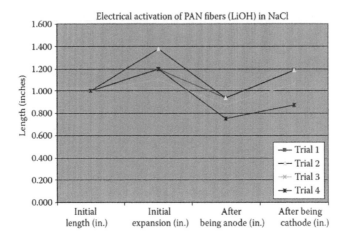

FIGURE 4.33   Electric activation of PAN fibers (LiOH) in NaCl.

TABLE 4.5
LiOH PAN in LiOH Solution

	Initial Length (in.)	Initial Expansion (in.)	After Being Anode (in.)	After Being Cathode (in.)
Trial 1	1.000	1.300	1.250	1.400
Trial 2	1.000	1.300	1.188	1.063
Trial 3	1.000	1.100	1.000	1.000

average reduced in size to 1.146 in. After acting as cathodes, the fibers on average increased in size to 1.15 in. We believed that the LiOH solution contained too much OH^- ions, so as to force the PAN maintaining the expanded state. However, at this moment, the true role of OH^- is not well understood. A counter-experiment with a low concentration of HCl as a solution is planned to investigate the effect of H^+ that potentially predetermines the contracted state but allowing expansion once OH^- ions are generated. Our previous PAN actuation mechanism described in Figure 4.24 cannot describe this observation fully.

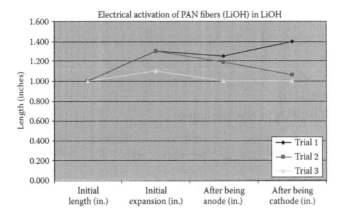

FIGURE 4.34   Electric activation of PAN fibers (LiOH) in NaOH.

TABLE 4.6
NaOH PAN in NaCl Solution

	Initial Length (in.)	Initial Expansion (in.)	After Being Anode (in.)	After Being Cathode (in.)
Trial 1	1.000	1.050	0.969	1.063
Trial 2	1.000	1.050	1.000	1.000
Trial 3	1.000	1.100	0.938	1.000

Fibers that had been boiled in sodium hydroxide did the worst with electrical activation (see Table 4.6 and Figure 4.35). Upon being placed in sodium chloride solution, they increased on average from 1 to 1.07 in. After acting as anodes, the fibers on average reduced in size to 0.97 in. Acting as cathodes, they brought about an average increase in size of 1.02 in.

Based upon these observations, it was decided to dynamically monitor the solution pH conditions and the generative force (or displacement) simultaneously. The proper setup is shown in Figure 4.36.

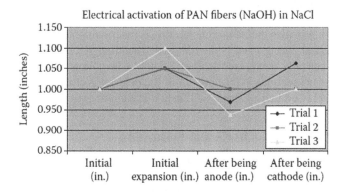

FIGURE 4.35 Electric activation of PAN fibers (NaOH) in NaCl.

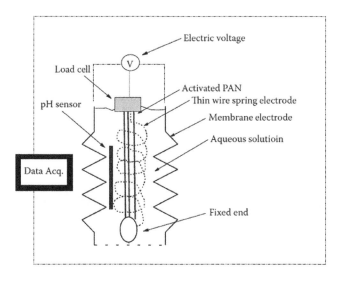

FIGURE 4.36 A PAN electrification configuration using a thin wire electrode and counter-membrane electrode.

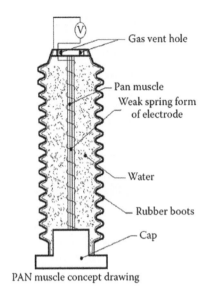

PAN muscle concept drawing

FIGURE 4.37 PAN actuator system design.

This configuration provides the membrane electrode functioning as a cathode and the thin conducting wire in a helical spring from being an anode shedding H^+. It can cause the PAN fiber to contract within the helically configured wire and flexible membrane, and the wire will also contract like a helical spring with the PAN fiber. Once the polarity is changed, the PAN fiber tends to expand, and the compressed helical spring will help it expand in a resilient manner.

4.3.8 PAN ACTUATOR SYSTEM DESIGN AND FABRICATION

The configuration in Figure 4.37 shows the updated design of the PAN actuator system. It can cause the PAN fiber to contract within the flexible membrane (bellows-type rubber boots) as shown in Figure 4.38. Once the polarity is changed, the PAN fiber tends to expand, and the compressed flexible membrane will help it expand in a resilient manner. The fabrication of this unit has been under way for performance testing.

4.3.9 PAN PERFORMANCE TESTING IN AN ACIDIC ENVIRONMENT

The objective of this experiment was to determine the contraction and elongation lengths of PAN fibers (cooked in 1-N LiOH for 30 min) during electrical activation in 0.1-M HCl. It was found that, while the fibers contracted to a much greater extent than fibers electrically activated in NaCl, they did not tend

FIGURE 4.38 Rubber boots.

FIGURE 4.39   Spring electrodes. Left: uncoated; right: gold coated.

to elongate at all. It was believed that the acid environment presets the condition as a contracted state, although the normality is low. In fact, the fibers generally contracted even more rather than elongating upon a change in polarity. It was concluded that, unless contraction of the fibers is the sole goal of the application, NaCl was again the best electrochemical solution in which to conduct electrical activation.

4.3.9.1   Procedures

PAN fibers were cooked at 230°C for 2 h. They were then boiled for 30 min in 1-N LiOH and stored in 1-N HCl. An apparatus was set up composed of a stainless steel anode, a 1.375-in. spring (gold coated; see Figure 4.39) with one 1-in.-length PAN fiber attached to it, a beaker filled with 150 mL of 0.1-N HCl solution and a 25-amp Midas Rectifier, which supplied the voltage.

The stainless steel anode was attached to the negative lead and held upright in the beaker of solution using a clamp. The spring with the fiber attached to the inside of it was attached to the positive lead. After the fiber and spring had been immersed in solution, the fiber was measured to see whether it had contracted at all.

The spring being served as the electrode and fiber was placed about 1 cm from the stainless steel anode. Note that the spring with the fiber was acting as the anode in this case. The voltage was then turned on to 5 V for a period of 1 min. The fiber was measured. Following this, the polarity was switched and the spring with the fiber became the cathode. Again, after 1 min at 5 V, the length of the fiber was measured.

4.3.9.2   Results

The PAN cooked in LiOH and used with HCl solution (see Table 4.7 and Figure 4.40) contracted far more than did the PAN cooked in LiOH and used with NaOH solution. (Table 4.8 and Figure 4.41

TABLE 4.7
Results of LiOH PAN Activated in 0.1-N HCl

	Initial Length (in.)	After Immersion in HCl (in.)	After Acting as an Anode (in.)	After Acting as a Cathode (in.)
Trial 1	1	1/2	3/8	1/4
Trial 2	1	2/5	3/8	7/16
Trial 3	1	1/2	7/16	13/32
Trial 4	1	4/5	11/16	1/2
Trial 5	1	4/5	3/4	5/8

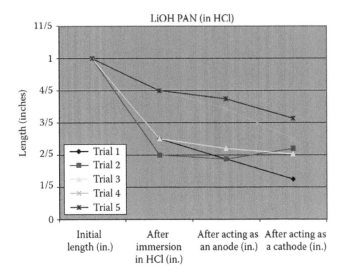

FIGURE 4.40   Results of LiOH PAN activated in 0.1-N HCl.

TABLE 4.8
Previous Results of LiOH PAN Activated in 0.1-N NaCl

	Initial Length (in.)	Initial Expansion (in.)	After Being Anode (in.)	After Being Cathode (in.)
Trial 1	1.000	1.200	0.938	1.188
Trial 2	1.000	1.380	0.938	1.188
Trial 3	1.000	1.100	0.875	1.063
Trial 4	1.000	1.200	0.750	0.875

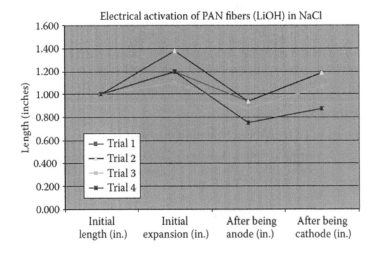

FIGURE 4.41   Previous results of LiOH PAN activated in 0.1-N NaCl.

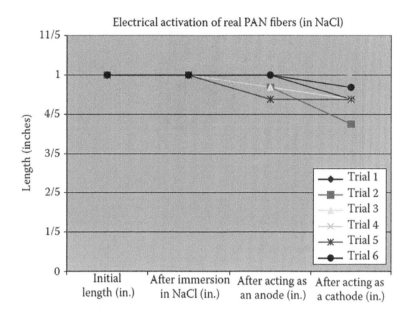

FIGURE 4.42 Electrical activation of real (pure) PAN fibers in 0.1-N NaCl.

are added to aid in this comparison.) However, the former displayed almost no characteristics of elongation. Additional test results are depicted in Figures 4.42 and 4.43.

4.3.10 PAN Film Casting Experiment

The objective of this experiment was to fabricate PAN in a film shape (as cast) to create a situation similar to an electrically controllable bender-type actuator. Figure 4.44 shows a PAN film after cross-linkage. Although it shows somewhat irregular shapes (due to bubble formation during solvent evaporation), it is anticipated that such a PAN will be smooth-film shaped after activation. This effort was undertaken. The Pt electrode deposition was attempted with a metal reduction process similar to the method attempted previously.

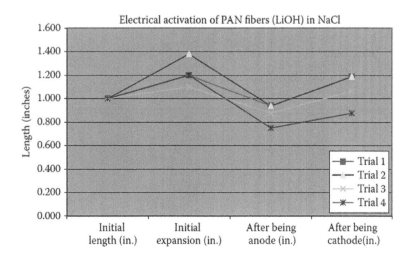

FIGURE 4.43 Electrical activation of regular PAN fibers in 0.1-N NaCl.

FIGURE 4.44   PAN film cast after cross-linking.

4.3.11   PAN Actuator System Design and Fabrication

The configuration in Figure 4.45 shows the fabricated PAN actuator system. The dimension is provided. It can cause the PAN fiber to contract within the flexible membrane (rubber boots). Once the polarity is changed, the PAN fiber tends to expand and the compressed flexible membrane will help it expand in a resilient manner. The fabrication of this unit was completed and it is now ready for performance testing. The instrumentation has been completed. Note that a leak-tight system is important.

First, we measured spring constant of the rubber boots by applying predetermined loads. The measurement gave the spring constant of $k = 0.01$ kg/mm. Inside the rubber boots, the following components were positioned as can be seen: the PAN muscle bundle (0.2 g, 15 strains), electrodes and a solution. Applying electrical currents through the electrodes can perform the system operation. The inner electrode (a circular shape) surrounds the PAN muscle bundle, and the other is attached to the boots' wall. The clearance between the boots' wall and the inner electrode is approximately 15 mm.

PAN fibers were cooked at 220°C for 90 min and boiled in 1-M LiOH for 30 min based upon the recent finding that LiOH boiling gives better performance in terms of elongation/contraction (Figure 4.46). The PAN fiber bundles were looped. The PAN muscle bundle (Figure 4.47) was placed between the top and bottom caps. PAN fibers were hung up from both hooks of the cap, then immersed into the water filling the insides of the rubber boots as shown in Figure 4.48.

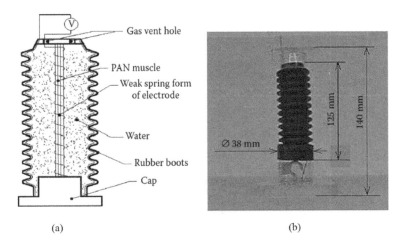

(a)   (b)

FIGURE 4.45   A concept drawing of (a) PAN muscle system and (b) fabricated model.

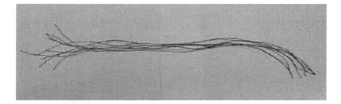

FIGURE 4.46 Annealed (cross-linked) PAN fibers.

FIGURE 4.47 Looped PAN fiber bundle.

FIGURE 4.48 Bottom and top caps. The bottom cap (left) places a magnetic stirrer that stirs the solution during the electric polarity change.

4.3.12 PAN Casting Experiment

A thin 1-mm PAN sheet was activated with high pH. Furthermore, nanocomposites of PAN/Pt via ion-exchange processes and a metal reduction process [Pt(II) → Pt(0)] was conducted. It appears that the Pt reduction is fast so that Pt particles precipitate in the solution rather than inside the PAN fibers (fast kinetics).

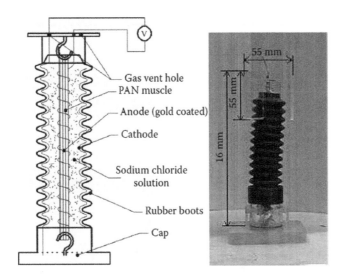

FIGURE 4.49 Revised design of PAN muscle system (left) and a fabricated model (right).

4.3.13 FURTHER PAN ACTUATOR SYSTEM DESIGN AND FABRICATION

Figure 4.49 shows the revised design of the fabricated PAN actuator system. First, to minimize the bending movement of the membrane boots, a guide set was added to the system. After inserting the cathode attached to the surface of the membrane boots, the spring constant was slightly increased to $k = 0.014$ kg/mm due to the added stiffness.

When the actuation occurs, the displacement of the membrane boots is measured. Subsequently, the force induced by the PAN muscle can be calculated. At high pH conditions, the PAN muscle usually turns into a gel, which exhibits an elastic behavior. Compared to raw PAN fibers, the mechanical properties of PAN fibers are weak (a modulus of 4–5 MPa and an ultimate stress of 1.5–2.0 MPa at elongated states; Schreyer et al., 2000). Therefore, it requires careful operation not to cause material damages or breakups.

We have gained a significant amount of experience for safe operation. In particular, any localized stress buildups should be eliminated. Much attention should be given to the connection between the cap and the PAN fibers. Both ends were tied up and fixed by epoxy to prevent loose bending.

Some difficulties have been found. The PAN fiber loop was broken several times during the initial shake-up testing of the system and caused the delay of the performance testing. Figure 4.50 depicts a loop of PAN fibers in a raw form and in an activated state.

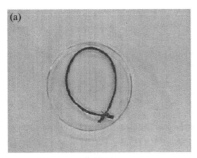
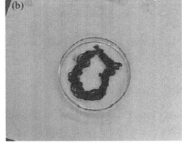

FIGURE 4.50 (a) Raw PAN fiber and (b) activated PAN fiber.

FIGURE 4.51 A PAN muscle system under an electric field: (a) initial and (b) after 15 min.

The PAN fibers were surrounded by a Cu-Zn spring-type anode coated by gold and immersed in a 0.1-M sodium chloride solution. We applied a typical operating voltage of 10 V to the system to operate it. Initially, we applied a D.C. current. A 2-mm displacement (equivalent to 28 gf) was maintained until 15 min had passed. When the time passed about 20 min, suddenly a large amount of current passed through the system, possibly due to the impurity buildups. The solution temperature was increased from 24°C (room temperature) to 35°C. After 20 min, the anode and the cathode was seemingly acting as an electric heat coil to boil the solution. It was also found that significant damage of the electrode (anode) occurs due to electrochemical corrosion. The gold plate was slipped off, and the anode surface was eroded. In this period, we were developing an improved gold-coating technique. Figure 4.51 depicts the electrical activation of active PAN fibers in an electrochemical cell subject to 7 V. To collect baseline data and to arrange criteria of electrical activation system design, we also carried out the PAN fiber contraction experiment in an acidic solution (1-M HCl; Figure 4.52).

From this experiment, we obtained 6-mm displacement that can be translated into a generative force of 84 gf (= 0.82 N). This information gives rise to the fact that the present system can produce an ultimate force of 0.82 N.

FIGURE 4.52 PAN fiber displacement in an HCl solution (1 M): (a) initial and (b) after 15 min positions.

FIGURE 4.53 Horizontal self-powered pH meter equipped with PAN fibrous muscle fibers and a resilient rotating cylinder.

4.4 PAN pH METERS

PAN fibers were used for a variety of applications including pH meters (Figures 4.53–4.56). In Figures 4.53 and 4.54, the fibrous PAN ionic muscles contract and expand and turn a dial gauge to indicate the pH of a solution.

Most conventional pH meters use some form of electric power source and are extremely sensitive to external environmental conditions such as temperature and pressure. This will limit the use of

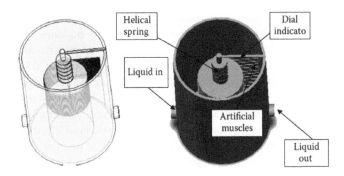

FIGURE 4.54 Vertical self-powered pH meters equipped with PAN fibrous artificial muscles and a resilient rotating cylinder.

FIGURE 4.55 Vertical pH meter assembly made from a parallel-type PAN fiber muscle and a specially designed container holding the fluid with unknown pH to be determined.

FIGURE 4.56   Linear pH meter using a graduated cylinder with calibrated PAN fiber bundle.

the meter to a controlled environment. Slight contamination of the probe tip or solution can cause erroneous readings. Although they can indicate up to several degrees of accuracy, their requirement for a power source makes it difficult to use them in remote field applications where there are often limited resources.

As shown in Figure 4.55, a simple cylindrical container made of acrylic was used to hold a PAN muscle arranged in parallel fibers much the same as pinnate muscles in the body. The ends of the fibers are secured to two glass rods of 1-mm diameter. Glass was used to avoid interfering with the corrosive fluids inside the container where the pH was unknown. The top and bottom caps of the container had slots and the muscle was wrapped around a core cylinder and put in tension via a retracting spring assembly located on the top cap assembly and secured to the center core of the container. By varying the pH of the solution, contraction or expansion of the muscle fiber assembly was initiated, causing the core drum to rotate. By a simple calibration and dial marking on the top plate, it was possible to get a crude pH reading in the range of 4–12. Although not entirely accurate readings, this assembly was perhaps the simplest pH meter that did not require a power source such as a battery to indicate the pH of a solution.

Another type of pH meter designed was a linear gauge type. This device was simpler in construction but required a larger container, such as a graduated glass cylinder that was carefully calibrated with various pH solutions to mark points along the cylinder indicating a range of pH of 2–12 (see Figure 4.56). For this type, a fiber bundle of the PAN attached from one end to a weight imposing a constant tension across the muscle was used. The muscle was suspended in pure water free to expand from one end and fixed from the top. When low pH (acidic) was applied, the muscle contracted and pulled the weight upward. A needle at the tip of the weight indicated the change in pH.

4.4.1 Skeletal Muscles Made with Fibrous PAN Artificial Muscles

Figures 4.57–4.59 depict the skeletal muscles of the forearm. Figure 4.58 represents an apparatus designed to test a biceps artificial muscle made from PAN fibers encapsulated in latex membrane. A multichannel peristaltic pump was used to direct fluid from each of the acid, base and deionized water reservoirs into and out of the muscle assembly. The microcontroller board incorporated a Motorola 6811 series processor to control opening and closing of appropriate solenoid valves, allowing pH solutions to flow when a command was executed.

FIGURE 4.57   Circulatory system assembly pumping pH solutions into a biceps PAN artificial muscle.

A PC interface was used to trigger the analog signals causing the valve sequence of closing and opening based on an up or down command requirement of the forearm. The fibers were accordingly contracted or relaxed to simulate the biological muscle contraction-relaxation. This system was essentially a closed-loop circulatory system with each fluid entering and leaving the muscle fiber separately. The deionized water was used as a rinse cycle between acid or base wash for contraction and relaxation, respectively. Platform-type spring loaded actuators were also designed and built for a series of experiments (Figure 4.60) including an electroactive PAN muscle experiment.

4.5   ELECTROACTIVE PAN MUSCLES

Artificial muscles made with PAN fibers are traditionally activated in electrolytic solution by changing the pH of the solution by the addition of acids and/or bases. This usually consumes a considerable amount of weak acids or bases. Furthermore, the synthetic muscle (PAN) has to be impregnated with an acid or a base and must have an appropriate enclosure or provision for waste collection after actuation.

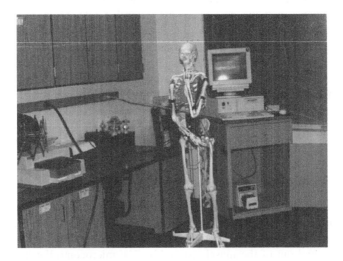

FIGURE 4.58   PAN biceps muscle shown in a test apparatus consisting of a multichannel pump, microcontroller board, solenoid valves and a desktop PC to activate a skeletal forearm.

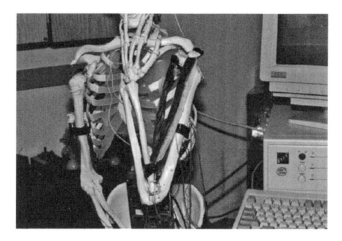

FIGURE 4.59 Close-up of the attachment of the fabricated biceps PAN muscle on a life-size human fore-arm skeleton showing the controller board housed in the pelvic bone area and solenoid valves attached to the humerus and ulna bones, respectively. We also designed and built a simple pH meter with PAN fiber bundles using the parallel-type packaging of the fibers as shown earlier (Figure 4.55). The picture shown is a rotary-type pH meter taking up a small space and fairly accurate for most cases.

This work introduces a method by which the PAN muscle may be elongated or contracted in an electric field. It is believed that for the first time, this has been achieved with PAN fibers as artificial muscles. In this new development, the PAN muscle is first put in close contact with one of the two platinum wires (electrodes) immersed in an aqueous solution of sodium chloride. Applying an electric voltage between the two wires changes the local acidity of the solution in the regions close to the platinum wires (Figures 4.60 and 4.61). This is because of the ionization of sodium chloride molecules and the accumulation of Na^+ and Cl^- ions at the negative and positive electrode sites, respectively.

This ion accumulation, in turn, is accompanied by a sharp increase and decrease of the local acidity in the regions close to either of the platinum wires. An artificial muscle, in close contact with the platinum wire, because of the change in the local acidity will contract or expand depending on the polarity of the electric field. This scheme will allow the experimenter to use a fixed flexible

FIGURE 4.60 Linear platform actuator for use in robotics.

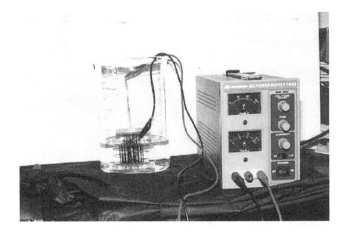

FIGURE 4.61 Electric PAN muscle apparatus showing gold-plated center rod as one electrode and gold-plated spring as the circumferential electrode.

container of an electrolytic solution, the local pH of which can be modulated by an imposed electric field while the produced ions are trapped to stay in the neighborhood of a given electrode.

This method of artificial muscle activation has several advantages. First, the need to use a large quantity of acidic or alkaline solutions is eliminated; second, the use of a compact PAN muscular system is facilitated for applications in active musculoskeletal structures. Third, the PAN muscles become electrically controllable, and therefore, the use of such artificial muscles in robotic structures and applications becomes more feasible (Figure 4.62). In this way, a muscle is designed such that it is exposed to either Na^+ or Cl^- ions effectively. In the following paragraphs, muscle contraction or expansion characteristics under the effect of the applied electric field are discussed.

As was discussed before, ionic polymeric gels are three-dimensional networks of cross-linked macromolecular polyelectrolytes that swell or shrink in aqueous solutions upon addition of alkali or acids, respectively. Reversible volumetric dilation and contraction of the order of more than 800% for PAN fibers have been observed in our laboratory. Furthermore, it has been experimentally observed that swelling and shrinking of ionic gels can be induced electrically. Thus, direct

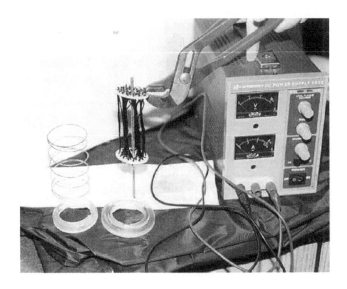

FIGURE 4.62 Exploded view of the electric PAN muscle apparatus.

computer control of large expansions and contractions of ionic polymeric gels by means of a voltage gradient appears to be possible.

These gels possess an ionic structure in the sense that they are generally composed of a number of fixed ions (polyions) pertaining to sites of various polymer cross-links and segments and mobile ions (counterions or unbound ions) due to the presence of an electrolytic solvent. Electrically induced dynamic deformation of ionic polymeric gels such as polyacrylic acid (PAA) plus sodium acrylate cross-linked with bisacrylamide (PAAM), poly(2-acrylamido-2-methyl-1-propanesulfonic acid) (PAMPS) or various combinations of chemically doped PAA plus polyvinyl alcohol (PAA-PVA) can be easily observed in our laboratory. Such deformation gives rise to an internal molecular network structure with bound ions (polyions) and unbound or mobile ions (counterions) when submerged in an electrolytic liquid phase.

In the presence of an electric field, these ionic polymeric networks undergo substantial contraction accompanied by exudation of the liquid phase contained within the network. Under these circumstances, there are generally four competing forces acting on such ionic networks: the rubber elasticity, the polymer–liquid viscous interactions due to the motion of the liquid phase, inertial effects due to the motion of the liquid through the ionic network and the electrophoretic interactions. These forces collectively give rise to dynamic osmotic pressure and network deformation and subsequently determine the dynamic equilibrium of such charged networks.

On the other hand, there are situations in which a strip of such ionic polymeric gels undergoes bending in the presence of a transverse electric field with hardly any water exudation. Under these circumstances there are generally three competing forces acting on the gel polymer network: the rubber elasticity, the polymer–polymer affinity and the ion pressure. These forces collectively create the osmotic pressure, which determines the equilibrium state of the gel. The competition between these forces changes the osmotic pressure and produces the volume change or deformation. Rubber elasticity tends to shrink the gel under tension and expand it under compression. Poly-polymer affinity depends on the electrical attraction between the polymer and the solvent. Ion pressure is the force exerted by the motion of the cations or anions within the gel network. Ions enter the gel attracted by the opposite charges on the polymer chain while their random motions tend to expand the gel like an ionic (fermionic) gas.

In the next sections, the cases of electrically induced contraction of PAN muscles are experimentally and theoretically described. Exact expressions are given relating the deformation characteristics of the gel to the electric field strength or voltage gradient, gel dimensions and other physical parameters of the gel.

4.6 ELECTROCHEMOMECHANICAL ACTUATION IN CONDUCTIVE POLYACRYLONITRILE (C-PAN) FIBERS AND NANOFIBERS

4.6.1 INTRODUCTION

Electrical activation of contractile ionic polymeric fibers dates back to the pioneering works of Kuhn (1949), Katchalsky (1949), Kuhn et al. (1950) and Hamlen et al. (1965). In recent years, Osada and Hasebe (1985), De Rossi et al. (1986), Chiarelli and De Rossi (1988), Chiarelli et al. (1989), Caldwell and Taylor (1990), Segalman et al. (1991), Umemoto et al. (1991), Shahinpoor et al. (1996, 1997a, 1997b) and Schreyer et al. (1999, 2000) have further contributed toward understanding of electrically controllable contractile ionic polymeric fibers. Figure 4.63 depicts a number of devices made in the early development of contractile fibrous synthetic chemomechanical muscles.

Activated PAN fibers that are suitably annealed, cross-linked and hydrolyzed are known to contract and expand when ionically activated with cations and anions, respectively. The change in length for these pH-activated fibers is typically greater than 100%, but up to 200% contraction/ expansion of PAN fibers has been observed in our laboratories at AMRI. They are comparable in

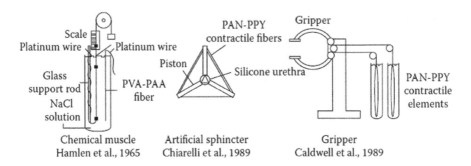

FIGURE 4.63 A number of early contractile synthetic chemomechanical muscles.

strength to human muscles (~20 N/cm^2) and have the potential of becoming medically implantable electroactive contractile artificial muscle fibers.

Increasing the conductivity of PAN by making a composite with a conductive medium such as a noble metal such as gold or platinum or graphite, carbon nanotube or a conductive polymer such as polypyrrole or polyaniline (C-PAN) has allowed for electrical activation of C-PAN artificial muscles when it is placed in an electrochemical cell. The electrolysis of water in such a cell produces hydrogen ions at a C-PAN anode, thus locally decreasing the cation concentration and causing the C-PAN muscle to contract. Reversing the electric field allows the C-PAN muscle to elongate. Typically, close to 100% change in C-PAN muscle length in a few minutes is observed when it is placed as an electrode in a 10-mM NaCl electrolyte solution and connected to a 20-V power supply. Nanofibrous versions of C-PAN or C-PAN nanofibers (C-PAN-Ns) obtained recently by electrospinning have allowed the response time to be reduced to a few seconds. These results indicate the potential in developing electrically activated C-PAN-N artificial nanomuscles and linear nanoactuators. Furthermore, these results present a great potential for using electroactive fiber bundles of C-PAN-Ns as artificial sarcomeres and artificial muscles for linear actuation as depicted in Figure 4.64.

Activated C-PAN contracts when exposed to protons, H$^+$, in an aqueous medium and elongates when exposed to hydroxyl ion, OH$^-$, in a strong alkaline medium. The length of activated PAN fibers can potentially more than double when going from short to long. A possible explanation for the contraction and elongation of activated C-PAN is the effect carboxylic acid groups (COO$^-$ and H$^+$) have on the molecular geometry. At high cationic concentration, all acid groups are protonated, potentially collapsing the network and contracting the polymer.

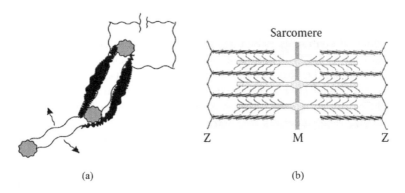

FIGURE 4.64 A possible configuration for the electroactive C-PAN-N artificial muscle in (a) an antagonist configuration to provide biceps and triceps (b) similar to the action of a sarcomere.

On the other hand, the presence of hydroxyl ions causes the protons to detach from the carboxylic groups and form water. The negatively charged carboxylic groups would then cause the cross-linked polymer network to expand by coulombic repulsion and attract the water molecules into the expanded network. These underlying mechanisms of expansion and contraction are considered later for the modeling of activated PAN fibers in an electrochemical cell in the presence of an imposed electric field.

4.6.2 PREPARATION OF IONICALLY ACTIVE PAN

Raw PAN fibers, which are composed of roughly 2,000 individual strands of PAN (each about 10 μm in diameter), are first annealed at 220°C for 2 h. The fibers can be bundled together at this point to form a PAN muscle. The PAN is then placed in a boiling solution of 1-N LiOH for 30 min, after which the PAN fibers become hyper elastic like a rubber band.

At this point, the PAN can be ionically activated. Flooding the activated PAN with a high concentration of cations such as H^+ induces contraction while the presence of hydroxyl ions (OH^-) causes the polymeric network to elongate the fibers. An assortment of such PAN muscles is shown in Figure 4.1.

PAN has been studied for more than a half century in many institutions and now is popularly used in the textile industry as a fiber form of artificial silk. The useful properties of PAN include the insolubility, thermal stability and resistance to swelling in most organic solvents. Such properties were thought to be due to the cross-linked nature of the polymer structure. However, a recent finding has shown that a number of strong polar solvents can dissolve PAN and, thus, has raised a question that its structure could be of the linear zigzag conformations (Umemoto et al., 1991; Hu, 1996) with hydrogen bonding between hydrogen and the neighboring nitrogen of the nitrile group. Yet, the exact molecular structure of PAN is still somewhat unclear. (We expect that it can be atactic with possibly amorphous phases in pure state.)

Also, it should be noted that, as received, PAN fibers are a somewhat different formulation from that of the pure PAN that is usually in a powder state. Generally, the PAN fibers used in the textile industry are often copolymerized with acrylamide (approximately 4–6% by weight) and manufactured through a spinning process. Therefore, changes in material properties of such PAN fibers are expected in a sense that it can be reorganized into a semicrystalline structure. Therefore, improved strength of the PAN fibers is a definite advantage over other types of ionic gels that are usually very weak.

Activated PAN fibers are highly elastic or hyperelastic like rubber bands. The term "activated PAN fibers" refers to the PAN fibers acting as the elastic fibers of which length varies depending upon the ionic concentration of cations in the solution. They contract at low cationic concentration and expand at high cationic concentration. At high cationic concentration, they appear to reject water out of the polymer network, resulting in shrinkage due to their elastic nature. Conversely, they elongate at low cationic concentration, attracting water from outside into the polymer network. In our laboratory, it has been observed that activated PAN fibers can contract more than 200% relative to that of the expanded PAN. The use of PAN as artificial muscles is promising since they are able to convert chemical energy to mechanical motion, possibly acting as artificial sarcomeres or muscles.

Other types of materials that have a capability to be electroactive are polyelectrolyte gels such as polyacrylamide (PAM), PAA-PVA and PAMPS. Under an electric field, these gels are able to swell and de-swell, inducing large changes in the gel volume. Such changes in volume can then be converted to mechanical work. Artificial sarcomeres or muscles made from PAN, unlike polyelectrolyte gels, have much greater mechanical strength than polyelectrolyte gels, thus having a greater potential for application as artificial sarcomeres and muscles (or soft actuators). Figure 4.65 shows mechanical strength of the activated PAN bundle in a contracted state and an elongated state. Different mechanical behavior of the ionically active PAN fibers clearly establishes the fundamental difference between the network structures in the presence of H^+ ions or protons as compared to the presence of hydroxyl OH^- ions.

The activation of PAN requires inducing cross-linking by the formation of pyridine rings by low-temperature annealing and, subsequently, converting nitrile groups to carboxylic acid groups by

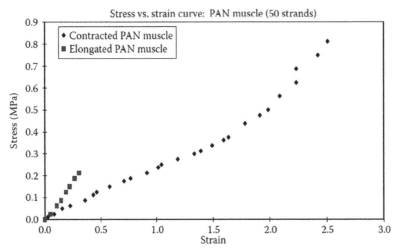

Stress vs. strain curve: PAN muscle (50 strands)

- ◆ Contracted PAN muscle
- ■ Elongated PAN muscle

Contracted muscle: Length under 0 load = 6.9 cm; maximum load applied = 650 gm (did not break); max length = 24.2 cm.
Elongated muscle: Length under 0 load = 15.5 cm; maximum load applied = 200 gm (broke under load); max length = 20.3 cm.
Estimated cross-sectional area of one strand = 1.57×10^{-7} meters squared; for 50 strands = 7.85×10^{-6} meters squared.

FIGURE 4.65 Normal stress–strain relationship for the contracted and the expanded state of PAN muscles.

saponification with sodium or lithium hydroxide. The degree of cross-linking depends on annealing temperature and time, which in turn determines the amount of free nitrile groups left to be converted to carboxylic acids during saponification.

A possible structure for activated PAN is given in Figure 4.66, as discussed by Umemoto et al. (1991) and Hu (1996). In Figure 4.67, scanning electron microscopy (SEM) micrographs show raw fibers as well as activated fibers. Typically, one strand is composed of about 2,000 fibers. Each fiber has an approximate diameter of 9 μm.

The strength of activated PAN and its ability to change length of up to 100% or more makes it an appealing material for use as linear actuators and artificial muscles. An attractive alternative is electrical activation. During the electrolysis of water, hydrogen ions are generated at the anode while hydroxyl ions are formed at the cathode in an electrochemical cell. Electrochemical reactions can then potentially be used to control the length of a PAN artificial muscle. This may be achieved by locating a PAN muscle near an electrode where the ions are generated or, if the conductivity of activated PAN can be increased, the PAN muscle can serve as the electrode.

The study reported here takes the second approach, where platinum is deposited on PAN fibers to increase conductivity, so that the muscle can serve as the electrode directly. This procedure of depositing Pt on the polymer and activating it in an electrochemical cell was initially demonstrated a few years ago by Hamlen et al. (1965) with a PVA–PAA copolymer. It resulted in about a 5% decrease in length, with contraction and elongation each taking about 12 min. However, our recent improvements to an electroactive nanofibrous version of PAN muscles or C-PAN-Ns have reduced

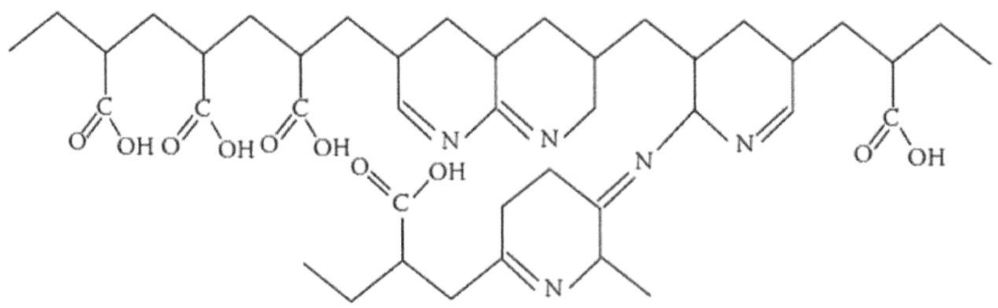

FIGURE 4.66 Possible structure of activated polyacrylonitrile.

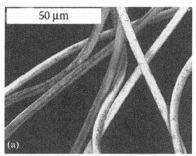
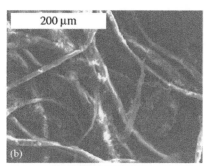

Raw PAN fibers with a small amount of platinum deposited

Contracted PAN fibers

Elongated PAN fibers

FIGURE 4.67 SEM micrographs that show (a) raw fibers, (b) contracted and (c) elongated states, respectively. It should be noted that SEM micrographs were taken for the dry samples. The elongated PAN fibers show that they contain a salt (possibly NaCl).

the activation time to a few seconds, with muscle strength approaching human muscles in the range of 20 N/cm^2 of muscle fiber cross section.

To activate as-received PAN fibers (Mitsubishi), first, they were oxidized in air at an elevated temperature range of 220–240°C. It is likely that the oxidation process makes a linear-like structure of as-received PAN fibers to a cross-linked structure of pyridine and cyano groups shown in Figure 4.65. The preoxidized PAN fibers show dark brown or black depending upon the level of cross-linking.

Second, they were saponified in an alkaline solution (2-N NaOH for 20–30 min). The elongation and contraction behavior of PAN fibers is interesting in the fact that hysteresis exists, changing upon pH values. In an approximate pH range of 3–10, the elongation process stays below the contraction process. This means that the equilibrium length of PAN fibers during contraction is larger than that of PAN fibers during elongation.

Based upon the Donnan theory of ionic equilibrium (Flory, 1953b), we believe that important forces arise from (1) induced osmotic pressure of free ions between activated PAN fibers and their environment; (2) ionic interaction of fixed ionic groups; and (3) the network itself. Among them, the induced osmotic pressure of free ionic groups could be the dominating force. Much study is needed to clarify this discussion.

Electrical activation of PAN fibers is performed in an electrochemical cell, shown in Figure 4.68. PAN fibers can be activated electrically by providing a conductive medium in contact with or within the PAN fibers. Such electrical activation can be made to have low overvoltage for hydrogen and oxygen evolution. At the anode, oxygen evolves via $2H_2O \Rightarrow O_2 + 4H^+ + 4e^+$ and the counterreaction at the cathode is $2H_2O + 2e^- \Rightarrow H_2 + 2OH^- + 4e^+$.

Upon being hydrogenated in the vicinity of the PAN anode, the decreased pH causes the PAN fibers to contract by the same effect as chemical activation. Also, reversing the polarity of D.C., elongation of PAN fibers is simply obtained.

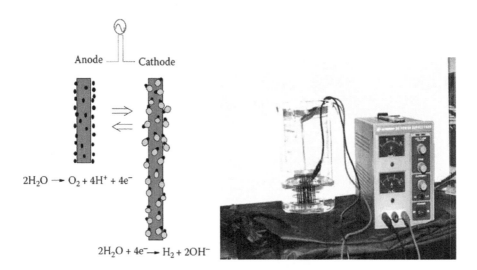

Anode   Cathode

$2H_2O \rightarrow O_2 + 4H^+ + 4e^-$

$2H_2O + 4e^- \rightarrow H_2 + 2OH^-$

FIGURE 4.68 Experimental setup for electrical activation of C-PAN artificial sarcomeres and muscles. It describes the operating principle of the C-PAN.

4.6.3 DIRECT METAL DEPOSITION TECHNIQUE

As a first trial, conductivity of PAN was increased by means of chemical deposition of platinum on PAN fibers. Raw PAN fibers were immersed in a tetraamineplatinum chloride monohydrate solution. The PAN fibers were then transferred to a reducing solution containing sodium borohydride. The solution was slowly heated to about 50–60°C with agitation at 100 rpm along with periodic additions of 5% $NaBH_4$ solution to reduce Pt metal. This process was repeated several times to seed the PAN fibers with Pt. After platinum deposition, the C-PAN fibers were activated by the method described previously.

The results of electrical contraction and elongation of a C-PAN platinum muscle are shown in Figure 4.69. As can be seen, the electric activation of C-PAN was successful to produce the

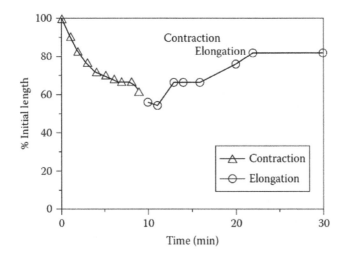

FIGURE 4.69 Electrical activation of muscle made up of fiber bundle of 50 C-PAN platinum fibers. Initial muscle length = 5.0 cm; number of fibers = 50; cell voltage = 20 V; current = 120 mA. Polarity of electrodes reversed at $t = 10$ min.

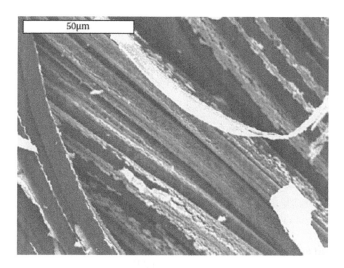

FIGURE 4.70   SEM micrograph shows platinum delamination over a number of cycles (right).

approximate mean rate of contraction of $L/L \approx 5\%/min$, while the approximate mean rate of elongation was $L/L \approx 3\%/min$. The anomaly in both contraction and expansion is believed to be due to the fact that the electroosmotic diffusion rate of H^+ cations and OH^- anions into and out of the C-PAN fiber bundle is not uniform. In this approach, we have found that delamination of Pt layer (see Figure 4.70) occurred over a repeated period of contraction and elongation of such C-PAN. As a result, a much higher overpotential was needed to produce the same contraction and elongation speed. In addition, it has been seen that the muscle expanded from the starting length but did not return to it, probably due to the resistive force induced by the Pt layer.

To address these problems, we have changed the electric activation scheme by using thin graphite fibers serving as an effective adjunct electrode circled around and combined with PAN and gold as a counterelectrode.

4.6.4   GRAPHITE AND GOLD FIBER ELECTRODE WOVEN INTO PAN MUSCLE
AS AN ADJUNCT ELECTRODE

The idea of using a graphite or gold fiber electrode was motivated by the fact that it could serve as an effective electrode producing necessary ions for contraction and elongation, respectively. Another advantage is its chemical/mechanical endurance in chemical environments during its activation process and contraction/elongation processes. Also, the problems encountered in the direct metal deposition technique could be adequately addressed. Figure 4.71 shows an SEM photograph of graphite fibers employed and the configuration used in this study.

In Figures 4.72 and 4.73, the electric activation of C-PAN fiber bundles with graphite fiber electrodes is shown. Significantly improved response times relative to the previously reported C-PAN platinum fibers can be observed. In fact, contraction and elongation as fast as $L/L \approx 10\%/min$ and $L/L \approx 5\%/min$, respectively, can be observed.

The efficient way of producing and diffusing the necessary ions (H^+ and OH^-) and their distribution over the C-PAN fibers is the key to producing fast-reacting C-PAN fibers. Thus, single-fiber response was also tested. For a single fiber comprising 2,000 strands of 10-μm diameter C-PAN-G fiber, the contraction time reduced by about 60% compared to the fiber bundle of 50 fibers, as shown in Figure 4.74.

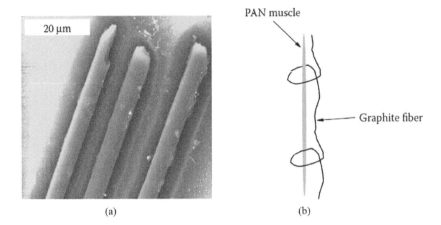

(a)   (b)

FIGURE 4.71   (a) SEM micrograph shows the graphite fibers used in this study. (b) Each fiber has a diameter of 6.4 μm and the configuration of the graphite electrode and PAN muscle.

The activation of a single strand of 10-μm fiber reduced the contraction time to a few seconds, as shown in Figure 4.75.

4.6.5   TOWARD NANOSCALE ARTIFICIAL MUSCLES AND MOTORS

Biological muscle is a magnificent nano- to micro- to macroscale actuating system with the capacity to perform diversified functions. According to Pollack (2001), the fundamental building blocks are microsized contractile units called *sarcomeres*, as shown in Figure 4.63. They contain three filament types: thick, thin and connecting. There has been a consensus that the muscle contraction is

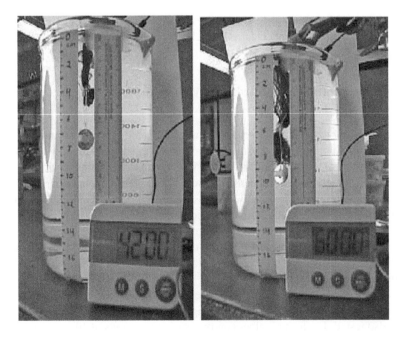

FIGURE 4.72   Variation of length of 100 C-PAN-G fibers in fiber bundle form with time in a 0.2-mN NaCl cell under an imposed electric field (1PAN-2G ratio, 100 fibers).

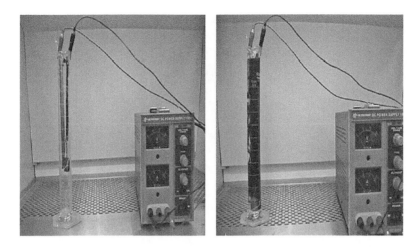

FIGURE 4.73 Variation of length of PAN-graphite muscle with time in a 0.2-mN NaCl cell under a voltage of 20 V (1PAN-2G ratio, 50 fibers).

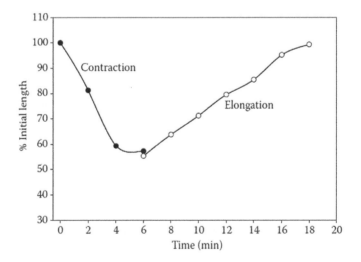

FIGURE 4.74 Electrical activation of C-PAN made with single PAN graphite fibers of 2,000 strands.

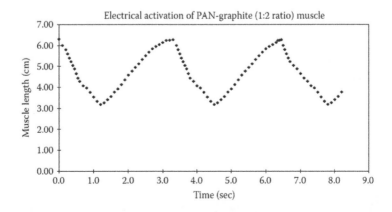

FIGURE 4.75 Variation of length of C-PAN-G strands of 10 μm in diameter with time in a 0.2-mN NaCl cell under a voltage of 20 V (1PAN-2G ratio in a special helically wound configuration).

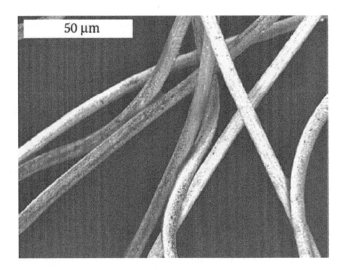

FIGURE 4.76   Conventional PAN fibers. The fiber diameter is 8–10 μm.

based upon thick and thin filaments' interaction. Connecting filaments link the thick filament to the end points of the sarcomere being served as molecular spring. Such filaments are polymeric materials: thin – monomeric *actin*, thick – *myosin* and connecting – vertebrates *titin* protein. With water, they form a gel-like lattice.

The textbook mechanism of muscle contraction is based upon the concept of "sliding filaments" where nanomotor (cross-bridge) rotation drives thin filament past thick. These nanosize cross-bridges, which are attached to the thin filament, swing and then detach in the presence of ATP, being ready for the next cycle. Such a mechanism is truly based upon a chemomechanical nanoscale motion that can ultimately cause biological muscle to contract in a collective manner to achieve macroscale motions with useful forces. Inspired by the fact that extremely fast response times can be attained by hierarchically moving toward smaller and smaller diameter fibers, just like the biological muscles, nanofibers of PAN were manufactured by electrospinning.

Conventional PAN fibers that are commercially available (Orlon) are shown in Figure 4.76. Normally, they are composed of 2,000 strands or microfibrils of 10 μm in diameter. They are used vastly for textile applications and in fact are known as artificial silk.

To manufacture nano-PAN fibers, a technique called "electrospinning" is employed. It typically produces fiber diameters in the tens or hundreds of nanometers, as shown in Figure 4.77, and can

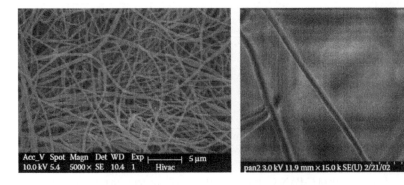

FIGURE 4.77   Spun PAN nanofibers. Average fiber diameter is approximately 300–600 nm.

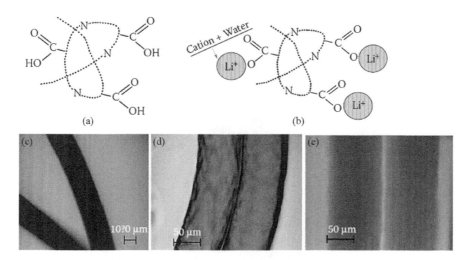

FIGURE 4.78 PAN elongation behavior explained by the osmotic behavior and PAN fibers in different states. (a) Neutral state; (b) under alkaline solutions. Therefore, if pure water is in contact with alkaline PAN, there will be an osmotic pressure driven water influx. (c) Oxidized PANs (prior to activation); (d) at low pH, contracted PAN (1-N HCl); (e) at high pH, expanded PAN (1-N LiOH).

offer new opportunities far beyond textiles to numerous other industrial, biomedical and consumer applications. In particular, as discussed before, once these PAN-Ns or strands are made conductive and used in an electrochemical cell for chemomechanical linear contractile transduction, they will be able to provide us with contraction response time comparable to biological muscles – that is, in the range of a few milliseconds.

Realizing that the response time of PAN artificial muscle is governed by the diffusional processes of ion–solvent interaction, the use of PAN-Ns or fibrils is promising for fabricating fast-response PAN artificial muscles. The contraction/elongation behavior explanation is based upon the exchange of counterions and solvent (in this case, water) into and out of activated PAN. Donnan equilibrium theory may possibly describe the situation properly (Figures 4.78–4.80).

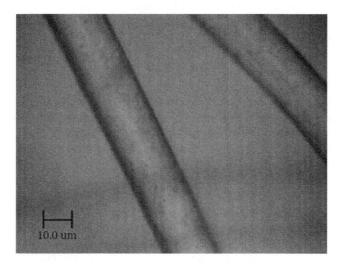

FIGURE 4.79 Raw PAN fibers.

FIGURE 4.80   After oxidation (cross-linked) – before activation.

If so, the swelling force may be identified by the net osmotic pressure difference associated with relevant ions. Also, the columbic force could play a role. The combination of such effects can describe the situation reasonably well. If we describe the kinetics of PAN fibers by using the diffusion-controlled slab-type gels, the contraction of PAN-Ns would be (Yoshida et al., 1996):

$$\frac{\Delta l}{\Delta l_0} = -1 - \sum_{n=0}^{\infty} \frac{8}{(2n+1)^2 \pi^2} \exp\left(-\frac{(2n+1)^2 t}{\tau}\right) \tag{4.2}$$

where the characteristic time, τ, is given by:

$$\tau = \frac{4 l_{ch}^2}{\pi^2 D} \tag{4.3}$$

where l_{ch} is the characteristic length (i.e., fibril diameter) and D is the overall diffusion coefficient of ions within the PAN-Ns. The implication of this equation is that the contraction or elongation kinetics should be dependent upon the length scale (i.e., nanofiber diameter), as we have observed experimentally. Therefore, the importance of our effort to fabricate PAN-Ns can be realized.

4.6.6   EXPERIMENT

The polymer solution used in the present electrospinning experiments was prepared using PAN purchased from Scientific Polymer Products. The electrospinning apparatus used a variable high-voltage power supply purchased from Gamma High Voltage Research. A 20-mL glass syringe with a Becton Dickinson 18 hypodermic needle was tilted at approximately 5 degrees from horizontal so that a small drop was maintained at the capillary tip due to the surface tension of the solution. The tip of the needle was filed to produce a flat tip.

A positive potential was applied to the polymer solution by attaching the lead to the high-voltage power supply directly to the outside of the hypodermic needle. The collection screen was a 10 × 10-cm^2 aluminum foil placed 20 cm horizontally from the tip of the needle as the grounded counter-electrode. The potential difference between the needle and the counterelectrode used to electrospin the polymer solution was 20 kV (electrical field strength, 1 kV/cm). The fiber diameter and polymer morphology of the electrospun PAN fibers were determined using SEM (Figure 4.81).

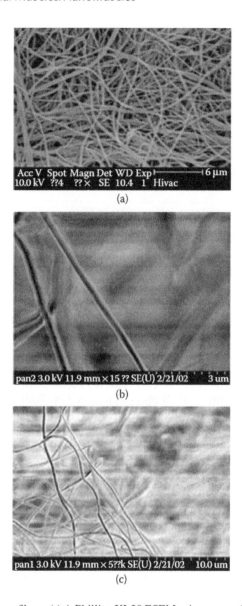

FIGURE 4.81 Spun PAN nanofibers. (a) A Phillips XL30 ESEM using an accelerating voltage of 10 kV was employed to take this SEM photograph; (b) ESEM image of the polyacrylonitrile nanofibers spun at 1 kV/cm; (c) PAN nanofibers (~300 nm diameter). Hitachi 4700 was used (an acceleration voltage of 3 kV).

The activation of PAN requires inducing cross-linking by the formation of pyridine rings by low-temperature annealing and, subsequently, converting nitrile groups to carboxylic acid groups by saponification with sodium or lithium hydroxide. The degree of cross-linking depends on annealing temperature and time, which in turn determine the amount of free nitrile groups left to be converted to carboxylic acids during saponification. Figure 4.82 shows a C13 NMR spectrum of saponified PAN fibers. Overall, it is anticipated that PAN should transit from a linear chain structure (or zigzag) to a planar structure that would be related to pyridine rings/cyano groups via cross-linking. Although the PAN fibers were tested at a dry state, it appears that the cyano group is weaker and the carboxyl group stronger after saponification. Generally speaking, the activated PAN (in wet state) can inherently carry cation (Na^+) and anion groups (pyridine rings/cyano groups).

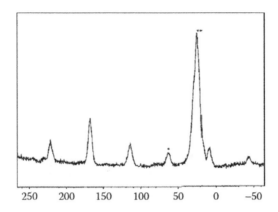

FIGURE 4.82 NMR spectrum of PAN processed fibers.

The PAN-Ns were suitably annealed, cross-linked and hydrolyzed to become "active". A key molecular structure of PAN – hydrogen bonding between hydrogen and the neighboring nitrogen of the nitrile group – exhibiting insolubility, thermal stability and resistance to swelling in most organic solvents is thought to be due to its cross-linked polymer structure. These results provide a great potential in developing fast-activating PAN muscles and linear actuators, as well as integrated pairs of antagonistic muscles and muscle "sarcomeres" and "myosin/actin"-like assembly with potential interdigitated control capabilities.

First, the effect of different cations (Li^+, Na^+ and K^+) on the contraction/elongation behavior of activated PAN fiber was determined. PAN fibers were tested in alkaline and acidic solutions of different normality to determine their optimum contraction and elongation properties. Interestingly, Li^+-based PAN fibers exhibited the largest elongation/contraction performance. PAN fibers were oxidized at 220°C for 90 min in air. In the first experiment, these fibers measured one inch in length and each was activated in boiling 1-N KOH for 30 min. The fibers were then soaked in distilled water for 30 min to obtain a base length. Then, several fibers were placed in each of 0.5-, 1- and 2-N KOH for 30 min and measured. Next, the fibers were again put into distilled water for 30 min and then measured. Following this, the fibers were soaked in 0.5-, 1- and 2-N HCl solutions (corresponding fibers from the alkaline, e.g., fibers from the 0.5-N KOH were placed in 0.5-N HCl) and measured. Also, this process was repeated using NaOH and LiOH for the boiling and alkaline-soaking media.

Fibers treated with LiOH had the largest increase in length following immersion in distilled water. Fibers soaked in all three media generally had the same decrease in length following immersion in the alkaline solutions, as also occurred following immersion in HCl. Especially noticeable with the fibers treated with LiOH was that greater displacement in the lengths occurred using the 2-N solutions. The lengths of fibers treated with NaOH were close to the same regardless of the normality of the solutions. In Figure 4.77, the importance of osmotic pressure is illustrated. It should be noted that the maximum displacement could be determined when conditions were switched from pure water to acidic conditions. These findings are important to describe casual pH hysteric behaviors reported previously.

4.6.7 NANOFIBER ELECTROSPINNING IN GENERAL

The electrospinning process is a variation of the better known and understood electrospraying technique. In the electrospinning process, a high electric field is provided between a highly viscous polymer solution held by its surface tension at the end of a capillary tube and a metallic target, as shown in Figure 4.83.

As the intensity of the electrical field increases, the surface of the liquid hemispherical drop (Figure 4.84(a)), suspended at equilibrium at the capillary tip, elongates to form a conical shape,

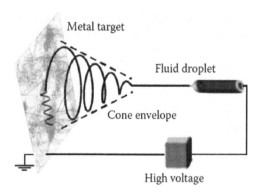

FIGURE 4.83 Schematic of the electrospinning process.

which is known as the *Taylor cone*. The "balancing" of the repulsive electrostatic force with the surface force of the liquid causes this distortion. When the electric field reaches a critical value (~0.5 kV/cm), the charge overcomes the surface tension of the deformed drop and a jet is produced, as shown in Figure 4.84(b). As a result of the low surface tension in low-viscosity solutions, the jet breaks apart into a series of droplets. This is the basis of electrospray technology. The electrically charged jet does not break apart for high-viscosity solutions, but instead undergoes a series of electrically induced bending instabilities ("necking") during its passage to the collection screen that result in hyperstretching of the jet stream.

This stretching process is accompanied by the rapid evaporation of solvent molecules that reduces the diameter of the jet in a "cone-shaped" volume. This is called the "envelope" cone (Figure 4.83). The as-spun dry fibers accumulate on the surface of the collection screen. This process results in a porous, nonwoven mesh of nanofibers. Polymer melts have also been processed into nanofibers and the metallic target used to quench the as-formed molten fiber mats. Replacing the metal target with a grounded coagulation bath target leads to instantaneous demixing of the polymer solution, which leads to the production of continuous nanofiber filaments.

The "under-researched" area of electrostatic fiber spinning is, thus, a highly versatile process that offers a potentially valuable tool for creating nanofiber structures. Since its discovery, this approach has remained little more than a laboratory curiosity because it was overshadowed by the important technological development of synthetic textile fibers (diameter >5 μm). In electrospinning of polymers, the polymer solution (melts or solution) is fed through a tube (glass) with a capillary opening. By applying an electric field between the capillary and the counterelectrode (collector), a pendant drop at the tip of the capillary is shaped into a conical protrusion often called Taylor cone. At the critical voltage (~0.5 kV/cm), the electrostatic force exerting on this cone overcomes the existing surface tension of the drop, ejecting the jet form of the cone toward the counterelectrode. When the electric field reaches a critical value, the charge overcomes the surface tension of the deformed drop,

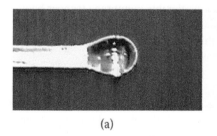

(a)

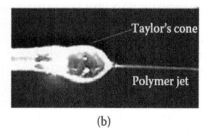

(b)

FIGURE 4.84 Photographs of viscous polymer solution suspended at a capillary tip with (a) no applied potential and (b) just above the critical voltage.

and a jet is produced. In general, the long chain molecules are oriented and entangled in the jet as the fiber solidifies. The electrically charged jet undergoes a series of electrically induced bending instabilities during its passage to the collection screen that results in hyperstretching of the jet stream.

Important process information and physical properties are:

- Polymer density (ρ)
- Polymer viscosity (η)
- Polymer surface tension (σ)
- Polymer electric conductivity (σ_e)
- Polymer dielectric permittivity ($\varepsilon = \varepsilon_r \varepsilon_o$)
- Capillary radius (R)
- Applied electric voltage (E)
- Capillary-to-target distance (H)
- Current density (current/area) (I)
- Volumetric flow rate (Q)

According to Senador and coworkers (2001), a general relationship between the critical voltage and other variables during jet formation can be written as:

$$E\sqrt{\frac{\varepsilon_e}{r\sigma}} = F\left[\left(\frac{\eta}{\sqrt{\rho r\sigma}}\right),\left(\frac{\sigma_e}{\varepsilon_e}\sqrt{\frac{\rho r^3}{\sigma}}\right),\left(\frac{h}{r}\right)\right] \tag{4.4}$$

where $F[\ ]$ is an undetermined function relating to the dimensionless groups. Also, the known expressions for the critical voltage are

$$E = 300\sqrt{20\pi\sigma r} \tag{4.5}$$

$$E = 0.863\sqrt{\frac{4\sigma}{e_0 r}} \tag{4.6}$$

Inspection of recast Equations (4.4) and (4.5) reveals the dimensionless critical voltage,

$$\frac{E}{h}\sqrt{\frac{e_e r}{\sigma}},$$

to be constant or a function only of the geometric ratio (h/r) with a power-law exponent of 0.5 or 1.

4.6.8 Fabrication of a PAN Actuator System

The configuration in Figure 4.85 shows the fabricated PAN actuator system. The dimension is provided. It can cause the PAN fiber to contract within the flexible membrane (rubber boots). Once the polarity is changed, the PAN fiber tends to expand and the compressed flexible membrane will help it expand in a resilient manner. The fabrication of this unit has been completed and is now ready for performance testing. The test results will be reported in the near future.

Note in Figure 4.85 that first we measured the spring constant of the rubber boots by applying predetermined loads. The measurement gave the spring constant of $k = 0.01$ kg/mm. Inside rubber boots, as can be seen, the following components are positioned: the PAN muscle bundle, electrodes and a solution. Applying electrical currents through the electrodes can perform the system operation. The inner electrode (a circular shape) surrounds the PAN muscle bundle and the other is attached to the boots' wall. The clearance between the boots' wall and the inner electrode is approximately 15 mm.

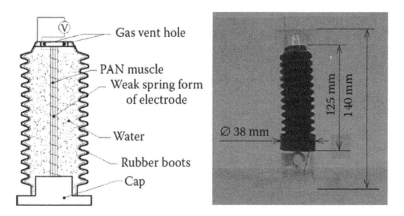

FIGURE 4.85 A PAN actuator system.

4.6.9 Contraction and Elongation Mechanism

According to Schreyer and colleagues (1999), a possible explanation for the contraction and elongation is based upon the carboxylic acid groups having the molecular geometry of activated PAN. At high cationic concentration, all carboxylic acid groups on activated PAN are likely to be protonated, thus potentially contracting the polymer chain through neutral charge of the acid groups and hydrogen bonding between neighboring carboxylic acid groups. At lower cationic concentration or higher anionic concentration, protons are likely to have been removed from the carboxylic acid groups, giving the group an overall negative charge. Negative charge repulsion between neighboring acid groups forces the polymer backbone to swell or expand. Other factors may affect the length of activated PAN, such as charges on pyridine rings. However, such electrostatic repulsions would prevail if the carboxylic ions are sole ions present but others exist. Therefore, the effect of fixed charges is expected to reduce the electrostatic repulsion forces significantly.

Another explanation is based upon the exchange of counterions and solvent (in this case, water) into and out of activated PAN and is illustrated in Figures 4.78 and 4.86. Donnan equilibrium theory

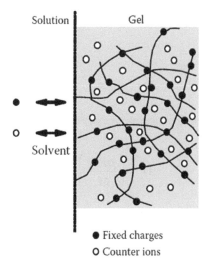

FIGURE 4.86 The exchange of counterions and surrounding solvent.

may possibly describe the situation properly. If so, the swelling force may be identified by the net osmotic pressure difference associated with relevant ions. Also, the coulombic force could play a role. The combination of such effects can describe the situation reasonably well. One key finding was that the degree of ionization governs the degree of swelling. In the next section, a mathematical model is presented for such electrochemically ion-induced contraction and expansion of PAN fibers and C-PAN-N microfibrils.

4.6.10 MATHEMATICAL MODELING OF CONTRACTION AND ELONGATION OF C-PAN FIBERS

A possible explanation for the contraction and elongation is based upon the carboxylic acid groups having the molecular geometry of activated PAN. At low pH concentrations, all carboxylic acid groups on activated PAN are likely to be protonated. This could potentially collapse the network by polymer-polymer affinity and contract the polymer chain through neutral charge of the acid groups and hydrogen bonding between neighboring carboxylic acid groups.

Based upon the Donnan theory of ionic equilibrium, the important forces arise from (1) induced osmotic pressure of free ions between activated PAN fibers and their environment; (2) ionic interaction of fixed ionic groups; and (3) the network itself. Among these sources, the induced osmotic pressure of free ionic groups could be the dominating force.

Electrical activation of PAN fibers is performed in an electrochemical cell such as shown in Figures 4.77 and 4.85. Note that, at the anode, oxygen evolves via $2H_2O \Rightarrow O_2 + 4H^+ + 4e^+$ and the counterreaction at the cathode is $2H_2O + 2e^- \Rightarrow H_2 + 2OH$. Upon being hydrogenated in the vicinity of the PAN anode, the decreased pH causes the PAN fibers to contract by the same effect as chemical activation. Also, by reversing the polarity of D.C., elongation of PAN fibers is simply obtained.

4.6.10.1 Basic Modeling

Three important working forces that drive contraction or elongation of PAN muscles were identified: rubber elasticity, proton pressure and polymer-polymer affinity. Rubber elasticity can be expressed by:

$$\Pi_r = -\frac{\left(\rho R T v_2^{1/3}\right)}{M_c} \tag{4.7}$$

where
  Π_r = contraction/elongation force by rubber elasticity
  v_2 = volume fraction
  T = absolute temperature
  ρ = density of unswollen polymer
  M_c = molecular weight
  R = gas constant

the proton pressure is

$$\Pi_e = \left(\frac{\rho R T f}{M_c}\right) v_2 \tag{4.8}$$

where f is a number of dissociated hydrogen ions per chain.
  The polymer–polymer affinity is

$$\Pi_p = \left(\frac{RT}{V}\right)\left[v^2 + \ln\left(1 - v_2\right) + x v_2^2\right] \tag{4.9}$$

where

V = molar volume of the solvent

x = Flory–Huggins parameter

The overall osmotic pressure is then

$$\Pi_t = \Pi_r + \Pi_e + \Pi_p. \tag{4.10}$$

In addition to these forces, it has been recognized that, as the diameter of fibers and number of strands increases, the induced force also increases (dimensional effect). Also, the temperature of the system will significantly contribute to the performance of the PAN artificial muscle.

The polymer–polymer affinity arising from the interaction between the polymer fibers and the solvent is believed to be an important driving force, although it is not yet clarified to explain the exact mechanism of PAN contraction–elongation behavior. One possible explanation is based on the carboxylic acid groups having the molecular geometry of activated PAN. At low pH concentrations, all carboxylic acid groups on activated PAN are likely to be protonated and contracting the polymer chain through neutral charge of the acid groups and hydrogen bonding between neighboring carboxylic acid groups. At high pH concentrations, protons have been removed from the carboxylic acid groups and give off negative charges. Negative charge repulsion between neighboring acid groups likely forces the polymer backbone to elongate. Electrical activation can be made for hydrogen and oxygen evolution. At the anode, oxygen evolves via $2H_2O \rightarrow O_2 + 4H^+ 4e^+$ and the counterreaction at the cathode is $2H_2O + 2e^- \rightarrow H_2 + 2OH^-$; the decreased pH causes the PAN fibers contract by the same effect as chemical activation. Elongation is simply obtained with reversing the polarity of D.C. while water diffuses into the PAN polymer network.

We also looked at the Donnan equilibrium carefully. If we assume that polymer and aqueous solution are composed of three parts – A, B and C – in turn from the anode side, the concentration of cation is changed abruptly across the boundary but uniform in any of A, B and C, the transport rates of cation, h, from A to B and from B to C are the same, and all ion-ion interactions are neglected, the cation concentration in each part can be expressed by,

$$C_A(t) = C_A(1 - ht) \tag{4.11}$$

$$C_B(t) - C_B(1 - ht) + C_A \frac{V_A}{V_B} ht(1 - ht) \tag{4.12}$$

$$C_C t(t) = C_C + C_B \frac{V_B}{V_C} ht + C_A \frac{V_A}{V_C} h^2 t^2 \tag{4.13}$$

where $C_i(t)$ is the cation concentrations in the ith species – namely, A, B and C, respectively. As the osmotic pressure π obeys van't Hoff's law, $\Delta\pi$ can be given as:

$$\Delta\pi = \pi_1 - \pi_2 = RT\left[C_B(t) - C_A(t)\right] - RT\left[C_B(t) - C_C(t)\right]$$

$$= RT\left[C_C(t) + C_B \frac{V_B}{V_C} ht + C_A \frac{V_A}{V_C} h^2 t^2 - C_A(1 - ht)\right] \tag{4.14}$$

where π_1, π_2 are the osmotic pressure of the anode and cathode side between gel and solution, respectively. R is the gas constant and T is the absolute temperature.

In what follows, a basic theory is presented for the contraction of PAN fibers in an electric field based on electrocapillary transport and electroosmotic dynamics.

4.6.10.2 Modeling

The objective of modeling was to understand the fundamental mechanism of PAN actuation principle, to predict the performance and to help the design and fabrication of the effective PAN system. Investigation had been carried out to study the effect of osmotic pressure when the electric field was applied between the electrodes. The electric field created proton movement toward the cathode, so as to cause the proton concentration gradient, which could change pH conditions at the vicinity of the electrodes and ion conjugate PAN fibers. Electrocapillary diffusion of ions within interstitial clusters within the ionic polymer was also studied. In this endeavor, the models described in the following sections were formulated.

4.6.10.3 More Detailed Mathematical Modeling PAN Fiber Contraction/Expansion

A possible explanation for the contraction and elongation is based upon the carboxylic acid groups having the molecular geometry of activated PAN. At low pH concentrations, all carboxylic acid groups on activated PAN are likely to be protonated, thus potentially contracting the polymer chain through neutral charge of the acid groups and hydrogen bonding between neighboring carboxylic acid groups.

Based upon the Donnan theory of ionic equilibrium (Flory, 1953b), it is believed that important forces arise from (1) induced osmotic pressure of free ions between activated PAN fibers and their environment; (2) ionic interaction of fixed ionic groups; and (3) the network itself. Among them, the induced osmotic pressure of free ionic groups could be the dominating force.

Electrical activation of PAN fibers is performed in an electrochemical cell such as those shown in Figures 4.85 and 4.87. PAN fibers can be activated electrically by providing a conductive medium in contact with or within the PAN fibers. Such electrical activation can be made to have low overvoltage for hydrogen and oxygen evolution. At the anode, oxygen evolves via $2H_2O \Rightarrow O_2 + 4H^+ + 4e^+$ and the counterreaction at the cathode is $2H_2O + 2e^- \Rightarrow H_2 + 2OH^-$. Upon being hydrogenated in the vicinity of the PAN anode, the decreased pH causes the PAN fibers to contract by the same effect as chemical activation. Also, by reversing the polarity of D.C., elongation of PAN fibers is simply obtained.

A possible explanation for the contraction and elongation is based upon the carboxylic acid groups having the molecular geometry of activated PAN. At high cationic concentration, all carboxylic acid groups on activated PAN are likely to be protonated, thus potentially contracting the polymer chain through neutral charge of the acid groups and hydrogen bonding between neighboring carboxylic acid groups. At lower cationic concentration or higher anionic concentration, protons are likely to have been removed from the carboxylic acid groups, giving the group an overall negative charge. Negative charge repulsion between neighboring acid groups likely forces the polymer backbone to elongate. Other factors may affect the length of activated PAN, such as charges on pyridine rings. However, it should be pointed out that such

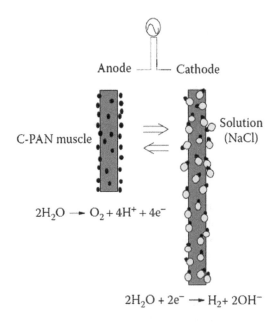

FIGURE 4.87 Experimental setup for electrical activation of PAN fibers.

electrostatic repulsions would prevail if the carboxylic ions were the sole ions present, but others do exist. Therefore, the effect of fixed charges is expected to reduce the electrostatic repulsion forces significantly.

Another explanation is based upon the exchange of counterions and solvent (in this case, water) into and out of activated PAN (illustrated in Figure 4.87). Donnan equilibrium theory may possibly describe the situation properly. If so, the swelling force may be identified by the net osmotic pressure difference associated with relevant ions. Also, the columbic force could play a role. The combination of such effects can describe the situation reasonably well. One key finding was that the degree of ionization governs the degree of swelling. However, much study is needed to further clarify the mechanism of PAN contraction/elongation behavior.

If the kinetics of PAN fibers is described by using the diffusion-controlled slab-type ionic gel model, the elongation would be

$$\frac{\Delta l}{\Delta l_0} = 1 - \sum_{n=0}^{\infty} \frac{8}{(2n+1)^2 \pi^2} \exp\left(-\frac{(2n+1)^2 t}{\tau}\right) \tag{4.15}$$

where the characteristic time, τ, is given by

$$\tau = \frac{4l_{ch}^2}{\pi^2 D} \tag{4.16}$$

The notation D is the overall diffusion coefficient of ions within PAN fibers. The implication of the preceding equation is that the contraction or elongation kinetics must be faster than that which we had observed in this study. In fact, a proper approximation of physical properties provides a time constant of $L/L \approx 5\%/s$ or less for both contraction and elongation. Therefore, it was decided to change the activation scheme into a much more compact form. One successful way was to deal with individual fibers rather than strands made with 2,000 fibers.

4.6.10.4 PAN Actuator System Design and Fabrication

Figure 4.88 shows the newly fabricated PAN actuator system. The new linear contractile PAN actuator system consisted of a much more flexible electrochemical cell membrane electrode (bellows-type

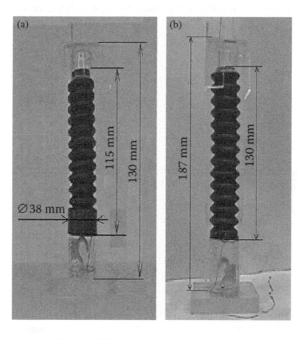

FIGURE 4.88 (a) Previous and (b) new PAN actuating systems.

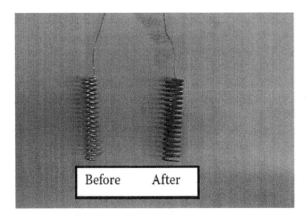

FIGURE 4.89   Erosion of the electrode used after the operation.

rubber boots) than that of the previous system. In this new configuration, a softer membrane (rubber boots) than that of the previous system was also adopted. As a result, the spring constant of these bellows-type rubber boots including the electrode was measured at $k_s = 0.01$ kgf/mm and turned out to be 0.004 kgf/mm less than the previously fabricated system. The PAN fiber length was also changed from 115 to 145 mm (a 26% increase). Detailed dimensions are provided in Figure 4.88.

To address the problem associated with the erosion and corrosion of the electrode (on the anode side), an erosion/corrosion-resistive beryllium copper springs (Small Parts Inc.) and a wire diameter of 0.0508 in. (1.29 mm) were chosen to serve as the anode that was also gold-plated (Figure 4.89). By doing this, it was anticipated that the electrodes would have long-term operation even under the action of electrochemical reactions caused by the imposed electric field. The operating voltage was lowered from 10 to 5 V, so as to prevent the quick temperature rise of the solution. The initial testing proceeded for 27 min and produced a 5-mm displacement (producing 50 gf) (Figure 4.90). It was noted that even after 27 min, the PAN muscle contraction was still in progress. The temperature of the saline electrolyte solution was maintained at about 50°C, which was significantly lower than that of the previous system. Nevertheless, minor erosion/corrosion of the electrode was still observed (Figure 4.89).

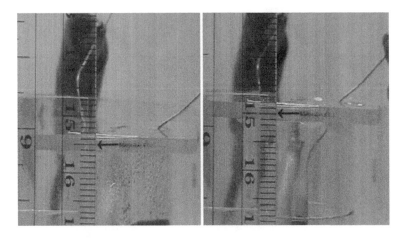

FIGURE 4.90   The new PAN system produced a 5-mm displacement (a generative force of 50 gf).

The following important issues were identified and left for the upcoming period of research and development to be addressed:

Electrode performance improvement: It is our belief that the electrode is a key component to properly design effective PAN-based linear contractile and electrically controllable artificial muscle systems. It was decided to consider different shapes, materials and manipulation of the electrode in the future.

Development of appropriate macramé of PAN fibers: We studied how to weave a bundle of PAN fibers for consolidation and their engineering properties that can affect the PAN muscle system. The performance improvement of the PAN muscle systems can be achieved by using an appropriate macramé of PAN fibers.

4.6.10.5 Further Modeling

Following the modeling presented in the second progress report (Appendix B of the first edition), three working forces seem to drive the contraction and elongation of PAN muscles. The osmotic pressure is a sum of rubber elasticity, proton pressure and polymer-polymer affinity. This can be expressed as

$$\Pi_t = \Pi_r + \Pi_e + \Pi_p = -\frac{\left(\rho R T v_2^{1/3}\right)}{M_c} + \left(\frac{\rho R T f}{M_c}\right) v_2 + \left(\frac{RT}{V}\right)\left[v^2 + \ln\left(1 - v_2\right) + x v_2^2\right] \qquad (4.17)$$

In this equation, the volume fraction, v_2, is usually given by V_0/V. Notations V_0 and V are the initial network volume and swollen volume, respectively. This volume-fraction ratio is considered a main driving means of PAN artificial muscles. Therefore, we attempted to measure the diameter change of the PAN muscle to estimate the value associated with the term V_0/V.

Comparing Figures 4.91–4.93, one can clearly observe that "chemical induction" is extremely effective in volume changes of the activated PAN fibers. This means that "electric activation" of PAN fibers in an electrochemical cell produces fewer dimensional changes at the present time compared with chemical activation. Such findings enable one to further improve the performance of the PAN system. (Note that the dimensional changes of the PAN fibers are 28.4 μm [electric activated] and 14.7 μm [chemically activated], respectively.) We also obtained micrographs of PAN fibers of "raw" and "Oxy-PAN", for basic analysis (Figures 4.94 and 4.95).

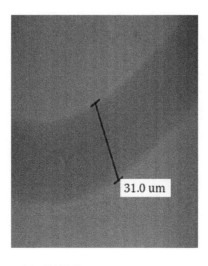

FIGURE 4.91 An expanded state of the PAN fiber.

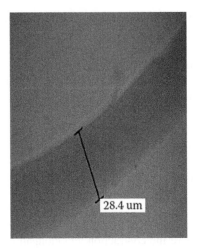

FIGURE 4.92   An electrically induced, contracted state of the PAN fiber.

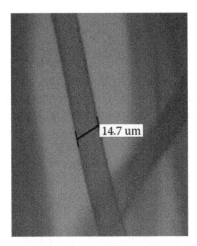

FIGURE 4.93   A chemically induced, contracted state of PAN fiber (at 2-N HCl).

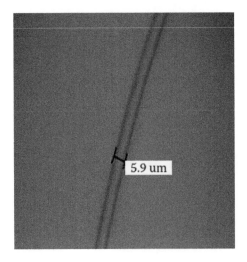

FIGURE 4.94   A micrograph of a PAN raw fiber.

FIGURE 4.95 A micrograph of an oxy-PAN (after heat treatment).

The fiber diameter of the "raw" PAN is approximately 6 μm. It expands to 7–7.5 μm after heat treatment (Oxy-PAN). An activated PAN (conditioned at 1-N LiOH) fiber expands to approximately 31 μm in diameter and shrinks approximately 3 μm in diameter upon electrical activation.

The diameter of a PAN fiber immersed in a 2-N HCl solution is approximately 15 μm. We also observed that the vividness of the PAN fiber color increases as the diameter decreases. The transport of water into and out of PAN dependent upon the environmental conditions and resultant expanding and/or contracting volume of PAN seemingly govern the overall properties of the PAN artificial muscle system.

Further electrical activation of the PAN linear contractile actuator system was performed. To record the detailed dynamic behavior of the PAN linear contractile actuator system, instrumentation was created with a load cell positioned at the proximity of the PAN actuator to conduct analytical studies on PAN fibers (NMR and DMA, or DSC) continue to improve the modeling and simulation of artificial muscle chemoelectrodynamics.

Three-strand braided configuration of a PAN muscle bundle was also fabricated, as strong PAN synthetic muscles fiber bundles, for testing in a newly designed electrochemical cell. The muscle fiber bundle was easy to handle but frictional interferences between woven strands restricted the PAN muscle movement. Linear contractile strains of about 17% have been achieved from woven PAN muscle fibers, generating force density of 20 gf/g of muscle fiber bundle.

Following the extensive modeling and simulation and realizing that the changes in volume caused by the electric activation may be the main driving factor of the PAN fibrous muscle system, a testing module was designed, using a tensile test machine (INSTRON 1011) to measure a pattern of force change with time variation during the actuation.

4.6.10.6 PAN Actuator Fabrication

To fabricate strong fiber bundles for linear contractile electrochemical synthetic muscle actuation, we embarked on making three-strand braided PAN muscle fiber bundles with each strand having five single fibers. Figure 4.96 depicts one such fabricated braided PAN muscle fiber bundle. To perform linear contractile stress-strain tests on such fiber bundles, both ends of braided PAN fiber bundle were solidified using epoxy glue. A minor change of the fabrication system was made for easier experimental measurements. Detailed dimensions are provided in Figure 4.97. The other parameter in the test environment remained the same as in the previous test environment, as reported in the previous progress reports.

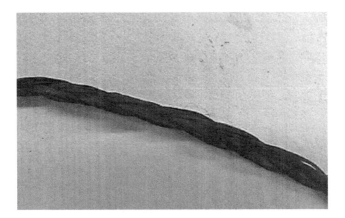

FIGURE 4.96   Braided PAN bundle in an expanded state.

Two kinds of PAN bundle were used. One was looped while another was kept straight, as shown in Figure 4.98(a) and (b). Both PAN muscle fiber bundles had the same number of strands (15 strands). The shape of the braided strands was still maintained after activation (1-M LiOH). The snap of each strand during the test manipulation was reduced. We noted that the braid restricted free movement of each fiber string, which disturbed expansion or shrinkage of PAN muscle. Straight PAN was tested to detect whether the looping caused negative effect on the muscle system.

The testing proceeded for 25 min and showed 2-mm displacement (producing a maximum of 20 gf) of PAN muscle fiber bundle system (Figure 4.99). It was observed that the volume change, which we thought was the main driving force, had its linear movement restricted and/or disturbed by the braided configuration. The straight braided fiber bundle system did not show any different behavior.

Small-scale PAN fiber bundle systems were also tested. More detailed data for a single PAN fiber were expected. Due to the small scale of the test rigs, fabrication was somewhat challenging compared to previously operated test rigs (Figure 4.100).

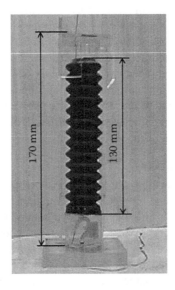

FIGURE 4.97   Dimensional changes of the electrochemical cell system.

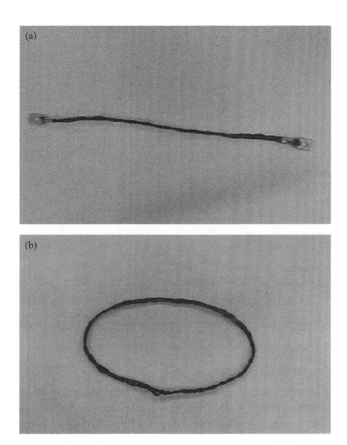

FIGURE 4.98 (a) Straight PAN fiber bundle muscle with end hooks and (b) looped PAN muscle.

4.6.10.7 Additional Modeling

As identified in previous research, volume change is considered a main driving force of PAN muscle. We worked on measuring the energy generated by the PAN muscle system, which helped to analyze how the volume change is converted into muscle forces. To achieve data of chemically induced

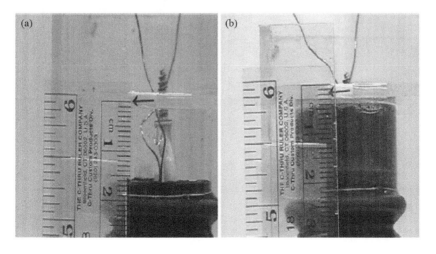

FIGURE 4.99 Braided PAN fiber actuator test: (a) initial and (b) after 25 min (right).

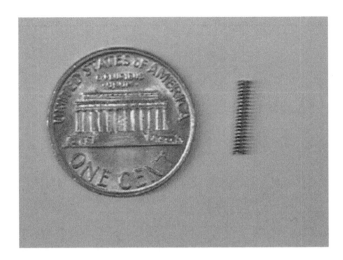

FIGURE 4.100   Spring for small-scale PAN fiber test.

force and electrically induced force, we used a tensile machine (INSTRON 1011). Figure 4.101 shows test configuration. Change of force with time variation was measured.

4.6.10.8   Small-scale PAN Actuator Fabrication

Figure 4.102 shows the fabrication of small-scale PAN muscle. A copper spring was used as a load (purchased from Small Parts Inc.). Outer diameter of the spring was 1.8 mm, the free length of the spring was 9.4 mm and the spring constant was 11.46 gf/mm. One single string of fiber was placed inside the spring and was glued to both ends on the small square of plastic plate to withstand and hold the possible contraction force. The PAN muscle fiber bundle was immersed in 1-M HCl solution with no significant displacement observed.

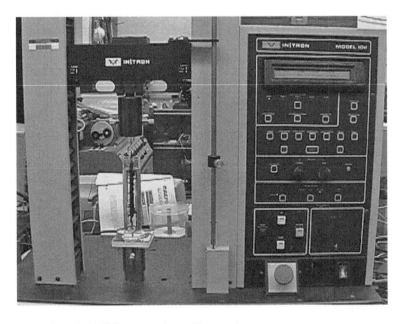

FIGURE 4.101   Small-scale PAN fiber test with resilient springs.

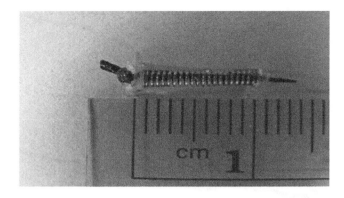

FIGURE 4.102 A fabricated small-scale PAN system.

We also measured the length change between elongation state and contraction state overall strain. Single-string (25-mm) fiber was taken and immersed in 1-M HCl. The fiber contracted to 11 mm. We worked on measuring displacement at each concentration change and generated force with varying pH levels.

Figure 4.103 depicts the change in length of the PAN fiber bundles before and after contraction. Figure 4.101 depicts the conductive gold-plated helical compression spring-loaded PAN fiber bundle synthetic muscles and electrode samples.

Three muscle samples were tested. Two of the samples had identical dimensions and the third one had larger dimensions. Two samples (acid activated and E-activated2, Figure 4.104) had 25 strands and 15.5-cm length; the other (red) had 30 strands (20% increased) and 17.5-cm length (16% increased). Electrodes of 7.5 and 10.5 cm, respectively, were used. One sample was soaked in 1-M HCl to compare results with an electric activation muscle system (the baseline). An input voltage of 5 V was used to operate the system. In acid solution, most of the reaction was done in a very short period of time when compared with the electrically activated fiber with the same dimensions, and more force was achieved than for the electrical system.

When the reaction was complete, the force was converged to 150 gf. The electrical system reached 40% of acid-activated force after 25 min of operation and the force was still increasing.

The larger dimension sample (red) reached 180 gf and 0.06-N force per a single string of PAN fiber and was achieved with a sharper gradient, which meant short response time. It also converged after 25 min of operation. It should be noted that the size of the electrodes affects the performance significantly. This finding was investigated more carefully. As the previous observations indicated, H^+ ions are created at the anode side, so a larger surface area of anode increased the possibility to

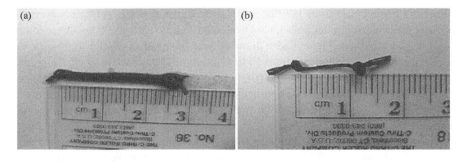

FIGURE 4.103 (a) Initial length and (b) contracted length.

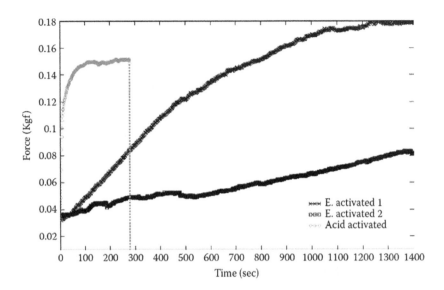

FIGURE 4.104 Force with time variations. Acid-activated: baseline condition (actuated at pH of 1-N HCl); E-activated2: sample 1 electrically actuated; E-activated1: sample 2 electrically actuated.

generate more H$^+$ to contact fiber with H$^+$ ion, which basically was to reduce the response time. Other than the geometric factor, fabrication is also an important consideration. Each string in the muscle bundle should have uniform stress and tension during the operation. The experimental setup is shown in Figure 4.101.

4.6.10.9 Small-scale PAN Actuator Fabrication and Testing

A single PAN fiber string was tested in the small-scale PAN muscle system. A 30-mm (1.18-in.) single PAN fiber string was prepared for force measurement. After the saponification process in 1-M LiOH solution, fiber was carefully connected to a 30-g load cell (transducer technique). Figure 4.105 shows the single-string fiber attached to the experimental setup composed of two screw bolts and a load cell.

The PAN muscle was placed in the 1-M HCl solution to contract the fiber. A Personal Daq/56(IOtech) data acquisition system was used to collect the force response. Within a few seconds, the PAN fiber stopped producing force and reached the steady state. A maximum force of 5 gf was achieved from a single PAN fiber. The fiber was then elongated by placing it in the 1-M LiOH solution. This process was repeated five times. The average value of the generative force was 5 gf per a single PAN fiber. The same fiber produced 5-mm displacement without load.

4.6.10.10 PAN Actuator System Testing

The previously fabricated cylinder vessel, which held the sodium chloride solution needed for electric activation, was modified. The cylinder diameter was changed from 31.5 to 94 mm. Distance between both electrodes also changed from 10 to 42.25 mm. The objective of enlarging the diameter was to minimize thermal effect caused from water temperature rising and electrode corrosion during electric activation. This new design helped to avoid possible cation or anion mismatches arising from the short distance between the two electrodes without losing the electric potential significantly.

Two thermocouples were also used to keep track of the temperature changes during the experiment. Figure 4.106 shows such a PAN actuator system. Each tested sample had the same dimension: a 175-mm length with 30 strands in a bundle having an approximate mass of 0.4 g.

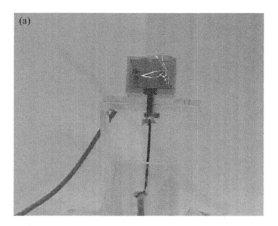

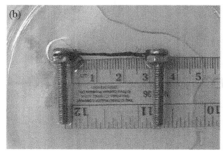

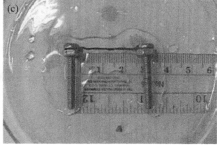

FIGURE 4.105 (a) Load cell setup to measure the generative force of PAN muscle. Single PAN fiber adhered to two screw bolts (b) before test and (c) after test.

Two voltages of 3 and 7 V were applied, respectively. Figure 4.107 depicts the forces generated. As can be seen, the generated forces increased nearly linearly as a function of time. The result was 90 and 51 gf for 7 and 3 V, respectively.

The gradients were 0.034 gf/s for 3 V with extremely small temperature change and 0.079 gf/s for 7 V with a 5°C temperature rising (from 25 to 30°C). Approximately 21- and 19-gf pretension was added in the 3- and 7-V experiments, respectively. Table 4.9 summarizes the test results. The test sample was composed of a 30-PAN-fiber bundle of 0.4 g with length of 175 mm.

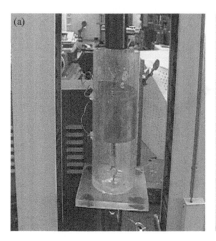

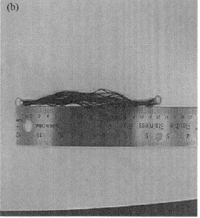

FIGURE 4.106 (a) The modified test setup and (b) fiber bundle.

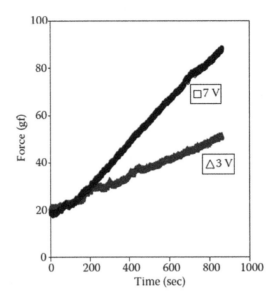

FIGURE 4.107   Force generation depending upon time with voltage change.

TABLE 4.9
Test Results

Applied Voltage (V)	Generative Force (gf)	Pretension (gf)	Time Dependency (gf/s)
3	51	21	0.034
7	90	19	0.079

Figure 4.108 shows behavior of PAN muscle when polarity is switched. The polarity was changed two times at 10 and 18 min. The force was increased at the rate of 0.05 gf/s, decreased at the rate of −0.04 gf/s with delay of time from the moment that the polarity was changed, and then resumed to increase at the rate of 0.07 gf/s, also with time delay. We applied 5 V to the system. Temperature was changed from 23 to 25°C. Discontinuity was observed at the moment that the polarity was changed.

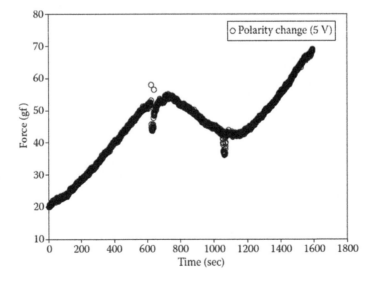

FIGURE 4.108   Force changes from switching polarity.

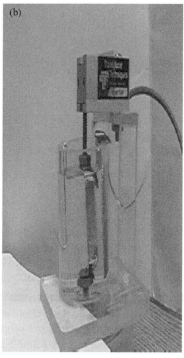

FIGURE 4.109   Load cell sensor to measure the force of PAN muscle: (a) front view and (b) side view (right).

Additional force testing was performed on an electroactivated PAN fiber bundle as depicted in Figure 4.109. Dimensions and other test environments were identical to the previous test. An aluminum foil electrode embraced a single PAN fiber. Aluminum foil was used because of easy fabrication. This figure shows the experimental apparatus. The PAN muscle was put under a 5-V electric field for 10 min. The force reached about 10 gf and reached steady state after 10 min of operation. The solution temperature had been raised from 26.7 to 29°C. Figure 4.110 shows the force curve depending upon time change.

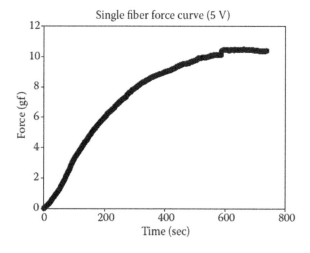

FIGURE 4.110   Single-fiber force curve under 5-V electric field.

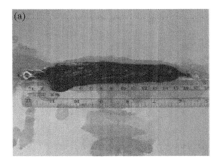
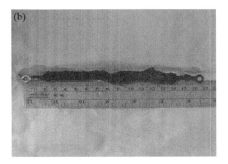

FIGURE 4.111   PAN fiber length (a0 before test: 170 mm; (b) after test 165 mm, under 5 V.

Three PAN muscles of 30-strand fiber bundles were further tested under 3, 5 and 7 V. One PAN muscle bundle with the same number of strands was tested under 5 V with polarity change. The polarity was changed every 5 min. Sodium chloride solution was stirred for 1 min to compensate the ion concentration difference near both the anode and cathode sides. This is thought to delay the contraction and elongation of the PAN muscle.

Dehydration of PAN muscle was observed on each fiber after the said operation. It is generally believed that it affects the mechanical property of PAN muscle and weakens the strength of the PAN fibers. The length of samples 1 and 2 was measured, but sample 3 was damaged during such measurements. It was assumed that the displacement of sample 3 was comparable with the displacements of samples 1 and 2. It was further assumed that the displacement of these samples is linearly related to force increase. Both samples contracted 5 mm after 20 min of operation. Figures 4.111 and 4.112 show length changes of such PAN muscle samples. Note that the electrically induced contraction was more pronounced in the lateral dimensions than the longitudinal direction.

Figures 4.113–4.115 depict the variation of force with time in electroactive PAN muscle fibers. Note that the force increases almost linearly with time increase and has an identical gradient, compared to all other tests.

Figures 4.116 and 4.117 show that both PAN muscle fiber bundles generate about 17 gf/mm gradients. Also note that, up to 5 V, the increase in voltage causes the force/displacement gradient slope to increase. These phenomena were investigated to see whether the variation in the slope of these curves was because of irregular fabrication of the PAN bundle or due to other causes or sources of manufacturing glitches.

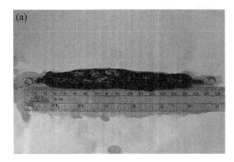
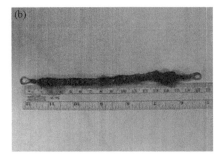

FIGURE 4.112   PAN fiber length (a) before test: 175 mm; (b) after test: 170 mm, under 7 V.

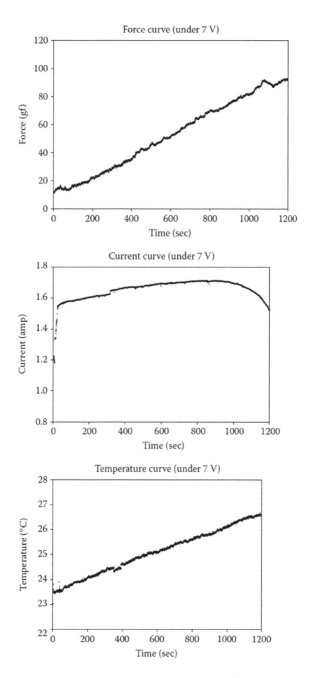

FIGURE 4.113 Force, current and temperature versus time, under 7 V.

Note that all these graphs show a time delay before any impending electrically induced contraction or force exertion. These delays were investigated. However, initial observations were that the process of electrolysis, which is responsible for the production of protons H^+ and hydroxyl ions OH^-, needed a voltage threshold of about 1.5 V before any impending electrolysis, and thus, some delays were to be expected.

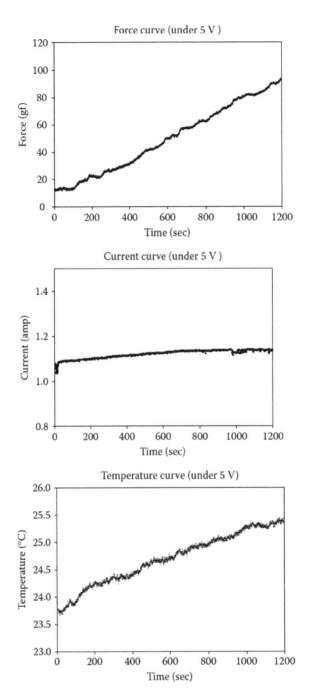

FIGURE 4.114 Force, current and temperature versus time, under 5 V.

Based on the experimental data and assuming that the displacement was linear, the amount of work of the PAN muscle, samples 1 (7 V) and 2 (5 V), was calculated to be about 200 gf/mm (2×10^{-4} kg/m) under 7 and 5 V, and 14 gf/mm under 3 V, respectively.

Figure 4.116 shows the force curve with polarity change. Figures 4.117 and 4.118 depict the variation of force versus displacements, respectively.

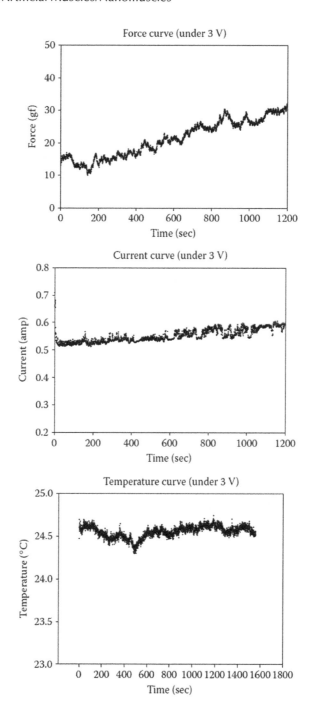

FIGURE 4.115 Force generation depending upon time, under 3 V.

4.6.11 ELECTROCAPILLARY TRANSPORT MODELING

Consider the gel fiber to be a swollen cylinder with outer radius r_o and inner radius r_i and assume the electric field to be aligned with the long axis of these cylindrical macromolecule ionic chains. Furthermore, we assume the polyions are evenly distributed along the macromolecular network at regular distance b. Thus, we employ the conservation laws – namely, conservation of mass and

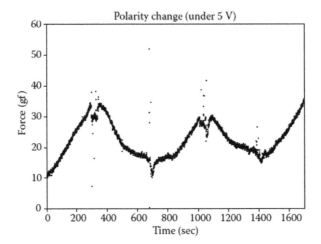

FIGURE 4.116   Force changes from switching polarity.

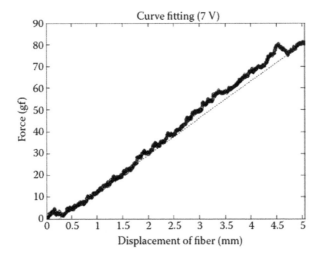

FIGURE 4.117   Force versus displacement, under 7 V.

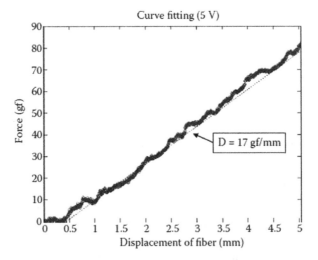

FIGURE 4.118   Force versus displacement, under 5 V.

momentum – to arrive at the following governing differential equation for the flow of counterions containing solvent into and out of the gel macromolecular network.

$$\rho \frac{dv}{dt} = \rho g + \rho^* E + \mu \nabla^2 v - \nabla_p \tag{4.18}$$

where

ρ = density of the liquid solvent, which is assumed to be incompressible
v = three-dimensional liquid velocity vector
∇ = gradient vector operator
∇^2 = Laplacian operator
g = local gravitational acceleration vector
μ = solvent viscosity
p = hydrostatic pressure
E = imposed electric field vector.

ρ^* is the charge density governed by the following Poisson's equation:

$$\rho^* = -D^* \nabla^2 \psi \tag{4.19}$$

where D^* is the dielectric constant of the liquid phase and ψ is governed by the following Poisson–Boltzmann equation:

$$\nabla^2 \psi = \left(\frac{4\pi n \varepsilon}{D^*} \right) \exp \left[-\frac{\varepsilon \psi}{kT} \right] \tag{4.20}$$

where

n = the number density of counterions
ε = their average charge
k = Boltzmann constant
T = absolute temperature.

The electrostatic potential in polyelectrolyte solutions for fully stretched macromolecules is given by the following equation, which is an exact solution to the Poisson–Boltzmann Equation (4.20) in cylindrical coordinates:

$$\Psi(r,t) = \left[\frac{kT}{\varepsilon} \right] \ln \left\{ \left[r^2 \left(r_0^2 - r_i^2 \right) \right] \sinh^2 \left[\beta \ln \left(\frac{r}{r_0} \right) - \tan^{-1} \beta \right] \right\} \tag{4.21}$$

where β is related to $\lambda = (\alpha \varepsilon^2 / 4\pi D \times bkT)$, where α is the degree of ionization (i.e., $\alpha = n/Z$, where n is the number of polyions and Z is the number of ionizable groups) and b is the distance between polyions in the network. Furthermore, $n = (\alpha \varepsilon^2 / 4\pi D \times bkT)$ and βs are found from the following equation:

$$\lambda = \frac{1 - \beta^2}{1 + \beta \coth \left[\beta \ln \left(r_0 / r_i \right) \right]} \tag{4.22}$$

Let us further assume that, due to cylindrical symmetry, the velocity vector $v = (v_r, v_\theta, v_z)$ is such that only v_z depends on r and further that $v_\theta = 0$. Thus, the governing equations for $v_z = v$ reduce to:

$$\rho \left(\frac{\partial v}{\partial t} \right) = f(r, t) + \mu \left(\frac{\partial^2 v}{\partial r^2} + \frac{1}{r} \frac{\partial v}{\partial r} \right) - \frac{\partial p}{\partial r} \tag{4.23}$$

Let us assume a negligible radial pressure gradient and assume the following boundary and initial conditions:

At $t = 0$, $r_i \leq r \leq r_0$, $v = 0$; at $r = r_i$, $\forall t, v(r_i) = 0$; and at $r = r_0$, $\forall t$, $\left(\dfrac{\partial v}{\partial r}\right)_{r=r_0} = 0$.

Furthermore, the function $f(r,t)$ is given by:

$$f(r,t) = n\varepsilon E(r,t) \left\{ \left[\frac{k^2 r^2}{2\beta^2}\right] \sinh^2\left[\beta \ell n\left(\frac{r}{r_0}\right) - \tan^{-1}\beta\right] \right\}^{-1} \tag{4.24}$$

where $k^2 = (r\varepsilon^2/DkT)$.

An exact solution to the given set of equations can be shown to be:

$$v(r,t) = \sum_{m=1}^{\infty} e^{-\left(\frac{\mu}{\rho}\right)\beta_m^2 \xi} A\left(\beta_m, \xi\right) d\xi \tag{4.25}$$

where β_m,s are the positive roots of the following transcendental equations:

$$\frac{J_0\left(\beta r_i\right)}{J_0\left(r_0\right)} - \frac{Y_0\left(\beta r_i\right)}{Y_0\left(\beta r_0\right)} = 0 \tag{4.26}$$

where J_0, Y_0, J'_0, Y'_0, are the Bessel functions of zero order of first and second kind and their derivatives evaluated at r_0, respectively, and

$$k_0\left(\beta_m, r\right) = N^{-(1/2)}\left\{\frac{J_0\left(\beta r\right)}{\beta_m J'_0\left(\beta r_0\right)}\right\} = N^{-(1/2)} R_0\left(\beta_m, r\right) \tag{4.27}$$

where

$$N = \left(\frac{r_0}{2}\right) R_0^2\left(\beta_m, r_0\right) - \left(\frac{r_i^2}{2}\right) R_0'^2\left(\beta_m, r_i\right), \tag{4.28}$$

$$A\left(\beta_m, \xi\right) = \left(\frac{1}{\mu\rho}\right) \int_{r_i}^{r_0} \varsigma\, k_0\left(\beta_m, \varsigma\right) f\left(\varsigma, \xi\right) d\varsigma. \tag{4.29}$$

Having found an explicit equation for $v(r, t)$, we can now carry out numerical simulations to compare the theoretical dynamic contraction of ionic polymeric gels in an electric field with those of experiments. To compare the experimental results and observations with the proposed dynamic model, a number of assumptions, simplifications and definitions are first made. Consider the ratio $W(t)/W(0)$, where $W(t)$ is the weight of the entire gel at time t, and $W_0 = W(0)$ is the weight of the gel at time $t = 0$, just before the electrical activation. Thus,

$$W(t) = W_0 - \int_0^t \int_{r_i}^{r_0} 2\pi\rho v(r,t)\, r\, dr\, dt \tag{4.30}$$

This can be simplified to

$$\left[\frac{W(t)}{W_0}\right] = 1 - W_0^{-1} \int_0^t \int_{r_i}^{r_0} 2\pi\rho v(r,t)\, r\, dr\, dt \tag{4.31}$$

The initial weight of the gel is related to the initial degree of swelling $q = V(0)/V_p$, where $V(0)$ is the volume of the gel sample at $t = 0$ and V_p is the volume of the dry polymer sample. Numerical simulations were carried out based on the assumptions that the cross section of the gel remains constant during contraction of the gel sample, and that

$$\varepsilon = e = 1.6 \times 10^{-19}\,\text{C} \tag{4.32}$$

$$T = 300\text{K},\ \alpha = 1,\ D = 80,\ \mu = 0.8 \times 10^{-3}\,\text{Pa s} \tag{4.33}$$

$$\rho = 1000\ \text{kg/m}^3, b = 2.55 \times 10^{-10}\,\text{m} \tag{4.34}$$

$$k = 1.3807 \times 10^{-23}\,\text{J/K},\ r_i = 6.08 \times 10^{-10}\,\text{m}, r_0 = r_i q^{(1/2)} \tag{4.35}$$

$$q = 25,\ 70,\ 100,\ 200,\ 256,\ 512,\ 750 \tag{4.36}$$

The initial length and cross section of the sample are, respectively, $\ell_0 = 1$ cm and $S = 1\ \mu\text{m}^2$, and the electric field is $E = 5.7$ V/cm.

The results of numerical simulation and experimental data are depicted in Figure 4.119 and show reasonable agreement with our experimental results, as well as the experimental results of Gong et al. (1994a, 1994b) and Gong and Osada (1994) shown in Figure 4.120.

4.6.12 OTHER ASPECTS OF PAN MUSCLE BEHAVIOR

The behavior of the PAN fibers may also be explained based on the molecular interaction forces. The main interaction forces may be considered the long-range electrostatic (coulomb) forces as well as the short-range van der Waals and double-layer forces. The combination of the van der Waals and double-layer interactions acting together is usually called the DLVO forces.

Figure 4.121 shows that PAN fibers can be woven into fabric forms and then activated as sheet-like or smart fabric-like artificial muscles. In this case the PAN fibers were suitably annealed, cross-linked and hydrolyzed to become "active". A cation-modification process was performed using

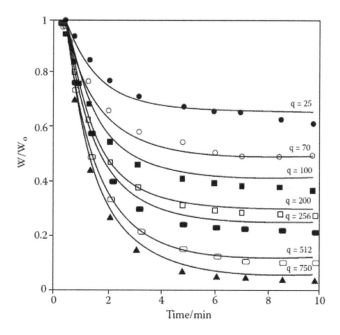

FIGURE 4.119 Computer simulation (solid lines) and experimental results (scattered points) for the time profiles of relative weight of the gel sample for various degrees of swelling q.

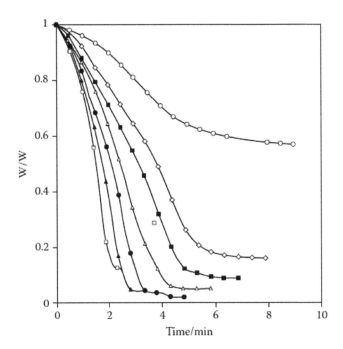

FIGURE 4.120 Experimental result taken from direct measurement of a sample PAMPS muscle. (Gong, J. P. and Y. Osada. 1994. In *Preprints of the Sapporo symposium on intelligent polymer gels*, 21–22; Gong, J. et al. 1994. In *Proceedings of the international conference on intelligent materials*, 556–564.)

KOH, NaOH and LiOH, respectively, for the boiling and alkaline-soaking media. It was found that the PAN fibers, regardless of whether activated in KOH, NaOH or LiOH, increased from their initial length after being activated and soaked in distilled water. Lengths then decreased after the fibers were soaked in the bases.

Fibers treated with LiOH had the largest increase in length following immersion in distilled water. Fibers soaked in all three media generally had the same decrease in length following immersion

FIGURE 4.121 Woven fabric forms of PAN muscles.

in the alkaline solutions as also occurred following immersion in HCl. Especially noticeable with the fibers treated with LiOH was that greater displacement in the lengths occurred using the 2-N solutions. It should be noted that the maximum displacement could be determined when conditions were switched between pure water to acidic conditions. The experimental observation reveals that the osmotic pressure is of great importance for the contraction/elongation behavior of PAN. It should be noted that PAN fibers have their capability of changing effective longitudinal strain more than 100% and their comparable strength to human muscle. Single microfibers of PAN of 10 μm in diameter have shown contraction/expansion linear strain of over 500% in our laboratories.

4.6.13 Force Generation with pH Difference

In this experiment, we were interested in the amount of ions sufficient to shrink the PAN muscle fiber. A single string of PAN fiber with 6.0-cm length before saponification was tested. After each force measurement, the fiber was saturated with 1-M LiOH and rinsed by a sufficient amount of distilled water. When the fiber reached equilibrium condition, different concentrations of HCl solution were sprayed on the fiber for the next measurement until the fiber reached steady state. For force measurement, a 30-g load cell (transducer techniques) was used.

Figure 4.122 shows the experiment's apparatus. In the first experiment, there is a step in force generation because of discontinuation of HCl supply. The maximum forces of PAN fiber in response to each concentration of HCl solution, which were 2.15 g for 0.01 M, 2.66 g for 0.1 M, 2.71 g for 0.5 M, 3.1 g for 1 M and 2 M = 2.84 g, were measured. There was 1-g force gap between 0.01- and 0.1-M solution, and when the concentration of solution was low, there was a time delay to reach the peak point. In high-concentration solution, such as 1 and 2 M, the force reached the maximum value almost immediately. Even in 0.01-M concentration of HCl, 72% of force, compared with the force generated in 1-M solution, the concentration of which is 100 times stronger, was observed.

Another force variation measurement along with pH difference using other acids such as sulfuric acid and nitric acid was ongoing to investigate whether the anion of acid solution affects the test

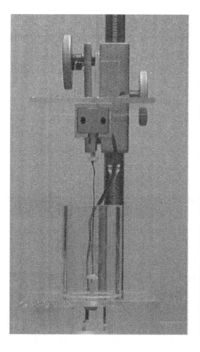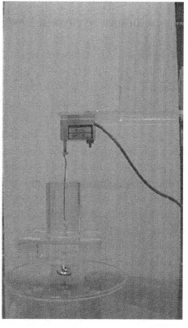

FIGURE 4.122 Electrochemical test setup.

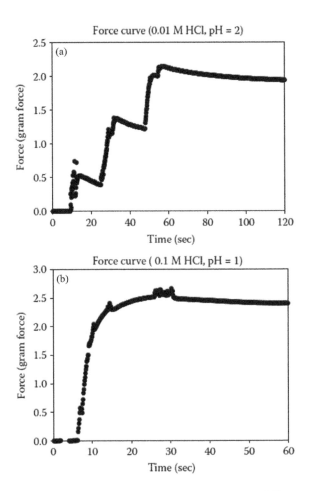

FIGURE 4.123   Force versus time curve with (a) 0.01-M HCl and (b) 0.1-M HCl.

result. The results of these experiments are depicted in Figure 4.123(a) and (b), as well as Figures 4.124(a) and (b) and 4.125.

To achieve kinetic energy, a PAN actuating system or a PAN muscle system depends on pH change of its environment. In an electrically controlled PAN muscle system, localized pH difference during the water electrolysis is the key parameter that actually makes the movement of PAN fiber. To reduce the response time, applying a diaphragm (ion-separation materials) between anode and cathode electrode was considered. Ionic diaphragms prevent ion transportation, which will increase concentration of ions rapidly and reduce the response remarkably. Furthermore, note that corrosion of anode electrodes occurs because H^+ ions chemically react with copper. The copper surface has been plated with gold to block direct contact between copper surface and H^+ ions. However, the plating is not dense enough to prevent chemical reaction. The ions penetrate, crack and corrode copper under the gold layer. To make a denser layer on the copper surface, use of appropriate plating material such as hexachloroplatinate (H_2PtCl_6) was also considered and tested.

To investigate the anion distribution for force production of PAN muscle, two kinds of acid were prepared: sulfuric acid and nitric acid. Force generation was measured under the same condition as the previous experiment. Normality was used to give equal amounts of hydrogen ion to the muscle system: 0.01, 0.1, 0.5 and 1 N of each solution was applied to the single PAN fiber. Like the previous experiment, a single fiber covered the whole experiment, and, after tensile work, 1-M LiOH solution was applied to release the muscle, followed by rinsing the fiber with distilled water.

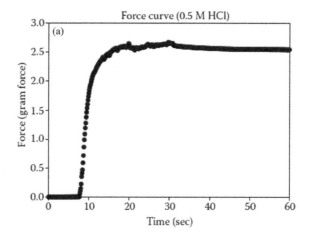

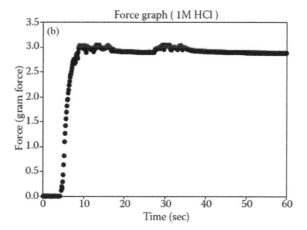

FIGURE 4.124   Force versus time curve with (a) 0.5-M HCl and (b) 1-M HCl.

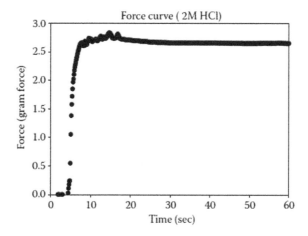

FIGURE 4.125   Force versus time curve with 2-M HCl.

TABLE 4.10

Maximum Force of Each Concentration

	Maximum Force of Each Concentration (gf)			
	0.01 N	**0.1 N**	**0.5 N**	**1 N**
HNO_3	6.17	7.45	7.63	7.28
H_2SO_4	5.1	7.06	7.13	6.92

Table 4.10 shows the maximum force of each concentration of acid. Force curves by acid concentration higher than 0.1 N had similar values and patterns. Under 0.01-N concentration, the force curve had a step increase due to discontinuous supply of acid solution (a 5-mL pipette was used to spray the solutions). The ion diffusion took longer than with a higher concentration because of the limited amount of ions carried within the 0.01-N solution. To confirm the differences between HCl and other acid solutions, 0.5 and 1 N of HCl were tested after two acid experiments were already completed. Maximum force of each concentration was 5.97 and 6.12 gf, respectively. This is approximately 1 gf less than that of sulfuric acid and nitric acid. However, the following experiments using 1 N nitric acid showed that the maximum of 5.0 gf fatigue or deterioration of PAN structure due to iteration load could affect the force produced. If the fatigue of the muscle system is taken into account, the effect of anion on the system would be small. These results are depicted in Figures 4.126–4.130.

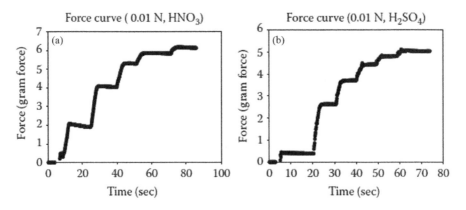

FIGURE 4.126   Force curve under 0.01 N of (a) HNO_3 and (b) H_2SO_4.

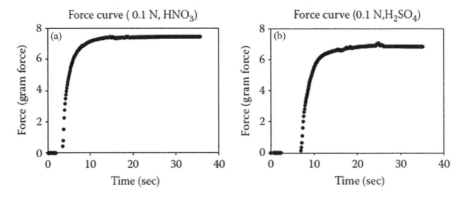

FIGURE 4.127   Force curve under 0.1 N of (a) HNO_3 and (b) H_2SO_4.

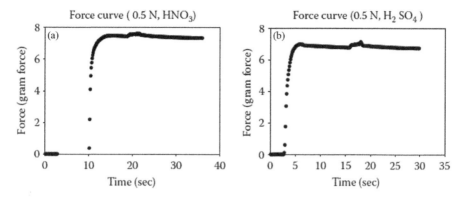

FIGURE 4.128 Force curve under 0.5 N of (a) HNO₃ and (b) H₂SO₄.

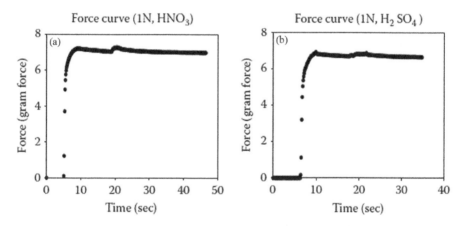

FIGURE 4.129 Force curve under 0.5 N of (a) HNO₃ and (b) H₂SO₄.

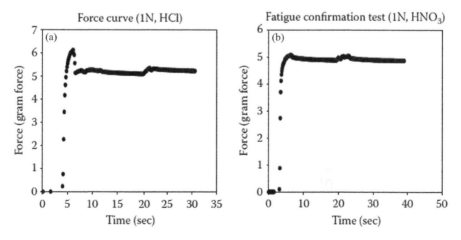

FIGURE 4.130 Fatigue confirmation test with (a) 1-N HCl and (b) 1-N HNO₃ after sulfuric and nitric acid test.

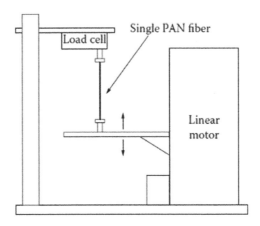

FIGURE 4.131 Single PAN fiber tensile machine setup.

4.6.13.1 Mechanical Property of Single PAN Fiber

A tensile testing machine to measure a single PAN fiber, as shown in Figure 4.131, was designed and built. A stand holds a load cell from the top and a single PAN fiber is slowly pulled downward by a linear motor (Razel™). This experiment was useful to achieve reliable mechanical property data of a single PAN fiber. Note that a test with a bundle of PAN fibers is very tricky and depends on the fabrication and assembly of fibers in the bundle.

4.6.14 EFFECTS OF ELECTRODE DETERIORATION ON FORCE GENERATION

The corrosion of electrodes has made significant impurities in the electrolytic solution and obstructed the process. Since these could be the cause of slow response time of PAN muscles, the material of electrodes was changed for the anode and the cathode sides. Electrodes were made of titanium (99.7%). Ruthenium dioxide and iridium dioxide were coated for the anode electrode with the sol-gel method. Both electrodes have mesh structure to increase contact surface and effectiveness. On top of that, an ion diaphragm is selected to prevent direct contact between hydrogen ions and hydroxide ions and enhance local pH concentration. Dimensions are shown in Figure 4.132. As shown in

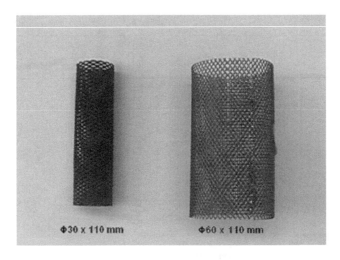

FIGURE 4.132 New electrodes for anode (left) and cathode (right).

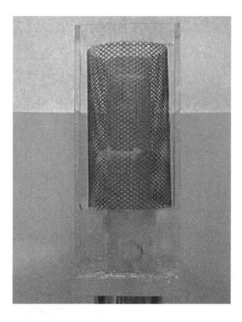

FIGURE 4.133   Test apparatus for new electrodes.

Figures 4.133 and 4.134, anode and cathode sides are isolated by the ion diaphragm. Figure 4.135 depicts the results of force generation by pH and voltage.

From HCl 2-M solution, 150 gf was obtained and, under 5-V electric field, it reached 100 gf for 20 min of operation. A 6 gf was obtained, which was more than the ones obtained in previous experiments. After the experiments, decolorization of some PAN fibers was found. This was caused by hypochlorite (HOCl) produced during the electrolysis of sodium chloride solution.

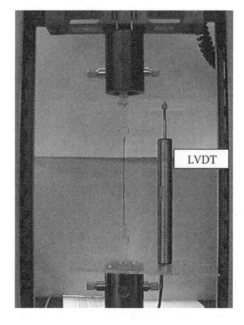

FIGURE 4.134   New electrodes for anode (left) and cathode (right).

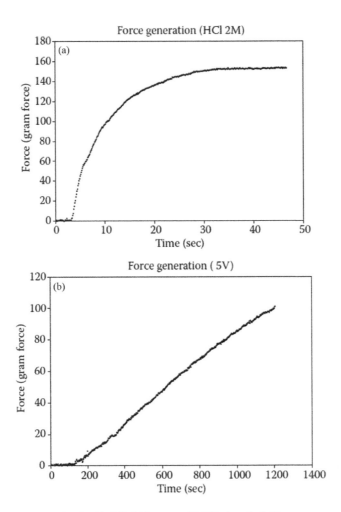

FIGURE 4.135   Force generation by (a) HCl 2 M versus (b) 5-V electric field.

In general, HOCl can cause appalling effects on a polymer structure that may retard or weaken the PAN muscle during the operation. In searching for other electrolytes to substitute, sodium chloride was examined to eliminate the sodium hypochlorite effect. Smaller diameter ($\sqrt{14}$ mm) titanium tubes with platinum coating for anode electrodes were also prepared and tested.

4.7   FIVE-FINGERED HAND DESIGN AND FABRICATION USING PAN FIBER BUNDLE MUSCLES

A robot hand using PAN muscle was fabricated. The skeletal bone of the hand was made of acrylic and grab-and-release motion was performed by a combination of PAN muscles and springs. When PAN fiber is shrunk, the counterpart extension spring stores energy, which makes the finger move back to its original position. Dimensions of one of the fingers are shown in Figure 4.136. Each finger section makes a circular motion about the pivot point until the cylinder of supporting muscle force meets the stopper of other finger section. Figure 4.137 depicts the solid design of the PAN five-fingered hand.

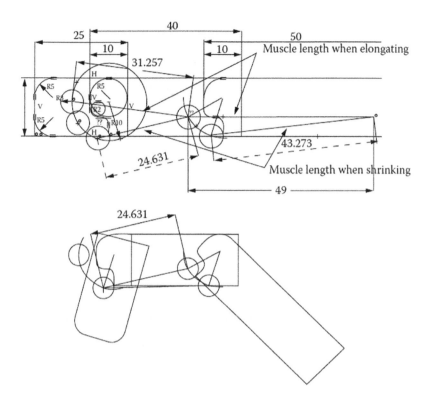

FIGURE 4.136 Dimensions of the robotic finger.

FIGURE 4.137 Schematics of PAN muscle hand, original position (upper) and in grab motion (bottom).

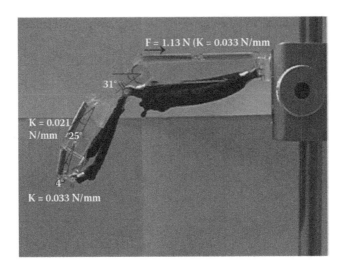

FIGURE 4.138   Initial position of a single PAN finger.

4.7.1   Fabrication of Five-fingered Hand Equipped with Fiber Bundle PAN Muscles

A prototype of a PAN muscle finger was made and shake-up was tested. PAN fingers in initial and grab motion are shown in Figures 4.138 and 4.139. Three different kinds of springs were prepared to practice turn-back motion. Detailed spring specification is provided in Table 4.11. The specific dimensions and spring combinations were optimized for better performance.

Each PAN finger except for the thumb consists of three knuckles, four PAN fiber muscles, two return springs and two tubes that deliver acid and basic solution used to shrink and elongate the fiber. The thumb consists of three knuckles, two PAN muscles, one return spring and two tubes. The finger body is made of acrylic with thickness of 3 mm. Figure 4.140 shows one PAN finger in a contracted state. Detailed dimensions of finger joint and assembly drawings are included.

The final configuration of a five-fingered hand equipped with five pairs of PAN fiber bundle muscles is shown in Figures 4.141(a) and (b) and 4.142(a) and (b). Figure 4.143 depicts the detailed line drawings pertaining to the designed and fabricated five-fingered hand equipped with contractile PAN fiber bundle ionic polymeric muscle design.

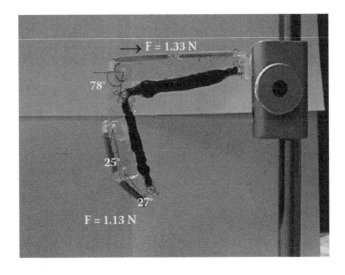

FIGURE 4.139   Grab position performance of a single PAN finger.

TABLE 4.11
Spring Specification

Outside Diameter (mm)	Wire Diameter (mm)	Max Load (N)	Free Length (mm)	Rate (N/mm)	Max Length (mm)	Initial Tension (N)	Material
4.775	0.355	2.962	25.400	0.033	109.219	0.146	SS
4.775	0.355	2.962	34.925	0.021	163.703	0.146	SS
4.775	0.355	2.962	57.150	0.012	298.450	0.146	SS

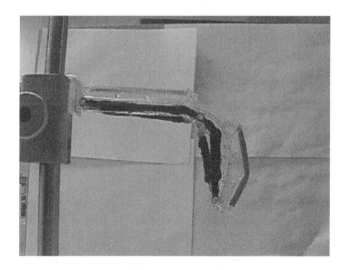

FIGURE 4.140 PAN fiber bundle muscles in action as a single finger.

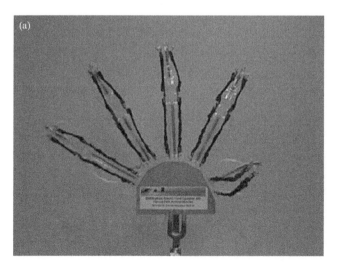

FIGURE 4.141 (a) Frontal and (b) back view of PAN five-fingered hand. *(Continued)*

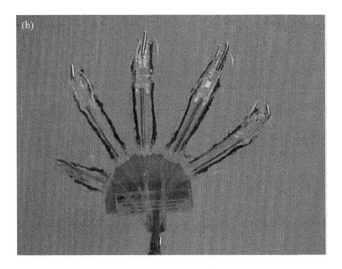

FIGURE 4.141   (b) back view of PAN five-fingered hand. *(Continued)*

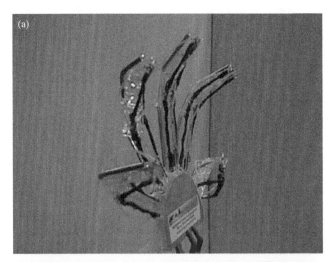

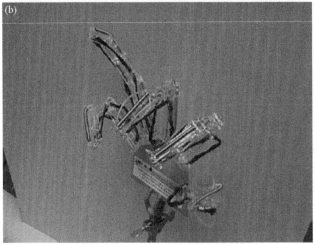

FIGURE 4.142   (a) Left side and (b) right side view of PAN five-fingered hand.

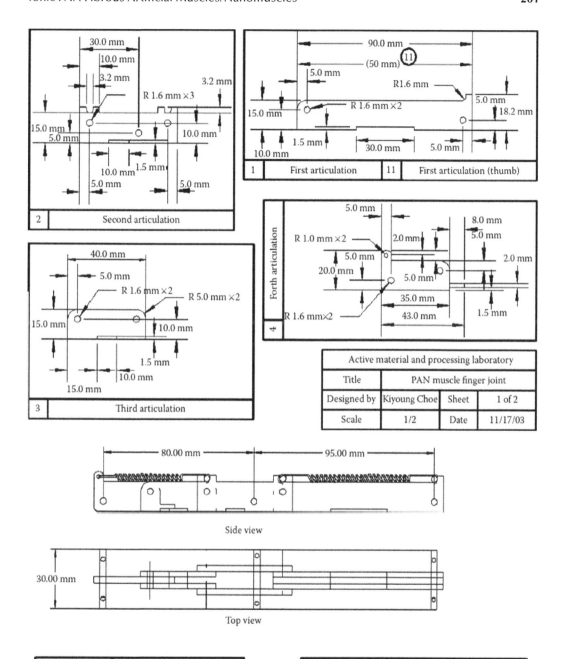

FIGURE 4.143 Five-fingered hand drawings.

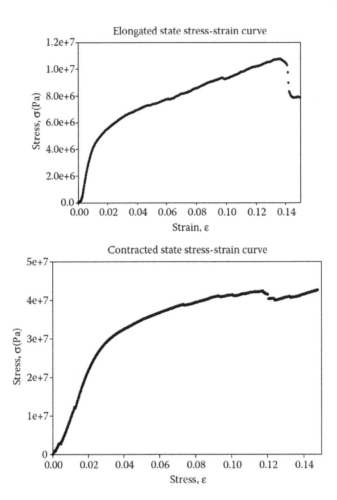

FIGURE 4.144   Stress–strain curve of elongation and contraction state of single PAN fiber.

4.7.2   ADDITIONAL MECHANICAL PROPERTY MEASUREMENT OF SINGLE PAN FIBER

Mechanical properties of PAN fiber muscle were measured. PAN fibers were oxidized at 220°C and then saponified in 1-M LiOH at 75°C for 30 min. Ten samples were tested for both elongated and contracted states. A typical stress–strain curve is shown in Figure 4.144. It was found that, above 80°C saponification temperature, fibers were completely transformed to a gel-like substance and mechanical properties became hard to measure. The contracted state and elongated state exhibited different mechanical properties. Table 4.12 summarizes the mechanical properties of each status.

TABLE 4.12
Mechanical Properties of Single PAN Fiber

	E (Young's Modulus) (MPa)	Yield Strength (MPa)
Contracted state	107	25.0
Elongated state	40	6.5

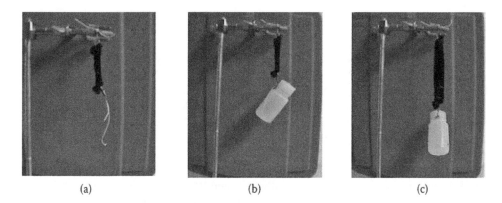

FIGURE 4.145   (a) A typical setup for a PAN fiber bundle; (b) oxidation or contracted mode of a PAN fiber bundle; (c) reduction or expanded mode of a PAN fiber bundle.

PAN fiber bundles' (of 50 fibers) typical contraction of more than 100% is shown in Figures 4.145(a)–(c).

4.8   MICRO-PAN FIBER OBSERVATION

Figure 4.146 shows the volume change of minute PAN fibers. Marked fibers in the circle mean the same fiber, which elongates and contracts under basic and acidic solution. In elongation status, the length and the diameter of the fiber were 350 and 40 μm, respectively. In contraction status, they were constricted to 120 and 14 μm, respectively.

After simple calculation, the volume of the minute PAN fibers was measured to be 4.0×10^{-4} mm^3 in swelling state and 1.8×10^{-5} mm^3 in contracting state. The volume change ratio (swelling status/contracting status) is about 2,380%.

It was determined from a previous experiment that a single fiber can generate 5 gf under 1-M HCl solution. If we assume that a single fiber has about 2,000–2,500 minute fibers and each fiber equally contributes to force generation, a minute fiber generates 2.0–2.5×10^{-3} gf. Observing volume change of the minute PAN fiber under applied electric field is also under consideration.

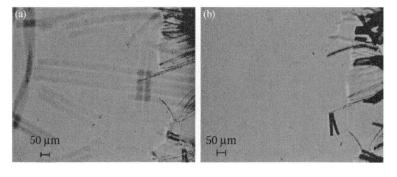

FIGURE 4.146   Minute PAN fiber change (a) before and (b) after applying 1-M HCl solution.

4.9 CONCLUSIONS

Described in this chapter are results obtained over a period of eight years on electroactive PAN as well as a model and experimental results that allow one to electrically control the actuation of active C-PAN fiber bundles, or C-PAN. Increasing the conductivity of PAN fibers by making a composite of them with a conductive medium such as platinum, gold, graphite or carbon nanotubes and conductive polymers such as polyaniline or polypyrrole has allowed for electric activation of PAN fibers when a (C-PAN) fiber bundle is placed in an electrochemical cell as an electrode. A change in concentration of cations in the vicinity of a C-PAN fiber electrode leads to contraction and expansion of C-PAN fibers depending upon the applied electric field polarity.

Typically, pH-activated PAN fibers exhibit more than 100% contraction or expansion in a few seconds. However, in cases of electrochemically activated PAN fibers, more than a 50% change in C-PAN length in a few seconds is observed in a weak electrolyte solution with tens of VDC power supply. Furthermore, C-PAN-Ns have been fabricated first by the electrospinning method and then ionically activated to have fixed-charge carboxylic groups. The use of such C-PAN-Ns is promising for fabricating fast response PAN artificial muscles. PAN-Ns were suitably annealed, cross-linked and hydrolyzed to become "active". Active PAN fibers are contractile synthetic muscles with strength comparable to biological muscles. These results provide a great potential in developing fast-activating C-PAN-N nanomuscles and linear actuators, as well as integrated pairs of antagonistic muscles and muscle sarcomere and myosin/actin-like assemblies.

Electrical activation of C-PAN artificial muscles is demonstrated by increasing the conductivity of PAN artificial muscles. The conductivity of PAN is increased by depositing a coat of metal on the fibers or interweaving it with conductive fibers such as graphite fibers. Electrochemical reactions are used to generate hydrogen ions or hydroxyl ions for the contraction and elongation, respectively, of these C-PAN muscles. Therefore, by increasing the conductivity of activated PAN, a PAN-based linear actuator can be electrically activated in an antagonistic manner suitable for industrial and medical applications. Increasing the conductivity of PAN fibers by making a composite of them with a conductive medium such as platinum, gold, graphite or carbon nanotubes and conductive polymers such as polyaniline or polypyrrole was shown to allow for electric activation of PAN fibers when a C-PAN fiber bundle is placed in a chemical electrolysis cell as an electrode. Typically, close to 50% change in C-PAN fiber length in a few seconds is observed in a weak electrolyte solution with tens of VDC power supply.

To decrease the response time of C-PAN, PAN-Ns were also successfully fabricated by the electrospinning method. As expected, the response time of C-PAN is governed by the diffusional processes of ion–solvent interaction; the use of such PAN-Ns is promising for fabricating fast-response PAN artificial muscles.

Experimental results provided a great potential in developing fast-activating C-PAN-N muscles and linear actuators, as well as integrated pairs of antagonistic muscles and muscle sarcomere and myosin/actin-like assembly.

5 PAMPS Ionic Polymeric Artificial Muscles

5.1 INTRODUCTION

Ionic polymeric gels are three-dimensional networks of cross-linked macromolecular polyelectrolytes that swell or shrink in aqueous solutions upon adding alkali or acids, respectively. Reversible dilation and contraction of the order of more than 800% have been observed in our laboratory for polyacrylonitrile (PAN) gel fibers. Furthermore, it has been experimentally observed that swelling and shrinking of ionic gels can be induced electrically. Thus, direct computer control of large expansions and ionic polymeric gels' contractions utilizing an electronic system with a voltage controller is possible. Some ionic gels such as poly(2-acrylamido-2-methyl propane sulfonic acid) or PAMPS actuate and sense well in imposed electrical fields.

Swelling and contraction of ionic polymeric gels by pH variations have been historically discovered and reported by Kuhn et al. (1948) and Katchalsky (1949), followed by a wealth of additional papers on pH activation of ionic gels by them and their students to the late 1960s. Such investigation on the pH response of ionic gels is still going strong, as can be witnessed by Aluru and his coworkers (De et al., 2002).

The first paper connected with the electrically controllable polymer gels' response was by Hamlen and coworkers on electrolytically active polymers published in *Nature* in 1965. In modern times, De Rossi et al. (1986) reported for the first time on contractile behavior of electrically activated mechanochemical polymer actuators. Segalman et al. (1991, 1992a, 1992b, 1992c) reported their investigation on electrically controlled polymeric gels as active materials in adaptive structures. The first papers on using electrically controllable ionic polymeric gel actuators and artificial muscles connected with swimming robotic structures were produced by Shahinpoor (1991, 1992, 1993a). The first patent (U.S. patent 5,250,167) in the world on electrically controllable polymeric gel actuators was reported in 1993 by Adolf et al.

These electrically controllable gel actuators possess an ionic structure because they are generally composed of several fixed ions (polyions) pertaining to sites of various polymer cross-links and segments and mobile ions (counterions or unbound ions) due to the presence of an electrolytic solvent. Electrically induced dynamic deformation of ionic polymeric gels, such as polyacrylic acid plus sodium acrylate cross-linked with bisacrylamide (PAAM), or PAMPS, or various combinations of chemically doped polyacrylic acid plus polyvinyl alcohol (PAA-PVA), can be easily observed in the laboratory. Such deformations give rise to an internal molecular network structure with bound ions (polyions) and unbound or mobile ions (counterions) when submerged in an electrolytic liquid phase. In the presence of an electric field, these ionic polymeric networks undergo substantial contraction accompanied by exudation of the liquid phase contained within the network.

Under these circumstances, there are generally four competing forces acting on such ionic networks: rubber elasticity, viscous interactions due to the liquid phase's motion, inertial effects due to the liquid's motion through the ionic network and electrophoretic interactions. These forces collectively give rise to dynamic osmotic pressure and network deformation and subsequently determine such charged networks' dynamic equilibrium.

On the other hand, there are situations in which a strip of such ionic polymeric gels undergoes bending in the presence of a transverse electric field with hardly any water exudation. Under these circumstances, there are generally three competing forces acting on the gel polymer network: rubber elasticity, polymer-polymer affinity and ion pressure. These forces collectively create the osmotic pressure, which determines the equilibrium state of the gel. The competition between these forces changes the osmotic pressure and produces the volume change or deformation. Rubber elasticity tends to shrink the gel under tension and expand it under compression. Polymer-polymer affinity depends on the electrical attraction between the polymer and the solvent. Ion pressure is the force exerted by the motion of the cations or anions within the gel network. Ions enter the gel attracted by the opposite charges on the polymer chain, while their random motions tend to expand the gel-like ionic (fermionic) gas.

Kuhn et al. (1948) originally reported on the possibility that certain copolymers can be chemically contracted or swollen like a synthetic muscle (pH muscle) by changing the pH of the solution containing them. As originally reported by Kuhn et al. (1950), a three-dimensional network consisting of polyacrylic acid can be obtained by heating a polyacrylic acid foil containing polyvalent alcohol such as glycerol polyvinyl alcohol. The resulting three-dimensional networks are insoluble in water but swell enormously in the water on the addition of alkali and contract on acids' addition. Reversible dilations and contractions of the order of more than 1000% have been observed for ionic gel muscles made with PAN fibers (Shahinpoor and Mojarrad, 1996). Chemically stimulated pseudomuscular actuation has also been discussed recently by Li and Tanaka (1989), De Rossi et al. (1986), and Caldwell and Taylor (1990). Hamlen et al. (1965) were the first to report that these gels' contractions and swelling can also be obtained electrically by placing fibrous samples of PAA-PVA in an electric field.

Tanaka and coworkers at MIT presented one of the earlier mathematical modelings of deformation of ionic gels in an electric field as early as 1978, proposing a phase transformation phenomenon responsible for such electrically induced contraction. Other experimental and theoretical investigations addressing the electrically induced contractile behavior of ionic polymeric gels have been presented by Osada and Hasebe (1985), De Rossi et al. (1986), Osada and Kishi (1989), Grodzinsky and Melcher (1974), Grimshaw et al. (1990a), Shiga and Kurauchi (1990), and Kishi et al. (1990). Shahinpoor has discussed applications of electrically activated ionic polymeric gel muscles to swimming robotic structures. These findings indicated that the gel's short-time response to the electric field is due to electrophoretic migration of unbound counterions in the gel and impingement of solvent ions on the gel samples' surfaces. The surplus or deficiency of such ions determines the osmotic pressure and free volume and therefore deformation of such gels. In the next section, the case of electrically induced nonhomogeneous deformation of transparent PAMPS and PAAM cylindrical lenses has been experimentally and theoretically described. Exact expressions relate the gel's deformation characteristics to the electric field strength or voltage gradient, gel dimensions and other physical parameters such as the resistance and the gel samples' capacitance.

5.2   PAMPS GELS

This chapter briefly concentrates on and describes the characteristics and application of PAMPS gel as electroactive artificial muscles capable of actuation and sensing. Although many experiments have been performed by applying this polyelectrolyte in swimming structures, the emphasis is placed on other applications, such as electrically controllable and active optical lenses. The reason is that, in simple engineering actuation configurations, such as swimming robotic structures in which PAMPS gel actuators are used as caudal fin structures (Shahinpoor, 1991, 1992), there is not enough structural strength in these materials in gel form. There exist several papers on the applications of PAMPS to robotic structures and, in particular, gel starfish by Osada et al. (1992), Shahinpoor (1992, 1993a, 1993b, 1993c, 1993d), Shiga et al. (1993), Osada et al. (1994), Gong et al. (1994a, 1994b), Okuzaki and Osada (1994c), Ueoka et al. (1997), Narita et al. (1998), and Otake et al. (2000, 2001, 2002).

These papers deal with weakly cross-linked PAMPS gel. This behavior's principle is based on an electrokinetic molecular assembly reaction of surfactant molecules on the polymer gel caused by electrostatic and hydrophobic interactions. Under an electric field, PAMPS gel undergoes significant and quick bending. The response could be controlled effectively by changing the surfactant molecule's alkyl chain length, the salt concentration and the current applied.

The results allow us to consider that cooperative complex formation between PAMPS gel and C_nPyCl is responsible for this effective chemomechanical behavior.

This copolymer is clear in color when synthesized, and similarly to other polyelectrolyte gels, it swelled in water several times its original volume. When placed in an electric field, strips of the PAMPS bent within a few seconds. Depending on the thickness of the gel synthesized, the response time for bending actuation varied. Thinner and smaller actuators had a faster response, whereas thicker samples were slower.

5.3 GEL PREPARATION

Here we present a very simple procedure to prepare an electroactive version of PAMPS gels. Electrically active PAMPS gels can be prepared by free-radical copolymerization. A 15 wt% aqueous solution (deionized [DI] water) of a reaction mixture with the desired molar ratio of a common monomer and a cross-linker density of about 1 mol% (of total monomers) should be bubbled with nitrogen for about 20 min to remove any oxygen in the reaction mixture. Then 35 μl of a 10 wt% ammonium persulfate (APS) solution should be added to the mixture. The final solutions should be filtered through a fine filter.

One portion of the resulting gels is washed in deionized water to remove the unreacted monomers and titrated with aqueous sodium hydroxide solution to determine the copolymer's composition. Another portion of the gels should be washed in a large amount of dilute sodium hydroxide solution for 20 days to neutralize the poly(methacrylic acid) and remove the unreacted monomers. The dilute sodium hydroxide solution should be changed every two days, and the final pH value should be kept above the neutral pH of 7.

Specifically, to make an electroactive PAMPS gel, make two mixtures and mix them quickly to form the gel. First, mix 0.5 g of tannish sodium acrylate powder with 2.13 g of whitish acrylamide powder in the presence of 0.08 g of a cross-linker N,N'-methylene-bisacrylamide $[(H_2C=CHCONH)_2CH_2]$ in 30 ml of DI water reaction mixture. Then, mix 0.156 g of N,N,N',N'-tetramethyl ethylene diamine (TMEDA) $[(CH_3)_2NCH_2CH_2N(CH_3)_2]$ with 0.02 g of APS. Put these two into 20 ml of DI water, shake well, and then add to the 30-mL reaction mixture immediately within few seconds. Otherwise, APS decomposes and never forms a gel of PAMPS. The mixture will take a few minutes to gel up. The resulting PAMPS gel is then highly electroactive.

5.4 PAMPS GEL APPLICATION

Since PAMPS gel appears colorless and is soft and compliant, there are many adaptive smart materials into which it can be incorporated. The following sections describe one application of PAMPS ionic gels investigated in our Artificial Muscle Research Institute. This particular hydrophilic gel has also been investigated extensively for use in drug delivery systems (DDS) by Osada et al. (1985, 1989, 1991, 1992, 1993, 1994, 1995, 2005). For example, Okuzaki and Osada (1994a) have described the electro-driven chemomechanical polymer gel as an intelligent soft material. In their work, weakly cross-linked PAMPS gel was synthesized, and the chemomechanical behaviors in the presence of N-alkyl pyridinium chloride (C_nPyCl, $n = 4, 12, 16$) were studied.

This behavior's principle is based on an electrokinetic molecular assembly reaction of surfactant molecules on the polymer gel caused by electrostatic and hydrophobic interactions. Under an electric field, PAMPS gel underwent significant and quick bending. The response could be controlled effectively by changing the surfactant molecule's alkyl chain length, the salt concentration, and the current applied. The results allow us to consider that cooperative complex formation between PAMPS gel and C_nPyCl is responsible for this effective chemomechanical behavior.

5.4.1 Adaptive Optical Lenses

Reversible change in optical properties of ionic polymeric gels, PAMPS, under an electric field's effect is reported. The shape of a cylindrical piece of the gel, with a flat top and bottom surfaces, changed when affected by an electric field. The top surface became curved, and the sense of curvature (whether concave or convex) depended on the applied electric field's polarity. The curvature of the surface changed from concave to convex and vice versa by changing the electric field's polarity. Using an optical apparatus, the curved surface's focusing capability was verified, and the focal length of the deformed gel was measured.

The effect of the applied electric field's intensity on the surface curvature and, thus, the gel's focal length is tested. Different mechanisms are discussed; either of them or their combination may explain the surface deformation and curvature. Practical difficulties in the test procedure and the potential of the electrically adaptive and active optical lenses are also discussed. These adaptive lenses may be considered smart adaptive lenses for contact lenses or other optical applications requiring focal point undulation. This research area has only recently been explored and is projected to be a major driver in future biotechnology and industrial optics applications.

5.4.2 Theoretical Model

To describe the electrically induced deformation of ionic polymeric gels, we consider the process of symmetrical deformation of a cylindrical segment of ionic polymeric gels in a transverse electric field. Suppose a cylindrical sample of an ionic gel of radius r and thickness $t^* = 2C$ is deformed into a convex lens (Figure 5.1) by an imposed voltage gradient across its radius r. It is assumed that initially, the gel is in a natural stress-free state equilibrated with a pH = 7.

Due to the presence of an electrical voltage gradient across the radius r of the gel, the cylindrical gel sample is deformed nonhomogeneously into a convex (Figure 5.2(a)) or a concave (Figure 5.2(b)) lens by a nonuniform distribution of fixed as well as mobile ions in the gel. Note that an electric field gradient may be imposed across the gel's body by charged surfactant molecules.

If the gel possesses a specific capacitance C_g and a specific resistance R_g, Kirchhoff's law can be written in the following form:

$$v = C_g^{-1}Q + R_g\dot{Q}, \text{ where } \dot{Q} = i \tag{5.1}$$

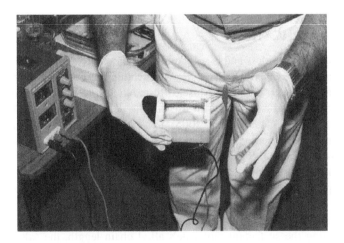

FIGURE 5.1 PAMPS muscle is shown in the bent state after applying a 30-V D.C. field. There are two gold-plated electrodes at each side of the Teflon (polytetrafluoroethylene, PTFE) container.

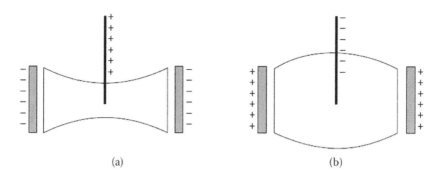

(a)   (b)

FIGURE 5.2   Deformation of PAMPS ionic gel cylinder due to an imposed radial voltage gradient.

where v is the voltage across the gel's thickness and Q is the specific charge (charge per unit mass) accumulated in the gel, $\dot{Q}$ being the current density i through the cylindrical gel sample across its radius. Equation (5.1) can be readily solved to yield

$$Q = C_g v \left[1 - \exp\left(-\frac{t}{R_g C_g} \right) \right],$$ (5.2)

assuming that at $t = 0$, $Q = 0$.

Equation (5.2) relates the voltage drop across the gel's radius to the charge accumulated, which eventually causes the gel to deform. Thus, Equation (5.2) will serve as a basis for the electrical control of gel deformations.

The imposed voltage gradient across the gel's radius forces the internal fixed and mobile ions to redistribute. The possible charge distribution of the gel is shown in Figure 5.3. A number of simplifying assumptions are made to mathematically model the nonhomogeneous deformation or bending forces at work in an ionic polyelectrolyte gel strip.

The first assumption is that the polymer segments carrying fixed charges are cylindrically distributed along a given polymer chain and independent of the cylindrical angle θ. This assumption is not essential but greatly simplifies the analysis. Consider the field of attraction and repulsion among neighboring rows of fixed or mobile charges in an ionic gel. Let r_i and Z_i be the cylindrical polar coordinates with the ith row as an axis such that the origin is at a given polymer segment.

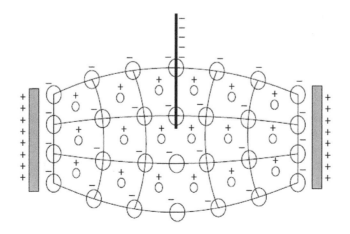

FIGURE 5.3   A possible charge redistribution configuration in ionic gels.

Let the spacing in the ith row be b_i, and let the forces exerted by the atoms be central and of the form cr^{-s} such that the particular cases of $s = 2$, $s = 7$ and $s = 10$ represent the Coulomb, van der Waals and short-range repulsive forces, respectively. Then it can be shown that the component of the potential electric field per unit charge at the point (r_i, Z_i) perpendicular to the row is represented in a series such that

$$R(r_i, Z_i) = \sum_{n=-\infty}^{\infty} \left\{ \frac{r_i}{\left[\left(Z_i - nb_i \right)^2 r_i^2 \right]^{(1/2)(s+1)}} \right\}, \tag{5.3}$$

This function is periodic in Z_i, of period b_i, and is symmetrical about the origin. It may therefore be represented in a Fourier series in the form

$$R(r_i, Z_i) = \left(\frac{1}{2} \right) C_{0i} + \sum_{m=1}^{\infty} C_{mi} \cos\left(\frac{2\pi m Z_i}{b_i} \right) \tag{5.4}$$

where

$$C_{mi} = \left(\frac{2}{b_i} \right) \sum_{n=\infty}^{\alpha_i} \int_0^{} \left\{ \frac{r_i \cos\left(2\pi m Z_i / b_i \right)}{\left[\left(Z_i - nb_i \right)^2 + r_i^2 \right]^{(1/2)(s+1)}} \right\} dZ_i \tag{5.5}$$

Evaluating the coefficients C_{mi}, $m = 0, 1, 2,\ldots\infty$ in Equation (5.5), for $m = 0$, then

$$C_{oi} = \frac{r_i \Gamma(1/2) \Gamma(1)}{b_i \Gamma(3/2)}, \tag{5.6}$$

and for $m > 0$

$$C_{mi} = \left(\frac{4r_i}{b_i} \right) \left(\frac{\pi m_i}{b_i r_i} \right)^{(1/2)s} \left[\Gamma(1/2) / \Gamma(1/2)(s+1) \right] K\left(\frac{1}{2} \right) s\left(\frac{2\pi m_i r_i}{b_i} \right) \cos\left(\frac{2\pi m_i Z_i}{b_i} \right) \tag{5.7}$$

where Γ and K are modified Bessel functions. In the remainder of this section, only the Coulomb types of attraction and repulsion forces will be considered to simplify these expressions. With this assumption, the expression for $R(r_i, Z_i)$ becomes

$$R(r_i, Z_i) = \left(\frac{2r_i}{b_i} \right) \left[r_i^{-2} + 4 \sum_{m=1}^{} mK_1\left(\frac{2\pi m r_i}{b_i} \right) \cos\left(\frac{2\pi m Z_i}{b_i} \right) \right] \tag{5.8}$$

Now recall that many molecular strands are generally cross-linked, oriented and entangled in an ionic polymer network. Assuming that in the presence of an imposed voltage gradient across the thickness of the cylindrical gel sample, the rows of fixed and mobile ions line up, the mean field can be obtained by superimposing the field corresponding to positive charges, namely,

$$R(r_i, Z_i) = \left(\frac{2r_i}{b_i} \right) \left[r_i^{-2} + 4 \sum_{m=1}^{\infty} mK_1\left(\frac{2\pi m r_i}{b_i} \right) \cos\left(\frac{2\pi m Z_i}{b_i} \right) \right] \tag{5.9}$$

and corresponding to negative charges, namely,

$$R\left(r_i, Z_i\right) = \left(\frac{2r_i}{b_i}\right)\left\{r_i^{-2} + 4\sum_{m=1}^{\infty} mK_1\left(\frac{2\pi mr_i}{b_i}\right)\cos\left[\left(\frac{2\pi mZ_i}{b_i}\right)\left(Z_i + \left(\frac{1}{2}\right)b_i\right)\right]\right\} \quad (5.10)$$

The resulting field for a pair of rows is

$$R\left(r_i, Z_i\right) = \left(\frac{16\pi}{b_i^2}\right)\left[\sum_{m=1,3,5...}^{\infty} mK_1\left(\frac{2\pi mr_i}{b_i}\right)\cos\left(\frac{2\pi mZ_i}{b_i}\right)\right] \quad (5.11)$$

The whole field due to the presence of N strands per unit length of the sample is then given by

$$R^*\left(r, Z\right) = \sum_{i=1}^{N} 16\pi b_i^{-2}\left[\sum_{m=1,3,5...}^{\infty} mK_1\left(\frac{2\pi mr_i}{b_i}\right)\cos\left(\frac{2\pi mZ_i}{b_i}\right)\right] \quad (5.12)$$

For simplicity, let us assume that all b_i, Z_i and r_i are equal to b, Z and r, respectively, where b is defined as the mean interior ionic distance in the gel so that, on average, $N = 2/b$. With this assumption, Equation (5.12) reduces to

$$R^*\left(r, Z\right) = 32\pi b^{-3}\left[\sum_{m=1,3,5...}^{\infty} mK_1\left(\frac{2\pi mr}{b}\right)\cos\left(\frac{2\pi mZ}{b}\right)\right] \quad (5.13)$$

The mean Coulomb attraction or repulsion force associated with the mean field $R^*(r, Z)$ is given by $F(r, Z)$ such that

$$F\left(r, Z = kq^2r^*R^*\left(r, Z\right)\right) \quad (5.14)$$

where k is a constant of proportionality, r^* is the mean radius of the gel strip and q corresponds to the total charge between a pair of adjacent ionic surfaces in the gel sample.

Thus, $q = (4/3)b^{-1}\rho\pi r^3 Q$, where ρ is the gel sample's average density. This force is repulsive (positive) or attractive (negative) according to whether like or unlike charges lie in the adjacent rows of charges. Experimental observations on the bending of ionic gels in the presence of a voltage gradient generally indicate no gross motion in the field's direction, suggesting that the force field $F(r, Z)$ along the long axis of the gel may be nonuniformly distributed. The nature of $F(r, Z)$, namely,

$$F\left(r, Z\right) = 32\pi kr^*q^2b^{-3}\left[\sum_{m=1,3,5...}^{\infty} mK_1\left(\frac{2\pi mr}{b}\right)\cos\left(\frac{2\pi mZ}{b}\right)\right] \quad (5.15)$$

suggests that even a one-term series solution gives rise to a possible solution, namely,

$$F_1\left(r, Z\right) = \left[32\pi kr^*q^2b^{-3}\right]\left[K_1\left(\frac{2\pi r}{b}\right)\cos\left(\frac{2\pi Z}{b}\right)\right] \quad (5.16)$$

This implies that one possible force configuration is when ions are located at the sample's outskirts with mobile ions located in the middle.

The solution for the force field given by Equation (5.16) can create a nonhomogeneous deformation field in the cylindrical sample. This allows the designer to robotically control such deformations in ionic gels utilizing a voltage controller or, conversely, measure the mechanical deformation of gels by the voltage produced due to such deformations. In this sense, the gel body becomes a large strain or deformation sensor.

5.4.3 Electrically Controllable Ionic Polymeric Gels as Adaptive Optical Lenses

Reversible change in optical properties of ionic polymeric gels, PAMPS and PAAM under the effect of an electric field is reported. The shape of a cylindrical piece of the gel, with a flat top and bottom surfaces, changed when affected by an electric field. The top surface became curved, and the sense of curvature (whether concave or convex) depended on the applied electric field's polarity. The curvature of the surface changed from concave to convex and vice versa by changing the electric field's polarity.

Using an optical apparatus, the curved surface's focusing capability is verified, and the focal length of the deformed gel is measured. A more exact method (thermal) to measure the focal length of the gel is also discussed. The effect of the applied electric field's intensity on the surface curvature and the gel's focal length is tested. Two different mechanisms are discussed based on ion mobility and interaction forces; either of them or their combination may explain the surface deformation and curvature.

Practical difficulties in the test procedure and the future potential of the electrically adaptive and active optical lenses are also discussed. These adaptive lenses may be considered smart adaptive lenses for contact lenses or other optical applications requiring focal point undulation.

5.4.4 Experimental Results: PAMPS

The experimental gel was cut in a cylindrical piece with a height of about 7 mm and a diameter of about 25 mm. It was placed in the middle of a copper ring with a diameter of about 40 mm. NaCl solution was added such that the gel and the copper ring were partially immersed in the electrolyte solution. A platinum wire of 1 mm in diameter was inserted into the center of the cylindrical gel. The copper ring and the platinum wire were connected to a power source's positive and negative terminals. A schematic of this setup is shown in Figure 5.4.

When the electric voltage was applied, liquid exudation started from the gel's upper surface, accompanied by swelling at the center. The swelling zone stretched toward the edges of the gel. The speed at which the swollen zone stretched toward the gel's edges varied with the applied electric voltage. A higher

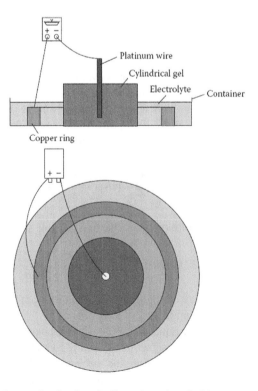

FIGURE 5.4 Experimental setup for the electrically activated optical lens.

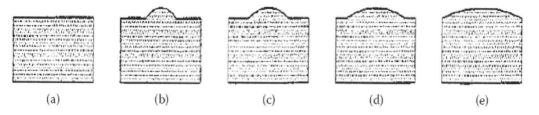

(a)   (b)   (c)   (d)   (e)

FIGURE 5.5 Stretching of the swollen zone due to the electric field.

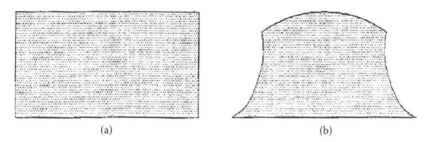

(a)   (b)

FIGURE 5.6 Observed deformation of the gel lens under the influence of the electric field.

applied electric voltage caused a higher speed of stretching out of the swollen zone toward the cylindrical gel's outer edges. Figure 5.5 indicates the stretching of the swollen zone as it was observed.

At all the applied electric voltages (10, 15, 20, 25 and 30 V), the swollen zone eventually reached the outer edge. However, this was achieved at different stretching speeds. The gel also was deformed at its circumferential wall, where it was immersed in the electrolyte. The gel was photographed before (cylindrical sample with a flat top surface) and after (deformed with the convex top surface) applying the electric voltage. These two configurations are indicated in Figures 5.5 and 5.6.

The electric field's polarity was reversed in a different test by connecting the platinum wire and the copper ring to the positive and negative terminals. This resulted in a concave upper surface. These observations were noticed for PAMPS and PAAM gels. The convex gel was placed on a stationary glass below which a screen was attached to a laboratory jack.

Figure 5.7(a) and (b) depict an experimental set showing how PAMPS optical lenses can be electrically controlled in terms of their power and focal length under the action of a small electric field.

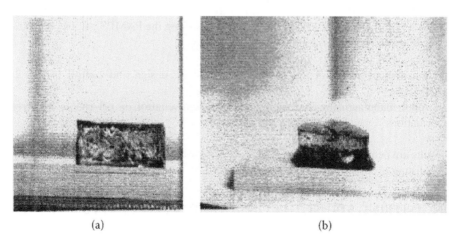

(a)   (b)

FIGURE 5.7 Actual experimental observation on the deformation of the adaptive optical lens under electrical control. (From Salehpoor, M. et al. 1996a. In *Proceedings of the SPIE conference on intelligent structures and materials*, 2716:36–45.)

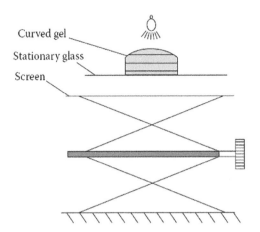

Curved gel
Stationary glass
Screen

FIGURE 5.8   Experimental setup for measuring the focal length of the gel lens.

The distance between the stationary glass and the screen could be adjusted. A light source was placed on top of the glass. With the light on and the convex gel on the stationary glass, the distance between the screen and the glass was adjusted to place the screen at the convex PAMPS gel's focal length. This distance was measured as about 65 mm for both PAAM and PAMPS lenses.

The setup for measuring the gel focal length is shown in Figure 5.8.

5.5   ELECTROACTIVE PAMPS GEL ROBOTIC STRUCTURES

As discussed before, the first use of electrically controllable ionic polymeric gel actuators and artificial muscles connected with swimming robotic structures was reported by Shahinpoor (1991, 1992, 1993a) and Shahinpoor and Mojarrad (1996). Figure 5.9 depicts such concepts using active fins such as the ones made with IPMCs. On the other hand, Mihoko Otake (Otake et al., 1999, 2000, 2001, 2002) should be given credit for making a starfish robot from electroactive PAMPS. These results can be found in her University of Tokyo doctoral dissertation and the subsequent papers from her. Her dissertation studies the modeling, design and control of deformable machines consisting of actively deformable materials referred to as artificial muscles. The main focus is to propose methods for deriving various shapes and motions of such machines, using PAMPS gel and its copolymer gel. However, Osada and coworkers (1992) should be given credit for the first crawling robot made with electroactive PAMPS gel.

In her dissertation, Otake described the mechanisms using the PAMPS gel, or "gel robots", and how to control them. Her dissertation includes the following:

1. A mathematical model of the gel to be applied for design and control of distributed mechanisms
2. Gel robot manufacturing and their driving system control of gel robots for dynamic deformations

The results are demonstrated for beam-shaped gels curling around an object and starfish-shaped gel robots turning over (Otake et al., 1999, 2000, 2001, 2002).

5.6   ENGINEERING STRENGTH CONSIDERATIONS ON PAMPS IONIC GELS

It is worthwhile to mention a few works on the engineering strength of PAMPS gel for practical applications. It is certainly true that PAMPS gels are rather soft gels and can easily break apart under tension or even compression or shear loadings. Whiting et al. (2001) have studied the shear

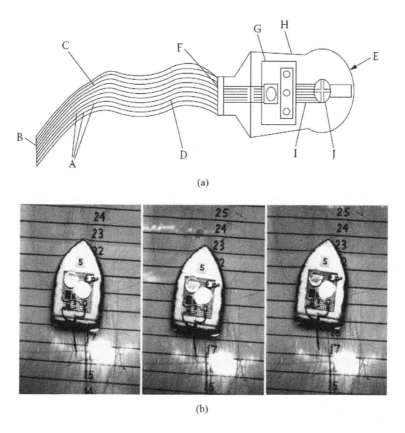

(a)

(b)

FIGURE 5.9   (a) Robotic swimming structure and (b)swimmer with muscle undulation frequency of 3 Hz (bottom). The scale shown is in centimeters.

strength and moduli of PAMPS for practical soft engineering applications. In their study, the shear modulus of two polyelectrolyte gels – poly(acrylic acid), a weak acid with small-side groups, and PAMPS gel, a strong acid with large-side groups – has been measured with and without application of an electric field across gel samples fully swollen with water.

According to their theoretical predictions, the shear modulus of PAMPS gels is inversely proportional to the swelling degree in the absence of an electric field. Under a steady electric field, the measured modulus of these gels was observed to reduce with time. They explain that this is due to the migration of the free counterions and associated hydration and added mass water toward the cathode, resulting in reduced contact between gel and rheometer due to exuded water. Theoretical prediction of G' and G'' in the presence of a thin slip layer is in good quantitative agreement with experimental observations. Preliminary measurements of the reduction of shear modulus under the pulsed electric field have also been obtained. They observed a spontaneous recovery of the initial modulus on all subsequent field applications before the continued reduction to successively lower values. When roughened platens were used, a stepwise variation in measured modulus was observed, with a slightly lower modulus being recorded in the presence of an applied electric field.

Yasuda et al. (2005) have studied the biomechanical properties of high-toughness double-network (DN) hydrogels. Using pin-on-flat wear testing, they have evaluated the wear property of four novel DN hydrogels composed of two kinds of hydrophilic polymers. The gels involved PAMPS-PAAm gel, which consists of poly(2-acrylamide-2-methyl-propane sulfonic acid) and polyacrylamide; PAMPS-PDAAAm gel, which consists of poly(2-acrylamide-2-methyl-propane sulfonic acid) and poly(N,N'-dimethyl acrylamide); cellulose-PDMAAm gel, which consists of bacterial cellulose and polydimethyl-acrylamide; and cellulose-gelatin gel, which consists of bacterial cellulose and gelatin. Ultrahigh molecular weight polyethylene (UHMWPE) was used as a control of clinically available material.

Using a reciprocating apparatus, a million cycles of friction between a flat specimen and ceramic pin were repeated in water under a contact pressure of 0.1 MPa. A confocal laser microscope was used to determine the depth and roughness of the wear's concave lesion. As a result, the PAMPS-PDMAAm gel (3.20 μm) was minimal in the five materials, while there was no significant difference compared to UHMWPE. There were significant differences between UHMWPE and one of the other three gels, the PAMPS-PAAm gel (9.50 μm), the cellulose-PDMAAm gel (7.80 μm), and the cellulose-gelatin gel (1302.40 μm). This study demonstrated that the PAMPS-PDMAAm DN gel has an amazing wear property as a hydrogel comparable to the UHMWPE. The PAMPS-PAAm and cellulose-PDMAAm DN gels are also resistant to wear to greater degrees than conventionally reported hydrogels. On the other hand, this study showed that the cellulose-gelatin DN gel was not resistant to wear.

5.7   IONIC GEL ROBOTICS

There has been some development on gel robotics by the University of Tokyo and Hokkaido University researchers (Otake et al., 1999, 2000a, 2000b, 2002a, 2002b, 2002c). They have proposed a framework for describing mollusk-type elastic robots' behaviors made of electroactive PAMPS polymer gel. Their numerical simulation and experimental results show that large deformations can be obtained with multiple electrodes in a planar configuration. They have designed a starfish-shaped gel robot that can flip over using a spatially varying electric field.

5.8   RESULTS AND CONCLUSIONS

Ionic polymeric gels are three-dimensional networks of cross-linked macromolecular polyelectrolytes that swell or shrink in aqueous solutions upon adding alkali or acids, respectively. Reversible dilation and contraction of the order of more than 800% have been observed in our laboratory for PAN gel fibers. Furthermore, it has been experimentally observed that swelling and shrinking of ionic gels can be induced electrically. Thus, direct computer control of large expansions and ionic polymeric gels' contractions utilizing an electronic system with a voltage controller is possible. Some ionic gels such as PAMPS actuate and sense well in imposed electrical fields.

6 Modeling and Simulation of IPMCs as Distributed Soft Biomimetic Nanosensors, Nanoactuators, Nanotransducers and Artificial Muscles

6.1 INTRODUCTION

Several earlier attempts have been made to analytically understand the sensing and actuation mechanisms at work in ionic polymer–metal composites (IPMCs), or ionic polymer conductor composites (IPCNCs) and ionic gels. Kuhn (1949), Katchalsky (1949) and Kuhn et al. (1950) originally reported that certain ion-containing copolymers might be chemically contracted or expanded like a synthetic muscle. According to their results, a three-dimensional ion-containing network or a polyelectrolyte, consisting of polyacrylic acid (PAA) and polyvinyl alcohol (PVA), could be obtained by heating a PAA containing polyvalent alcohol such as glycerol or PVA. The resulting three-dimensional networks were insoluble in water but swelled enormously (more than 400%) in water upon addition of an alkali and contracted enormously (more than 400%) on acids' addition. However, the processes of swelling and contraction took days to complete and were very slow. Kuhn made the early attempts; his student, Katchalsky, and coworkers started in 1948 and continued to the mid-1960s. Their strategy was based on the degree of ionization of the network and its effect on its swelling or contraction.

Grodzinsky, Melcher, Yannas, and coworkers out of MIT were the first to present coherent theories on the electromechanics of deformable, charged polyelectrolyte biomembranes as early as 1973, Figure 3.41(a), and their efforts have continued to the present time through the works of Grimshaw et al. (1989, 1990). Their approach has been based on the effect of movement of charged species, electrodiffusion phenomena, dissociation of membrane charge groups, intramembrane fluid flow, electroosmotic drag and mechanical deformation of membrane matrix – all due to charge redistribution within the polymeric network.

In the early 1980s, another MIT researcher, Toshio Tanaka, and his coworkers started a flurry of research publications on polymer gels' collapse in an electric field. They essentially treated the mechanism of collapse as a phase transition phenomenon.

In the mid-1980s, another pioneer, Danilo De Rossi, and coworkers out of the University of Pisa in Italy presented a series of papers on the determination of mechanical parameters related to the kinetics of electroactive swelling polymeric gels. They were the first to discuss biological tissues' analogs for mechanoelectrical transduction: tactile sensors and muscle-like actuators. Their contribution has continued to the present time.

DOI: 10.1201/9781003015239-6

As early as the mid-1980s, another pioneer in this area from Japan, Yoshihito Osada and coworkers presented possible theories of electrically activated mechanochemical devices using polyelectrolyte gels and mechanism and process of chemomechanical contraction of polyelectrolyte gels under an electric field. Their contributions have continued to the present time.

By the late 1980s and early 1990s, new contributors to the mechanisms of actuation and sensing of polyelectrolytes and ionic polymers had appeared. Shahinpoor (1991, 1992) discussed conceptual design, kinematics and dynamics of swimming robotic structures using active polymer gels. He presented a set of ion transport equations and continuity, conservation of momentum and conservation of energy equations involving an imposed electric field. Segalman et al. (1991, 1992a, 1992b, 1992c, 1993, 1994) presented a series of papers on modeling and numerical simulation of electrically controlled polymeric muscles as active materials used in adaptive structures. Further, they presented a finite element simulation of a polyelectrolyte gel disk's two-dimensional collapse, considering neo-Hookean constitutive equations for the polymer network elasticity.

Shahinpoor (1993b) further presented a nonhomogeneous, large-deformation theory of ionic polymeric gels in electric and pH fields. Attempts to formulate a microelectromechanical theory for ionic polymeric gels as artificial muscles for robotic applications were initiated in a series of papers by Shahinpoor (1993b, 1993c, 1993d, 1994a, 1994b, 1994c, 1994d, 1994e, 1994f, 1995d, 1995e, 1999b, 2000b, 2002e). Shahinpoor also presented a continuum electromechanics theory of ionic polymeric gels as artificial muscles for robotic applications. Shahinpoor and Osada (1995a) presented a theory on the electrically induced dynamic contraction of ionic polymeric gels based on electrocapillary and electroosmotic forces.

Shahinpoor et al. (1998a) presented the first review paper on IPMCs such as biomimetic sensors, robotic actuators and artificial muscles. Pierre-Gilles de Gennes and coworkers (2000) presented the first phenomenological theory for sensing and actuation in ionic polymer–metal nanocomposites (IPMNCs). Asaka and Oguro (2000) discussed the polyelectrolyte membrane's bending-platinum composites by electric stimuli. They presented a theory on actuation mechanisms in IPMNCs by considering the electroosmotic drag term in transport equations.

Nemat-Nasser and Li (2000) presented modeling on the electromechanical response of IPMNCs based on the electrostatic attraction/repulsion forces in them. Later, Nemat-Nasser (2002) presented a revised version of their earlier paper and stressed the role of hydrated cation transport within the clusters and polymeric networks in IPMCs.

Nemat-Nasser and Wu (2003) have presented a discussion on the role of backbone ionic polymer and, in particular, sulfonic versus carboxylic ionic polymers, as well as the effect of different cations such as K^+, Na^+, Li^+ and Cs^+ and some organometallic cations on the actuation and sensing performance of IPMCs.

Tadokoro (2000) and Tadokoro et al. (2000, 2001) have presented an actuator model of IPMNC for robotic applications based on physicochemical phenomena.

On the characterization front, to understand the underlying mechanisms of sensing and actuation, Mojarrad and Shahinpoor (1997a) presented plots of blocked force versus time for an ionic polymer bender subjected to sinusoidal, triangular, square and sawtooth waveforms. Shahinpoor and Mojarrad (1997b) showed displacement versus frequency for a 2-V periodic input.

The first report of ionic polymer transducer sensing was published by Sadeghipour et al. (1992), who created a Nafion™-based accelerometer. They fabricated a wafer-like cell that applied pressure across the thickness of a piece of Pt-plated Nafion (an ion-exchange membrane product of DuPont), and they measured the voltage output. The cell was approximately 2 in. diameter, and its sensitivity was on the order of 10 mV/g. An interesting feature of their work is that the Nafion was not hydrated. Before use, it was saturated with hydrogen under high pressure. The load was also applied across the polymer's thickness, while most other ionic polymer transducer research has been performed using cantilevered benders.

Shahinpoor (1995a, 1996c) and Shahinpoor and Mojarrad (1997a, 1997b, 2002) reported the discovery of a new effect in ionic polymeric gels – namely, the ionic flexogelectric effect in which flexing or loading of IPMNC strips created an output voltage like a dynamic sensor or a transducer converting mechanical energy to electrical energy. Consequently, Mojarrad and Shahinpoor (1997b) investigated displacement sensing by measuring the output voltage versus the applied tip displacement for a cantilevered ionic polymer transducer and observed that the output depended on the orientation of the transducer with respect to the electrodes.

Motivated by the idea of measuring pressure in the human spine, Ferrara et al. (1999) applied pressure across the thickness of an IPMC strip while measuring the output voltage. Their experiment was repeated with a maximum stress of almost 900 kPa, and the results were similar. More recently, Henderson and colleagues (2001) performed an experimental frequency-domain analysis of the output voltage with a cantilevered bender's tip displacement input. Their purpose was to evaluate the suitability of ionic polymer transducers for use in near-D.C. accelerometers. The transducer used in the experiment was allowed to dry in typical atmospheric conditions for approximately one month before testing.

They observed a sensitivity of approximately 50 mV/m for an 11×29-mm^2 cantilever and concluded that ionic polymer transducers might be a useful technology for low-frequency accelerometer applications. Recent studies of Aluru and coworkers on ionic polymer gels (De et al., 2002) are also relevant to this book.

Concerning the modeling of IPMNC biomimetic sensing and actuation, it must be emphasized that most of the models proposed for IPMNCs can be placed in one of three categories: physical models, black box models and gray box models.

For the physical models, researchers select and model the set of underlying mechanisms they believe to be responsible for the electromechanical response and subsequent deformation (actuation) of electrical output (sensing).

For the black box models (also called empirical models and phenomenological models), the physics is only a minor consideration, and the model parameters are based solely on system identification. The gray box models employ a combination of well-established physical laws and empirically determined parameters with several physical interpretations.

In the following section, we first present a general continuum model for ionic polymeric gels swelling and deswelling. We then embark on presenting several theoretical models for the deformation, flexing and bending behavior of IPMNCs (and IPCNCs).

6.2 CONTINUUM ELECTRODYNAMICS OF IONIC POLYMERIC GELS' SWELLING AND DESWELLING

6.2.1 BASIC FORMULATION

Theoretical models describing the dynamic behavior of the expansion and contraction of ionic polymeric gels present challenging problems. This section describes how weighted residuals can be a good approach to solving the two-dimensional governing system of equations using finite element analysis. The solvent's imbibition/expulsion modulation by an ionic gel disk is studied as an example. Formulation of a continuum mathematical model that accurately describes the gel's dynamic behavior requires properly accounting for the swelling or contraction or deswelling of the charged ionic polymeric network, fluid transfer into and out of the substructures nanoclusters and the coupled effects between the two phenomena. Also, large deformation kinematics must be used.

The initial modeling of Tanaka and coworkers (1982) and more recent modeling of De Rossi et al. (1986) consider only infinitesimal elasticity, which is only appropriate for small deformation of gels in quasistatic equilibrium situations. The interaction between the solvent and actuator requires two internal state variables to describe the system completely. Grimshaw et al. (1990a) have offered complete dynamic model descriptions (Segalman et al., 1991, 1992a, 1992b, 1992c, 1993; Segalman and Witkowski, 1994; Brock et al., 1994a, 1994b).

Theoretical predictions based on a finite element analysis solution scheme were first investigated by Segalman and colleagues (1991, 1992a, 1992b, 1993) (the SWAS theory). They solved the one-dimensional, dynamic analysis problem of a gelled sphere imbibing solvent. A slightly modified form of the theoretical model presented in Segalman et al. (1991, 1992a, 1992b, 1992c, 1993) and Segalman and Witkowski (1994) is used in this analysis. Modifications include representing the solvent concentration in terms of a mass fraction instead of a density value. In this formulation, the mass transport relationships and the elasticity equations are derived in a Lagrangian, or convected, framework.

The swelling of the gel is determined by the rate of solvent absorbed. Assuming no volume change of mixing, this condition becomes:

$$\frac{D\alpha^3}{Dt} = \frac{\alpha^3}{(\rho_s - c)} \frac{Dc}{Dt} \tag{6.1}$$

where

c = solvent concentration (mass of solvent per unit volume of swollen polymer)
α^3 = volumetric swell of the gel relative to some reference state
ρ_s = density of the pure solvent.

The time derivative here is the "material derivative" – the derivative of states associated with particles rather than position.

The velocity of the solvent is that of the gel plus the differential velocity due to diffusion. The diffusion is driven by osmotic pressures that are functions of two internal coordinates or states: solvent concentration, c, and the mass of H^+ ions per unit volume of the swollen polymer, H. Equivalent measures of these quantities are the mass fractions of gel, which are solvent and hydrogen ions, respectively. These two quantities and the displacement components of the gel are used as primary variables in this exposition. The solvent concentration can be represented in terms of a mass fraction by:

$$c = \frac{\rho_p \xi_s}{\left(1 - \left(\xi_s / \rho_p\right)\left(\rho_s - \rho_p\right)\right)} \tag{6.2}$$

where

ρ_p = density of the pure polymer
ξ_s = solvent mass fraction.

There are two independent diffusion equations for a system of three components (polymer, solvent and protons), each depending on the gradients of, at most, two of the components. The isothermal diffusion equation describing the evolution of the solvent mass fraction ξ_s is:

$$\left(\frac{\rho_s}{\rho_s - c}\right)\frac{D\xi_s}{Dt} = \nabla.\left[D_{sp}\left(\xi_s, \xi_H\right)\nabla\xi_s + D_{H_s}\left(\xi_s, \xi_H\right)\nabla\xi_H\right] \tag{6.3}$$

where the terms D_{ij} are diffusivities, ξ_H is the mass fraction of the H^+ ions and ∇ represents spatial gradient. Because the preceding evolution equation is written in a frame moving with the gel, the convective term appears differently from an Eulerian formulation. Derivation of this equation requires exploitations of the continuity Equation (6.1). The transport of H^+ is not only similar to that of solvent but also involves a source term:

$$\frac{D\xi_H}{Dt} = \nabla.\left[D_H\left(\xi_s, \xi_H\right)\nabla\xi_H\right] + \frac{D\xi_s}{D_t}\xi_H\frac{\partial c/\partial_s}{\rho_s - c} + \xi_{\bar{H}} \tag{6.4}$$

where $\xi_{\bar{H}}$ accounts for the rate of creation or neutralization of H. Note that $\xi_{\bar{H}}$ is a tunable parameter that may be varied through electrical or chemical means. Also, note the modification to the convective terms resulting from the equations' formulation in a frame that convects with the gel.

The diffusion relationships are taken from the Flory–Huggins theory. The hydrogen diffusion coefficient is represented as:

$$D_H = D_H^T \left(\frac{1 - \xi_p}{1 + \xi_p} \right)^{1/2} \approx 10^{-5} \xi_s^2 \tag{6.5}$$

The diffusion coefficient for the solvent in the polymer is given as:

$$D_{sp} = \frac{L_{sp}}{\xi_p^{1/2}} \left[(1 - 2_X) - \frac{1}{3N} \left(\frac{\xi_0}{\xi_p} \right)^{2/3} + \frac{K_a}{K_a + 53\xi_H} \right] \tag{6.6}$$

The hydrogen diffusion coefficient through the solvent is formulated as:

$$D_{H^s} = \frac{-L_{sp}}{\xi_p^{1/2}} \frac{K - a\xi_p}{(H + K_a)^2} \tag{6.7}$$

The nomenclature for material parameters used here is introduced by Flory (1953a, 1953b) and Grimshaw et al. (1990a).

The stress relationships for large deformation elasticity require the use of large deformation strain quantities. The deformation gradient $F(t)$ is defined as $F(t) = \partial x_g / \partial X_g$, where x_g, X_g are the particle locations in the deformed and unstrained states, respectively. In this problem, it is useful to factor the deformation gradient into its unimodular part and a part representing isotropic swell:

$$F(t) = F_{\text{uni}}(t) \cdot (\alpha I) \tag{6.8}$$

where

$a(t) = (\det[F(t)])^{1/3}$
α^3 = volumetric swell of the gel

For a solvent concentration-dependent neo-Hookean type solid, the Cauchy stress, S, resulting from a given deformation, is:

$$S(t) = G(c) \left[I - E(t)^T \cdot E(t) \right] - pI \tag{6.9}$$

where

$$\mathbf{E}(t) = F_{\text{uni}}(t)^{-1} \tag{6.10}$$

G is the shear modulus and p is a Lagrange multiplier dual to the incompressibility constraint on the swollen polymer. Because of the gel/solvent system's assumed incompressibility, the preceding equation presents stress only up to an unknown pressure. Constitutive modeling of rubber-like materials is discussed with good clarity in Segalman et al. (1992a, 1992b, 1992c, 1993) and Segalman and Witkowski (1994).

The incompressibility condition on the swollen polymer is simply a statement that the material's volume is not a function of the imposed pressure.

The conservation of momentum for the gel is:

$$P_g \ddot{x}_s = \nabla \cdot S + \rho_g f_b \tag{6.11}$$

where f_b contains all local body forces, such as gravitational or electromagnetic forces.

6.2.2 COMPUTER SIMULATION OF SYMMETRIC SWELLING AND CONTRACTION OF POLYELECTROLYTE GELS

The numerical problem is solved in a fully Lagrangian sense: All field variables are expressed as functions of time, t, and of the original particle location, X_g. The vector quantity X_g effectively labels the particles of gel. This approach is natural for large deformation problems, for which the finite element mesh will undergo significant deflection.

The governing differential equations are transformed into algebraic equations through a standard Galerkin finite element formalism (Segalman et al., 1993; Segalman and Witkowski, 1994, for instance). However, all interpolation is over a material manifold rather than over space because of the Lagrangian formulation. For instance, the configuration field of the gel is interpolated:

$$\hat{b}\left(t, X_g\right) = \sum_{i=1}^{\text{Nodes}} x_{gi}(t) \phi_i^I\left(X_g\right) \tag{6.12}$$

where $x_{gi}(t)$ is the location of particle X_{gi} (attached to node i) at time t. The other fields are also represented as linear combinations of appropriate basis functions:

$$\hat{\xi}_s\left(t, X_g\right) = \sum_{i=1}^{\text{Nodes}} \xi_{s,i}(t) \phi_i^s\left(X_g\right)$$

$$\hat{\xi}_H\left(t, X_g\right) = \sum_{i=1}^{\text{Nodes}} \xi_{H,i}(t) \phi_i^H\left(X_g\right)$$

$$\hat{p}\left(t, X_g\right) = \sum_{i=1}^{\text{Nodes}} p(t) \phi_i^s\left(X_g\right) \tag{6.13}$$

The preceding approximations are substituted into the governing equations, and the residuals are integrated concerning the appropriate basis function over the current volume of the gel. This process is reasonably standard. However, care should be taken to use the chain rule in evaluating spatial derivatives:

$$\nabla(.) = \frac{\partial(.)}{\partial X_g}\left[\frac{\partial X_g}{\partial X_g}\right]^{-1} \tag{6.14}$$

The basis functions for pressure must be one order lower than the basis functions for displacement. The LBB condition is satisfied, or else some additional constraints must be imposed on the pressure field. (Both methods have been employed successfully in the computer code.) Integration by parts must be done in those equations involving second derivatives in the basic fields to accommodate low-order basis functions.

A sequential solution strategy was chosen to solve the system of equations. First, the mass transport equations were solved, and then these results were piped into the elasticity relationships to calculate the resulting expansion/contraction. The following are the resulting weak forms of the governing equations:

$$\int_v \phi_i^s\left(X_g\right)\dot{\xi}dV = -\int_v \nabla\phi_i^s\left(X_g\right)\cdot\left[D_{sp}\nabla\hat{\xi}_s + D_{Hs}\nabla\hat{\xi}_H\right]dV - \int_v \nabla\phi_i^s\left(X_g\right)\hat{\xi}_s - \frac{\hat{c}}{\rho_c - \hat{c}} \quad (6.15)$$

$$\int_v \phi_i^H\left(X_g\right)\hat{\xi}_H dV = -\int_v \nabla\phi_i^H\left(X_g\right)\cdot\left[D_H\nabla\hat{\xi}_H\right]dV$$

$$-\int_v \phi_i^H\left(X_g\right)\hat{\xi}_s\hat{\xi}_H \frac{\partial c/\partial\xi_s}{\rho_c - \hat{c}}dV + \int_v \phi_i^H\left(X_g\right)\hat{\xi}_H dV \quad (6.16)$$

As mentioned in the previous section, the convective term in the preceding two equations is modified, reflecting the convection implicit in the Lagrangian finite element formulation. Boundary integrals associated with natural boundary conditions have been neglected in the preceding, where specified boundary conditions would override their contributions. The continuity equation is dual to the pressure field, and the resulting discretization is:

$$\int_v \phi_i^p\left(X_g\right)\left[\frac{\partial\hat{a}^3}{\partial t} - \hat{a}^3\hat{\xi}_s\frac{\partial c/\partial\xi_s}{\rho_c - \hat{c}}\right]dV = 0 \quad (6.17)$$

The discretized form of the momentum equation also requires integration by parts because the stress equation is substituted:

$$\sum_k \ddot{u}_k(t)\int_v \phi_i^u\left(X_g\right)\phi_k^u\left(X_g\right)\rho_g dV = \int_{\partial v}\phi_i^u\hat{S}\cdot dA - \int_{\partial v}\nabla\phi_i^u\hat{S}\cdot dV + \int_{\partial v}\phi_i^u f_b\left(X_g\right)\rho_g dV \quad (6.18)$$

This equation is used to derive the pressure field.

6.2.3 Gel Contraction/Shrinkage Example Based on the Continuum-Diffusion Model

The numerical analysis solution scheme is demonstrated in a gel disk example (see Figure 6.1). An equilibrated gel is subjected to changes in its environmental conditions to shrink by discharging

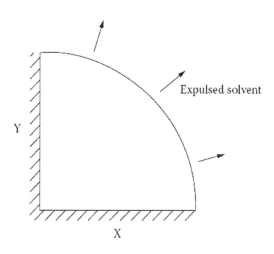

FIGURE 6.1 Gel disk swelling problem.

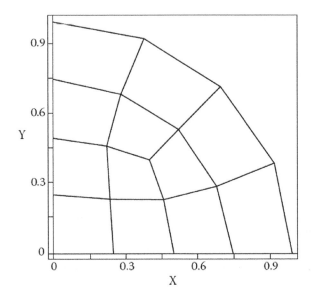

FIGURE 6.2   Finite element mesh on gel geometry.

solvent to attain its new equilibrium. Because of symmetry, only a quarter of the gel disk is modeled. The finite element mesh having 11 elements is shown in Figure 6.2.

The physical parameters and state variables are chosen to resemble typical values for a polyacrylamide gel system in the problem presented. Initially, the gel is assumed to be equilibrated at a pH of 8 with a corresponding equilibrium solvent mass fraction equal to 0.989. The gel's pH is then changed to 3 (with a corresponding solvent mass fraction equal to 0.775). This will cause the gel to expel solvent and shrink. Physical model parameters were derived from experimental observations ($K_a = 6 \times 10^{-5}$, $\times = 0.2$, $L_{sp} = 1 \times 10^{-5}$ and $N = 4$).

Figure 6.3 shows the displacement of the outer edge of the disk as a function of time. As expected, the initial shrinkage is the fastest, and then it slows until the new equilibrium dimension is reached.

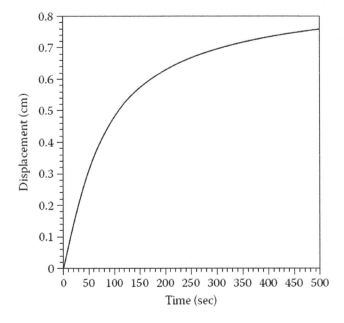

FIGURE 6.3   Outer edge radial displacement as a function of time.

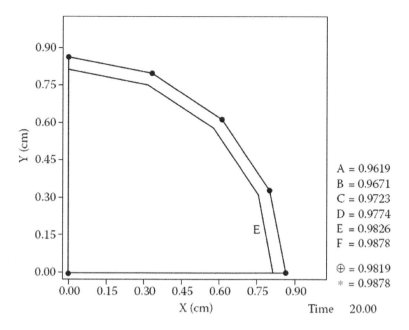

FIGURE 6.4 Solvent mass fraction at time = 20 s.

The solvent concentration gradients for three different time values (20, 200 and 500 s) are shown in Figures 6.4–6.6.

The change in equilibrium conditions causes the gel to expel solvent through its outer edge. Therefore, solvent leaves first from the outer edge region and moves from the inner region to the outside edge, causing concentration gradient rings. These rings cause internal stresses, producing tension and compression fields within the gel. As the gel nears equilibrium, these mechanical forces dissipate, and the gel becomes "stress-free" (see Figures 6.7–6.9).

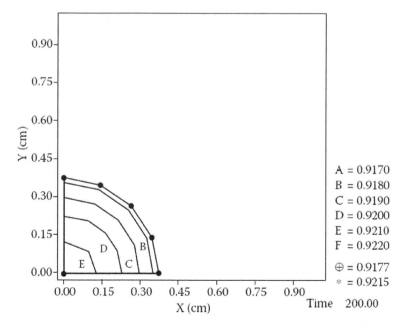

FIGURE 6.5 Solvent mass fraction at time = 200 s.

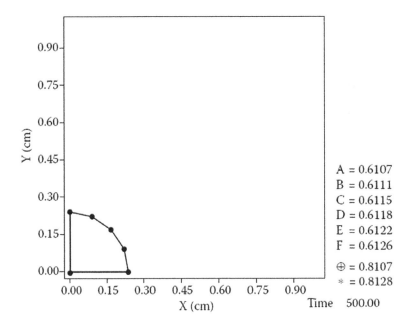

FIGURE 6.6   Solvent mass fraction at time = 500 s.

6.3   CONTINUUM-DIFFUSION ELECTROMECHANICAL MODEL FOR ASYMMETRIC BENDING OF IONIC POLYMERIC GELS

6.3.1   ANALYTICAL MODELING

An electromechanical continuum theory is presented for the dynamic deformation and, in particular, ionic polymeric gels in the presence of an imposed electric field. The proposed theory is based on some recent experimental results obtained in our laboratory for the deformation of ionic polymeric gels – in particular, PAA plus sodium acrylate cross-linked with bisacrylamide.

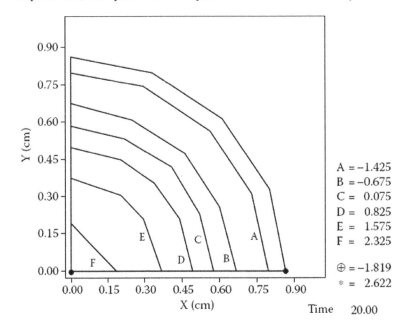

FIGURE 6.7   Pressure at time = 20 s.

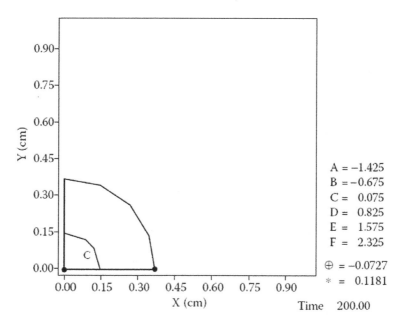

FIGURE 6.8 Pressure at time = 200 s.

The proposed model takes into account the electro-osmosis, electrophoresis and ionic diffusion of various species. It further considers the spatial distributions of cations and anions within the gel network before and after applying an electric field. The model will then derive exact expressions relating the deformation characteristics of the gel as a function of electric field strength or voltage gradient; gel dimensions and physical gel parameters such as diffusivities of cations D_{GM} and anions D_{GP}; elastic modulus E; temperature T; charge concentration of cations, C_{GM}; charge concentration of anions, C_{GP}; resistance R_g; and capacitance C_g of the gel. Thus, direct electrical and computer control of these polymeric ionic gels' expansion and contraction is possible because ionic polymeric

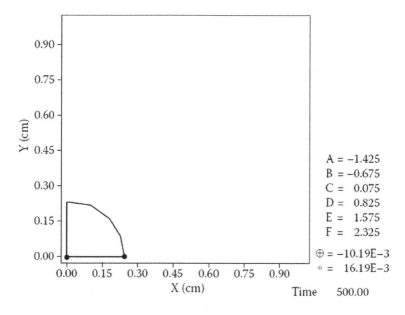

FIGURE 6.9 Pressure at time = 500 s.

gels are electromechanical in nature. Because they can convert electrical and chemical energy to mechanical energy, they may become of particular importance to some unique applications in engineering and medical professions.

To control the large deformation behavior of ionic polymeric gels electrically by a computer, it is necessary to develop a model to microelectrodynamically simulate the large deformations of ionic polymeric gels and subsequently be able to computer-control such large deformations for the design of practical devices and applications. The proposed model's technical objectives are to provide a computational tool to design, simulate and computer-control the electrically induced large expansion and contraction of ionic polymeric gels as smart materials and artificial muscles for various engineering applications. These novel applications will include smart or adaptive structures, bionic robots, artificial muscles, drug-delivery systems, large motion actuators and smart material systems. The modeling is based on formulating a macroscopic theory for the large deformation of ionic polymeric gels in the presence of an electric field.

The proposed theoretical development considers the spatial distributions of cations and anions within the gel network before and after applying an electric field. The model will then derive exact expressions relating the deformation characteristics of the gel as a function of electric field strength or voltage gradient; gel dimensions and physical parameters, such as diffusivities of cations D_{GM} and anion D_{gp}; elastic modulus E; temperature T; charge concentration of cations C_{GM} and anions C_{GP}; resistance R_g; and capacitance C_g of the gel.

In the following sections, a model is presented for the deformation of a hydrated and swollen ionic gel strip chemically plated by a conductive medium and placed in a transverse electric field gradient.

Suppose a long, thin, straight strip of an ionic gel of length l_g, width w_g and thickness t_g is bent into a curve strip by the presence of a transverse electric field across its thickness t_g. Figure 6.10 depicts the schematics of large bending deformation of a strip of IPMNC and the basic actuation configuration under an electric field.

This load can easily be applied by redistribution of fixed and mobile charges in an ionic gel due to an electric field. It is further assumed that, initially, the gel is in a natural bending-stress-free state equilibrated with a pH = 7.

For a discussion on the effect of pH on swelling of gels or the relationship between the strain in a gel and the pH of its environment, see Shahinpoor (1993c), Tanaka et al. (1982), and Umemoto

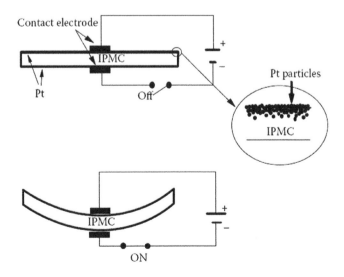

FIGURE 6.10 Bending of an ionic gel strip due to an imposed electric field gradient.

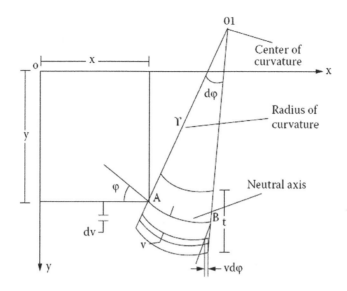

FIGURE 6.11 Geometry of microbending for an elastic strip of ionic gels.

et al. (1991). Now, referring to Figure 6.11, note that the strain ε (defined as the ratio of the actual incremental deformation (δl) to the initial length (l_0) is given by:

$$\varepsilon = \varepsilon_c + \kappa_E \eta = \lambda - 1 \qquad (6.19)$$

where

ε_c = strain along the neutral central line (line going through the centers of areas of all cross sections)

K_E = curvature of the strip due to an electric field

λ = stretch

η = a cross section variable (Figure 6.11).

Let us now assume an electrical field across the thickness t of the gel strip such that on the positive side of the anode side, the strain is ε_+, and on the negative or the cathode side, the strain is ε_-.

Assuming that the IPMNC strip is cationic and bends toward the anode electrode, it is obvious from Equation (6.19) and symmetry that:

$$\varepsilon_+ = \varepsilon_c - \kappa_E C^* \qquad (6.20)$$

$$\varepsilon_- = \varepsilon_c - \kappa_E C^* \qquad (6.21)$$

$C^* = (1/2)t_g$, such that $-C^* \leq \eta \leq C^*$, and t_g is the gel strip's thickness. Now, from Equations (6.19)–(6.21), it is clear that

$$\varepsilon_+ - \varepsilon_- = -2\kappa_E C^* \qquad (6.22)$$

Note that $\varepsilon = \lambda - 1$, where λ or stretch is defined as the ratio of the current length of an element of the gel to the same element's length in its natural undeformed state. Thus, it is clear that

$$\lambda_+ = \lambda_- = -2\kappa_E C^* \qquad (6.23)$$

where K_E is the total curvature of the gel strip due to the pressure of a voltage gradient across the gel strip's thickness.

It is well known that the axial stress σ in polymers is well modeled by the following (neo-Hookean constitutive model) equation (Truesdell and Noll, 1965; Beatty, 1987):

$$\sigma = \left[\frac{Y(C_s,\text{pH},T)}{3}\right](\lambda - \lambda^{-2}) \tag{6.24}$$

Y is Young's modulus of elasticity of the ionic gel and is a function of the concentrations of solvent C_s and pH of the gel and the absolute temperature T.

Assume that, due to the presence of an electrical voltage gradient across the gel's thickness t_g, the gel strip is bent into a curved strip by a nonuniform distribution of fixed and mobile ions (cations) in the gel. Accordingly, an electric field gradient may be imposed across the thickness of the gel by charged molecules. Note from Figure 6.11 that the gel acts like an electrical circuit with resident capacitors and resistors. No inductive properties may be attributed to ionic gels at this stage of the investigation. However, it is well established experimentally (Shahinpoor and Kim, 2000a, 2000b, 2000c, 2001g, 2002d, 2002h) that an ionic gel possesses a cross-capacitance, C_g, and a cross-resistance, R_g. Note that Kirchhoff's law can be written for a gel strip in the following form:

$$V = C_g^{-1}Q + R_g\dot{Q}, \leftrightarrow \dot{Q} = i \tag{6.25}$$

where V is the voltage across the thickness of the gel, and Q is the charge accumulated in the gel, $\dot{Q}$ being the current i through the gel strip across its thickness. Equation (6.25) can be readily solved to yield:

$$Q = C_g V\left[1 - \exp\left(\left(-\frac{t}{R_g C_g}\right)\right)\right] \tag{6.26}$$

assuming that, at $t = 0$, $Q = 0$. Equation (6.26) relates the voltage drop across the gel's thickness to the charge accumulated, which eventually contributes to the deformation of the gel strip. Thus, Equation (6.26) will serve as a basis for the electrical control of gel deformations. The imposed voltage gradient across the gel's thickness forces the internal fixed and mobile ions to redistribute, as shown in Figure 6.12.

The deformation characteristics of ionic polymer gels by electric fields have been theoretically modeled by Shiga and Kurauchi (1990), Shiga (1997) and Doi et al. (1992). Both formulations relate the change in osmotic pressure to the change of volume of the gel samples. In their experiments, the

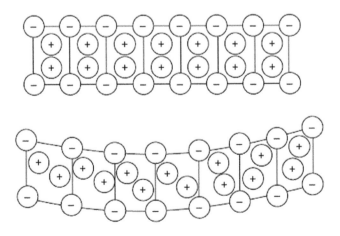

FIGURE 6.12   A possible charge redistribution configuration in ionic gels.

gel sample is not chemically plated but rather placed in an electrolyte solution in the presence of a pair of cathode/anode electrodes.

In particular, since an ionic gel's deformation because of an imposed electric field's influence is due to redistribution and shift of ions in the ionic gel, the change in osmotic pressure, Π, is associated with ion redistribution and concentration gradients should be considered. Flory's theory of osmotic pressure (1953a, 1953b, 1969) described an ionic polymer gel strip's bending deformation in an electric field. According to Flory's theory, the equilibrium volume V_g of an ionic gel is determined by:

$$\Pi_{\text{Network}}\left(V_g\right) + RT \sum_i \left(C_{i,g} - C_{i,s}\right) = 0 \tag{6.27}$$

where

$\Pi_{\text{Nework}}\left(V_g\right)$ = osmotic pressure of a neutral gel
R = universal gas constant
T = absolute temperature

$C_{i,g}$, $C_{i,s}$ determine the ionic concentrations of species i in the gel and the outer solution, respectively, subscript i stands for mobile cations M^+, anions P^-, hydrogen ion H^+, and hydroxyl ion OH^-.

Equation (6.27) implies that if the gel has no charge, its equilibrium volume is determined by the competition between the attractive forces due to polymer-polymer affinity and the network elasticity.

If the gel also has electrical charges, there will be strong ionic concentration gradients on the gel's boundaries and the aqueous solution. These contribute to the osmotic pressure and thus cause the ionic gel to swell or shrink accordingly.

Let us assume that the gel's osmotic pressure on the anode side is Π_1 and Π_2 denotes that of the cathode side. The difference $\Delta\Pi = (\Pi_1 - \Pi_2)$ causes the strip to bend such that the amount of bending is dependent on the difference between the stretches λ_+ and λ_- on the anode side and the cathode side, respectively. Then, based on Equation (6.24), the stretches λ_+ and λ_- are related to the state of uniaxial stresses $\sigma_+ = \Pi_1$ and $\sigma_- = \Pi_2$, such that one has:

$$\Pi_1 = \left(\frac{Y_+}{3}\right)\left(\lambda_+ - \lambda_+^{-2}\right) \tag{6.28}$$

$$\Pi_2 = \left(\frac{Y_-}{13}\right)\left(\lambda_- - \lambda_-^{-2}\right) \tag{6.29}$$

Y_+ and Y are the corresponding elastic Young's moduli at the anode and the gel strip's cathode sides, respectively. The idea here is to find direct relations for Π_1 and Π_2 so that λ_+ and λ_- can be found from Equations (6.28) and (6.29). One can then relate the resulting bending curvature κ_E to the difference $\lambda_+ - \lambda_-$ by Equation (6.23).

It is assumed that the spatial correlation in ionic redistribution or ion-ion interaction is negligible. The Donnan type of equilibrium holds in the presence of an electric field. As discussed by Doi and coworkers (1992), the magnitude of the perturbation due to ion-ion interaction compared to diffusional interaction is of the order of 10^{-3} for the usual strength (10 V/cm) of the imposed field. The polymer gel and the surrounding aqueous solution are divided into three regions: the anode side A, the gel G and the cathode side C. We further consider systems with only one kind of mobile cation M^+. The imposed electric field forces the cations to move toward the cathode, causing abrupt concentration gradients across the boundaries. These cation concentrations are then given by:

$$C_{A,M}\left(t\right) = C_{A,M}\left(0\right)\left(1 - h_{GA,M}t\right) \tag{6.30}$$

$$C_{G,M}(t) = C_{G,M}(0)(1 - h_{GC,m}t) + C_{A,M}(0)\left(\frac{V_A}{V_G}\right)h_{GA,M}t - C_{A,M}(0)\left(\frac{V_A}{V_G}\right)h_{GA,M}h_{GC,M}t^2 \quad (6.31)$$

$$C_{C,M}(t) = C_{C,M}(0) + C_{G,M}(0)\left(\frac{V_G}{V_G}\right)h_{GC,M}t + C_{A,M}(0)\left(\frac{V_A}{V_C}\right)h_{GA,M}h_{GC,M}t^2 \quad (6.32)$$

where

$h_{GA,M}$, $h_{GC,M}$ = cation transport rates across the G–A, G–C boundaries, respectively
V_A, V_g and V_C = volumes of A, G and C parts
t = time of exposure to the electric field.

For complete derivations of such ionic transport equations and the appearance of quadratic terms due to the Donnan equilibrium assumption, see Shiga and Kurauchi (1990), Shiga et al. (1993), and Shahinpoor (1993b, 1993c, 1993d, 1994a, 1994b, 1994d, 1994e, 1994f). The corresponding osmotic pressures $\Pi_{1,M}$ and $\Pi_{2,M}$ on the anode and the cathode boundaries of the gel strip due to redistribution of cations are then given by

$$\Pi_{1,M} = RT\left[C_{G,M}(t) - C_{A,M}(t)\right] \quad (6.33)$$

$$\Pi_{2,M} = RT\left[C_{G,M}(t) - C_{C,M}(t)\right] \quad (6.34)$$

Similarly, the anions P^- will be redistributed by migrating toward the anode. Since the whole system is neutral, some water resolves into H^+ and OH^- to satisfy these cations and anions' redistributions by migrating toward the cathode and the anode electrodes, respectively. Some water will be riding with the cations and anions as hydrated water and cause a corresponding stress on the cathode and anode sides. Therefore, the contribution to the total pressures on the anode and the cathode sides due to these charge redistributions can similarly be calculated such that

$$\Pi_1 = \Pi_{1,M} + \Pi_{1,P} + \Pi_{1,H} + \Pi_{1,OH} - \Pi_{hw} \quad (6.35)$$

$$\Pi_2 = \Pi_{2,M} + \Pi_{2,P} + \Pi_{2,H} + \Pi_{2,OH} - \Pi_{hw} \quad (6.36)$$

where Π_{hw} is the stress magnitude due to migration of hydrated water riding on the cations as they migrate toward the cathode. It is assumed that, due to symmetry, the magnitude of this stress, because it is compressive and negative in the anode side and tensile and positive in the cathode side, is the same in the anode and the cathode sides. In summation form, Equations (6.35) and (6.36) can be written as:

$$\Pi_1 = RT\sum_i\left[C_{G,i}(t) - C_{A,i}(t)\right] - \Pi_{hw}, \, i = M, P, H^+, OH^- \quad (6.37)$$

$$\Pi_2 = RT\sum_i\left[C_{G,i}(t) - C_{C,i}(t)\right] - \Pi_{hw}, \, i = M, P, H^+, OH^- \quad (6.38)$$

where

$$C_{A,i}(t) = C_{A,i}(0)(1 - h_{GA,i}t) \quad (6.39)$$

$$C_{G,i}(t) = C_{G,i}(0)(1 - h_{GC,i}t) + C_{A,i}(0)\left(\frac{V_A}{V_G}\right)h_{GC,i}t - C_{A,i}(0)\left(\frac{V_A}{V_G}\right)h_{GV,i}h_{GC,i}t^2 \quad (6.40)$$

$$C_{C,i}(t) = C_{C,i}(0)(1 - C_{G,i}(0))\left(\frac{V_G}{V_C}\right)h_{GC,i}t + C_{A,i}(0)\left(\frac{V_G}{V_C}\right)h_{GA,i}h_{GC,i}t^2, i = M, P, H^+OH^- \quad (6.41)$$

Due to neutrality, the following relations also hold:

$$C_{k,M}(t) = C_{k,\mathrm{OH}}(t), k = A, G, C \tag{6.42}$$

$$C_{k,P}(t) = C_{k,H}(t), K = A, G, C \tag{6.43}$$

Thus, Equations (6.37) and (6.38) simplify to

$$\Pi_1 = RT \sum_j \left[C_{G,j}(t) - C_{A,j}(t) \right] - \Pi_{hw}, \; j = M, P \tag{6.44}$$

$$\Pi_1 = RT \sum_j \left[C_{G,j}(t) - C_{C,j}(t) \right] - \Pi_{hw}, \; j = M, P \tag{6.45}$$

As discussed before, these stresses will be related to the induced stretches λ_+ and λ_- on the anode and the cathode sides, respectively, by Equations (6.28) and (6.29). The resulting cubic equations are then solved for λ_+ and λ_- to calculate the induced curvature K_E such that:

$$\kappa_E = \left[\frac{(\lambda_+ - \lambda_-)}{t_g} \right] \tag{6.46}$$

where t_g is the thickness of the gel strip. The resulting cubic equations are

$$\lambda_+^3 + \left(\left(\frac{3\Pi_1}{Y_+} \right) \right) \lambda_+^2 - 1 = 0 \tag{6.47}$$

$$\lambda_-^3 + \left(\left(\frac{3\Pi_2}{Y_-} \right) \right) \lambda_-^2 - 1 = 0 \tag{6.48}$$

Let us convert the cubic Equations (6.47) and (6.48) to a reduced form by substituting $\lambda_+ = x_+ - (\Pi_1/Y_+)$ and $\lambda_- = x_- - (\Pi_2/Y_-)$, respectively, in Equation (6.47) and (6.48) to obtain:

$$x_+^3 + 3p_+ x_+ + 2q_+ = 0 \tag{6.49}$$

$$x_-^3 + 3p_- x_- + 2q_- \tag{6.50}$$

where

$$p_+ = (\Pi_1)^2 Y_+^{-2}, \; q_+ = (\Pi_1)^3 Y_+^{-3} - \left(\frac{1}{2} \right) \tag{6.51}$$

$$p_- = (\Pi_2)^2 Y_-^{-2}, \; q_- = (\Pi_2)^3 Y_-^{-3} - \left(\frac{1}{2} \right) \tag{6.52}$$

The discriminants for Equations (6.49) and (6.50) are, respectively,

$$D_+ = -p_+^3 - q_+^3 = -(\Pi_1)^6 Y_+^{-6} - \left[\left(\frac{\Pi_1}{Y_+} \right)^3 - \left(\frac{1}{2} \right) \right]^3 \tag{6.53}$$

$$D_- = -P_-^3 - q_-^3 = -(\Pi_2)^6 Y_6^{-6} - \left[\left(\frac{\Pi_2}{Y_-} \right) - \left(\frac{1}{2} \right) \right]^3 \tag{6.54}$$

For Equations (6.49) and (6.50) to have all three real solutions, D_+ and D_- have to be positive definitive. If D_+ and D_- are negative, then there exists only one real root and two complex conjugate roots, which will be unacceptable. Applying the Tartaglia-Cardan formula to Equations (6.49) and (6.50) yields

$$x_{+1} = u_+ + v_+, x_{+2} = \varepsilon_1 u_+ + \varepsilon_2 v_+, x_{+3} = \varepsilon_2 u_+ + \varepsilon_1 v_+ \tag{6.55}$$

and

$$x_{-1} = u_- + v_-, x_{-2} = \varepsilon_1 u_- + \varepsilon_2 v_-, x_{-3} = \varepsilon_2 u_- + \varepsilon_1 v_- \tag{6.56}$$

where

$$\varepsilon_{1,2} = -\left(\frac{1}{2}\right) \pm \left(\frac{\sqrt{3}}{2}\right), \tag{6.57}$$

$$u = \left[-q_+ + \left(q_+^2 + p_+^3 \right)^{1/2} \right]^{1/3} \tag{6.58}$$

$$v = \left[-q_+ - \left(q_+^2 + p_+^3 \right)^{1/2} \right]^{1/3} \tag{6.59}$$

$$u_- = \left[-q_+ + \left(q_-^2 + p_-^3 \right)^{1/2} \right]^{1/3} \tag{6.60}$$

$$v_- = \left[-q_- - \left(q_-^2 + p_-^3 \right)^{1/2} \right]^{1/3} \tag{6.61}$$

The possible curvatures with a real value are then calculated as:

$$\kappa_E = \left(\frac{(\lambda_+ - \lambda_-)}{2C^*} \right) = \left(\frac{1}{2C^*} \right) \left[(x_+ - x_-) - \left((\Pi_1 / Y_+) - (\Pi_2 / Y_-) \right) \right] \tag{6.62}$$

where

$$x_+ - x_- = (u_+ - u_-) + (v_+ - v_-) \tag{6.63}$$

Now, from Equations (6.26), (6.39)–(6.41), (6.44), (6.45), (6.62) and (6.63), it is clear that the nonhomogeneous force field and the curvature in bending of ionic gels can be electrically controlled using an imposed voltage V across the gel. Note that in case of complete symmetry in bending of a strip of length l_g, width w_g, and thickness t_g, and only one kind of cation and fixed anions, $Y_+ = Y_- = Y$,

$$\kappa_E = \left(\frac{1}{2YC^*} \right) (\Pi_2 - \Pi_1) \tag{6.64}$$

or

$$\kappa_E = \left(\frac{1}{2YC^*} \right) \left\{ RT \sum_j \left[C_{C,j}(t) - C_{A,j}(t) \right] + 2\Pi_{hw} \right\}, \ j = M, P \tag{6.65}$$

$C_{C,J}(t)$ and $C_{A,J}(t)$ are the average molal charge concentrations in gram-moles per cubic meters on the cathode side and the gel strip's anode side, respectively. Note that, due to migration of cations toward the cathode side, the difference between the average total charges on the cathode side and the anode side is simply the accumulated charge Q given by Equation (6.26) and is also given by:

$$Q = \frac{1}{2} n^* e A_v \left(l_g w_g t_g \right) = \sum_j \left[C_{C,t}(t) - C_{A,j}(t) \right], \ j = M, P \tag{6.66}$$

or

$$\frac{2Q}{\left(n^{*}eA_{v}\left(l_{g}w_{g}t_{g}\right)\right)} = \sum_{j}\left[C_{C,t}\left(t\right) - C_{A,j}\left(t\right)\right], \; j = M, P \tag{6.67}$$

where

n* = valence of the cations
e = 1.602192E–19 C is the charge of an electron
A_{v} = 6.022E23 is Avogadro's number.

Thus, from Equations (6.26), (6.65) and (6.67), a simple expression for the voltage-induced local curvature in gels (also in IPMNCs) can be found:

$$\kappa_{E} = \left(\frac{1}{2YC^{*}}\right)\left\{RT\left[\frac{2Q}{\left(n^{*}eA_{v}\left(l_{g}w_{g}t_{g}\right)\right)}\right] + 2\Pi_{hw}\right\} \tag{6.68}$$

Note that Π_{hw} is the pressure generated due to the migration of hydrated water with the cations and can be obtained by noting that:

$$\Pi_{hw} = \left(\frac{Y}{3}\right)\left(\lambda_{hw} - \lambda_{hw}^{-2}\right) \tag{6.68a}$$

where λ_{hw} is the contribution to the stretch on the cathode side by the migrated hydrated water. Furthermore, it is related to the displaced volume V_{hw} of the migrated hydrated water such that:

$$V_{hw} = \left(\frac{1}{2}\right)l_{g}t_{g}w_{g}\lambda_{hw} \tag{6.68b}$$

or simply

$$\lambda_{hw} = \left(\frac{2V_{hw}}{l_{g}t_{g}w_{g}}\right) \tag{6.68c}$$

Furthermore, V_{hw} is the volume of migrated hydrated water of mass M_{hw} given by:

$$M_{hw} = \left(\frac{18}{A_{v}}\right)N_{hn}\left(\frac{Q}{n^{*}e}\right) \tag{6.68d}$$

where (Q/n*e) is estimated to be the total number of migrated cations and N_{hn} is the cations' hydration number. Note that V_{hw} can now be related to V_{hn} by the water density ρ_{hw} such that $V_{hw} = (M_{hw}/\rho_{hw})$. From Equations (6.68b)–(6.68d), one obtains the following equation for λ_{hw}:

$$\lambda_{hw} = \frac{36N_{hw}Q}{\left(n^{*}\rho_{hw}eA_{v}l_{g}t_{g}w_{g}\right)} \tag{6.68e}$$

Note that in Equation (6.68e), the unit of density, ρ_{hw}, is grams per cubic meter and the units of l_{g}, t_{g} and w_{g} are in centimeters. Thus, the additional stress due to migrated hydrated water is given by

$$\Pi_{hw} = \left(\frac{Y}{3}\right)\left(\lambda_{hw} - \lambda_{hw}^{-2}\right) \tag{6.68f}$$

Based on Equations (6.26), (6.65), (6.68e) and (6.68f), the equation for the total curvature of the ionic polymer strip becomes

$$\kappa_E = \left(\frac{1}{2YC^*}\right)\left\{RT\left[2C_gV\left[1-\exp\left((-t/R_gC_g)\right)\right]\Big/n^*eA_v\left(l_gw_gt_g\right)\right]+2\Pi_{hw}\right\} \quad (6.69)$$

Note that units in Equation (6.68) are consistent – namely, that $[2Q/(n^*eA_v(l_gw_gt_g))]$ has the unit of gram-moles per cubic meter. Thus, that unit times RT gives the units of pressure in pascals, which are then multiplied by $(1/2YC^*)$, which leaves the units of $(1/m)$ because Y also has the units of pascals.

Note that the electric field, $\underset{\sim}{E}$, is given by $\underset{\sim}{E} = V/2C^*$ and thus Equation (6.69) further simplifies to

$$\kappa\underset{\sim}{E} = \underset{\sim}{E}\left(\frac{2C_gRT}{n^*eA_v\left(l_gw_gt_g\right)Y}\right)\left[1-\exp\left(\left(-\frac{t}{R_gC_g}\right)\right)\right]+\left(\frac{\Pi_{hw}}{YC^*}\right) \quad (6.70a)$$

If the curvature at time $t = 0$ is denoted by κ_0, then Equation (6.70a) is further generalized to:

$$\kappa\underset{\sim}{E} - \kappa 0 = \underset{\sim}{E}\left(\frac{2C_gRT}{n^*eA_v\left(l_gw_gt_g\right)Y}\right)\left[1-\exp\left(\left(-\frac{t}{R_gC_g}\right)\right)\right]+\left(\frac{\Pi_{hw}}{YC^*}\right) \quad (6.70b)$$

Equations (6.70a) and (6.70b) are simple expressions for the time-dependent curvature of the strip in an average and approximate fashion and indicate that the induced curvature is directly proportional to the imposed electric field, $\underset{\sim}{E}$; is a nonlinear function of the capacitance of the gel strip, C_g, and cross-resistance, R_g; and is inversely proportional to the strip volume $V_g = (l_gw_gt_g)$ and the Young modulus of elasticity of the strip Y.

These observations are in complete harmony with experimental results on the bending of IPMNC strips. Figures 6.13–6.15 display several simulations for curvature κ_E versus capacitance C_g, electric field $\underset{\sim}{E}$, resistance R_g, and time t.

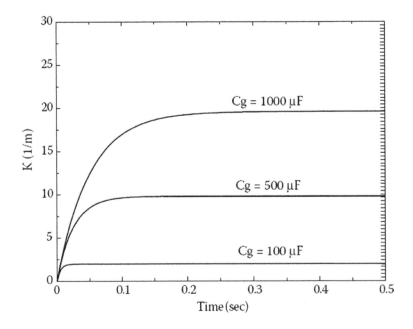

FIGURE 6.13 Variation of curvature versus cross-capacitance C_g and time t.

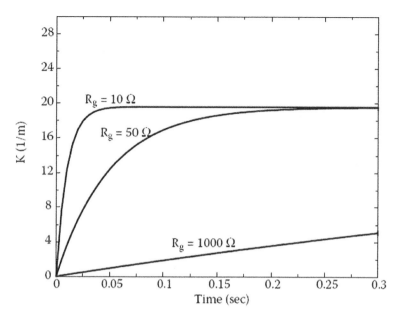

FIGURE 6.14 Variation of curvature versus cross-resistance R_g and time t.

Note that in Figure 6.13 and subsequently in Figures 6.14 and 6.15, the following values were used for the parameters:

$E = 20,000$ {electric field imposed, V/m = J/(C m)}
$C_g = 1000E-6$ {capacitance, C/V, F = C/V, 200 μF}
$R = 8.314$ {gas constant, (Pa m^3)/(g mol K)}
$T = 300$ {absolute temperature, K}
$V_g = (l_g w_g t_g) = 2E-8$ {sample volume, m^3}
$l_g = 2$ cm

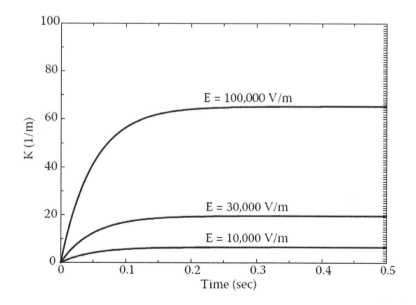

FIGURE 6.15 Variation of curvature versus cross-electric field E and time t.

$w_g = 0.5$ cm
$t_g = 0.2$ mm
$Y = 100E6$ {modulus of the sample, Pa}
$n = +1$ {valance charge}
$e = 1.602192E-19$ {an electron charge, C}
$A_v = 6.022E23$ {Avogadro's number, 1/mol}
$R_g = 100$ {resistance, 100 Ω = V/A; also C = A*s}

In a cantilever mode, the maximum tip deflection δ_{max} can be shown to be approximately related to the absolute value of the curvature $\left| {}^{\kappa}{}_{E} \right|$ by:

$$\left| {}^{\kappa}E \right| = \frac{2\delta_{max}}{l_g^2 + \delta_{max}^2} \tag{6.71}$$

When combined with Equation (6.70a), this results in the following equation relating the maximum tip deflection of an IPMNC bending strip to the imposed electric field, the time t, the strip cross-capacitance, the strip cross-resistance, the strip volume and the strip modulus of elasticity Y:

$$\frac{2\delta_{max}}{l_g^2 + \delta_{max}^2} = \underset{\sim}{E}\left(\frac{2C_g RT}{n^* e A_v \left(l_g w_g t_g\right) Y} \right)\left[1 - \exp\left(\left(-\frac{t}{R_g C_g}\right)\right)\right] + \left(\frac{\Pi_{hw}}{YC^*}\right) \tag{6.72}$$

Figures 6.16–6.18 depict the variations of maximum deflection versus the electric field E, the average cross-capacitance C_g, and the average cross-resistance R_g for the same values of the parameters used in Figures 6.13–6.15.

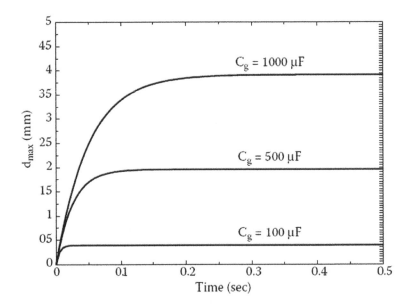

FIGURE 6.16 Variation of maximum tip deflection δ_{max} versus average cross-capacitance C_g and time t.

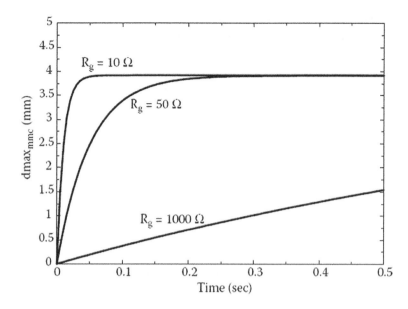

FIGURE 6.17   Variation of maximum tip deflection δ_{max} versus the average electric field $\underset{\sim}{E}$ and time t.

6.4   CONTINUUM MICROELECTROMECHANICAL MODELS

6.4.1   THEORETICAL MODELING

Following Shahinpoor (1994f) and Shahinpoor and Osada (1995b), we consider gel fiber bundles, the strands of which are in the form of a swollen cylinder with outer radius r_o and inner radius r_i. We assume the electric field is aligned with the long axis of these cylindrical macromolecular ionic chains. Further, we assume the ions are evenly distributed along with the macromolecular network at a regular distance of b. We concentrate on the flow of liquid solvent containing counterions into and out of the

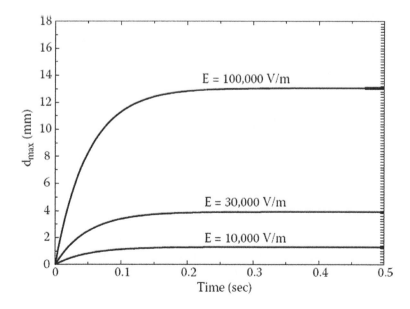

FIGURE 6.18   Variation of maximum tip deflection δ_{max} versus average cross-resistance R_g and time t.

gel network resembling electrocapillary flows. Thus, we employ the conservation laws – namely, conservation of mass and momentum – to arrive at the following governing differential equation for the flow of counterions containing solvent into and out of the gel macromolecular network:

$$\rho \frac{dv}{dt} = \rho g + \rho^* E + o\nabla^2 v - \nabla p \tag{6.73}$$

where

ρ = density of the liquid solvent, which is assumed to be incompressible
v = three-dimensional liquid velocity vector
∇ = gradient vector operator
∇^2 = Laplacian operator
g = local gravitational acceleration
μ = solvent viscosity
p = hydrostatic pressure
E = imposed electric field vector
ρ^* = charge density governed by the following Poisson's equation (Lifson and Katchalsky, 1954):

$$\rho^* = -D\nabla^2 \phi \tag{6.74}$$

where D is the dielectric constant of the liquid phase and ψ is the local electrostatic potential. Furthermore, ψ is governed by the following Poisson–Boltzmann equation (see Lifson and Katchalsky, 1954):

$$\nabla^2 : 5\left(\frac{4\pi n\varepsilon}{D}\right)\exp\left[-\frac{\varepsilon\psi}{kT}\right] \tag{6.75}$$

where

n = number density of counter-ions
ε = their average charge
k = Boltzmann constant
T = absolute temperature.

According to Lifson and Katchalsky (1954), the electrostatic potential in polyelectrolyte solutions for fully stretched macromolecules is given by the following equation, which is an exact solution to the Poisson–Boltzmann equation in cylindrical coordinates:

$$\phi(r,t) = \left[\frac{kT}{\varepsilon}\right]\ell n\left\{\left[\frac{r^2}{(r_0^2 - r_i^2)}\right]\sinh^2\left[B\ell n\left(\frac{r}{r_0}\right)_- \tan^{-1}\beta\right]\right\} \tag{6.76}$$

where β is related to λ, which is a dimensionless parameter given by:

$$\lambda = \left(\frac{\alpha^* \varepsilon^2}{4\pi DbkT}\right) \tag{6.77}$$

where α^* is the degree of ionization (i.e., $\alpha^* = a = n/Z$, where n is the number of ions and Z is the number of ionizable groups) b is the distance between polyions in the network. Furthermore, $\left[\alpha^*/\pi(r_0^2 - r_i^2)b\right]$ and betas are found from the following equation:

$$\lambda = \left\{\frac{1 - \beta^2}{1 + dCoth\left[\beta\ell n(r_0/r_i)\right]}\right\} \tag{6.78}$$

Let us further assume that, due to cylindrical symmetry, the velocity vector $v = (v_r, v_\theta, v_z)$ is such that only v_z depends on r and $v_\theta = 0$. Thus, the governing equations for $v_z = v$ reduce to:

$$\rho\left(\frac{\partial v}{\partial t}\right) = f(r,t) + \mu\left[\frac{\partial^2 v}{\partial^2 r} + r^{-1}\frac{\partial v}{\partial r}\right] - \frac{\partial p}{\partial r} \tag{6.79}$$

Let us assume a negligible radial pressure gradient and assume the following boundary and initial conditions:

At $t = 0$, $r_i \le r \le r_o$, $v = 0$; at $r = r_i$, $\forall t$, $v(r_i) = 0$; and $r = r_o$, $\forall t,$ $\left(\dfrac{\partial v}{\partial r}\right)_{r=r_o} = 0.$

Furthermore, the function $f(r,t)$ is given by:

$$f(r,t) = n\varepsilon E(r,t)\left\{\left[\kappa^2 r^2/2\beta^2\right]\sinh^2\left[\beta\ln\left(\frac{r}{r_o}\right) - \tan^{-1}\beta\right]\right\}^{-1} \tag{6.80}$$

where $\kappa^2 = (n\varepsilon^2/DkT)$.

An exact solution to the given set of equations can be shown:

$$v(r,t) = \sum_{m=1}^{\infty} e^{-(\mu/\rho)\beta_m^2 t}\kappa_0\left(\beta_m r\right)\int_0^t e^{-(\mu/\rho)\beta_m^2\xi}A\left(\beta_m,\xi\right)d\xi \tag{6.81}$$

where β_m,s are the positive roots of the following transcendental equation:

$$\frac{J_{0(\beta r_i)}}{J_0'\left(\beta r_0\right)} - \frac{Y_{0(\beta r_i)}}{Y_0'\left(\beta r_0\right)} = 0 \tag{6.82}$$

where J_0, Y_0, J_0' and Y_0' are the Bessel functions of zero order of first and second kind and their derivatives evaluated at r_o, respectively, and

$$A\left(\beta_m,\xi\right) = \left(\frac{1}{\mu\rho}\right)\int_{r_i}^{r_0}\varsigma\kappa_0\left(\beta_m,\varsigma\right)f\left(\varsigma,\xi\right)d\varsigma \tag{6.83a}$$

$$N = \left(\frac{r_0^2}{2}\right)R_0^2\left(\beta_m,r_0\right) - \left(\frac{r_i^2}{2}\right)R_0'^2\left(\beta_m,r_i\right) \tag{6.83b}$$

$$\kappa_0\left(\beta_m,r\right) = N^{-(1/2)}\left\{\frac{J_0\{\beta r\}}{\beta_m J_0'\left(\beta r_0\right)} - \frac{Y_0\left(\beta r\right)}{\beta_m Y_0'\left(\beta r_0\right)}\right\} = N^{-(1/2)}R_0\left(\beta_m,r\right) \tag{6.83c}$$

6.4.2 Numerical Simulation

Having found an explicit equation for $v(r, t)$, we can now carry out numerical simulations to compare ionic polymeric gels' theoretical dynamic contraction in an electric field with those of experiments. To compare the experimental results and observations with the proposed dynamic model, a number of assumptions, simplifications and definitions are first made. Consider the ratio $W(t)/W(0)$, where $W(t)$ is the weight of the entire gel at time t, and $W_0 = W(0)$ is the weight of the gel at time $t = 0$, just before the electrical activation. Thus,

$$W(0) - W(t) = \int_0^t \int_{r_i}^{r_0} 2\pi\rho v(\mathrm{r},\mathrm{t}) r d r d t \qquad (6.84)$$

This can be simplified to

$$\left[\frac{W(t)}{W(0)} \right] = 1 - W_0^{-1} \int_0^t \int_{r_i}^{r_0} 2\pi\rho v(r,t) r d r d t \qquad (6.85)$$

The initial weight of the gel is related to the initial degree of swelling $q = V(0)/V_p$, where $V(0)$ is the volume of the gel sample at $t = 0$ and V_p is the volume of the dry polymer sample. Numerical simulations were carried out based on the assumptions that the cross section of the gel remains constant during contraction of the gel sample, and that:

$\varepsilon = e = 1.6 \times 10^{-19}$ C, T = 300K
$\alpha = 1, D = 80$
$\mu = 0.8 \times 10^{-3}$ Pa s
$\rho = 1000$ kg/m^3
$b = 2.55 \times 10^{-10}$ m
$k = 1.3807 \times 10^{-23}$ J/K
$r_i = 6.08 \times 10^{-10}$ m
$r_0 = r_i q^{(1/2)}$
$q = 25, 70, 100, 200, 256, 512, 750$

The sample's initial length and cross section are, respectively, $\ell_0 = 1$ cm and $S = 1$ mm^2. Electric field strength, E, = 5.8 V/mm.

The numerical simulation results have been compared with Gong et al.'s experimental results (1994a, 1994b) and Gong and Osada (1994) for comparable cases, and good agreement is observed, as depicted in Figure 6.19.

Also, Asaka and Oguro (2000) present a bending IPMNC strips model that closely resembles the preceding formulation. In brief, they consider the water flux, J, of hydrated cations and cationic migration toward the cathode side. Note that an electric field's imposition causes the hydrated cations to migrate by electrophoresis, electro-osmosis and capacitive or ionic current. Thus, a material flux can be defined such that

$$J = -k_m \frac{\partial P_m}{\partial x} + \frac{\phi k j}{F} \qquad (6.86)$$

where
    k_m = water permeability constant, according to them
    P_m = fluid pressure
    κ = water transference coefficient in the gel network
    ϕ = water molar volume
    j = current density
    F = Faraday constant (96,500 C/(kg mol))

They, then, consider a linear stress–strain law – rather than hyperelastic neo-Hookean, as was proposed in our formulation – for the bending deformation of the strip in the form of $p_m = E_m \varepsilon$,

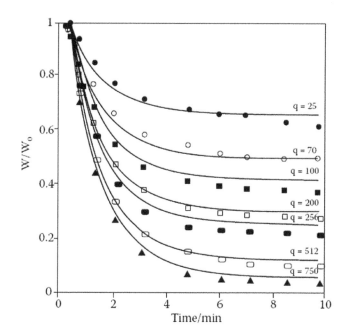

FIGURE 6.19 Computer simulation (solid lines) and experimental results (scattered points) for the time profiles of relative weight of the gel sample for various degrees of swelling, q.

where p_m is the swelling or osmotic stress, E_m is the elastic modulus and ε is the linear strain. They relate the strain ε to the hydration H and equilibrium hydration H_{eq} by:

$$3\varepsilon = \frac{H - H_{eq}}{1 + H_{eq}} \tag{6.87}$$

This appears to have an unnecessary factor of 3. They consider the inertial effect of being absent (slow motion) and assume conservation of momentum in the form of:

$$\frac{\partial P_m}{\partial x} = \frac{\partial p_m}{\partial x} \tag{6.88}$$

They solve the preceding equations to arrive at an expression for the curvature ($1/R$) such that:

$$\frac{1}{R} - \frac{1}{R_0} = \frac{\left(2d^3/9\right)\left(1/\left(1 + H_{eq}\right)\right)\left(\phi kj/FD_m\right)E_m\left[1 - \exp\left[-\left((1/2d)\pi\right)^2\right]D_m t\right]}{Q} \tag{6.89}$$

where
2d = thickness of the strip
$D_m = k_m E_m/[3(1 + H_{eq})]$ is the diffusion coefficient ($\sim 4.5 \times 10^{-6}$ cm^2/S)
Q = stiffness (i.e., the product of the elastic modulus and the area moment of inertia of the bending strip).

Their presentation further gets more empirical and unclear. However, it is interesting to note that Expression (6.89) for the curvature is similar to our expression but involves the current density j, which is unknown. The reader is referred to their paper for further explanation and graphical representation of their solutions. Figure 6.20 depicts the numerical simulation of these equations

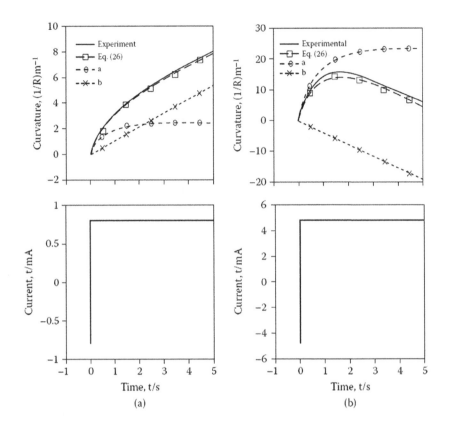

FIGURE 6.20 Numerical simulation of Equation (6.89) by Asaka and Oguro (Asaka and Oguro, 2000. *J. Electroanal. Chem.* 480:186–198.)

according to Asaka and Oguro (2000). As can be seen, the trends that they observe are very similar to the trends observed experimentally and simulated theoretically by the proposed model in this chapter.

6.5 MICROELECTROMECHANICAL MODELING OF ASYMMETRIC DEFORMATION OF IONIC GELS

To describe electrically induced deformation of ionic polymeric gels with no water exudation, we consider the process of bending strips of ionic polymeric gels in a transverse electric field.

Suppose a long, thin, straight strip of an ionic gel of length $l_g L$, thickness t_g and width w_g is bent into a curve-strip (Figure 6.10) by an imposed voltage gradient across its thickness t_g. It is assumed that, initially, the gel is in a natural bending-stress-free state equilibrated at pH = 7. Now, referring to Figure 6.11, note that the strain ε defined as the ratio of the actual incremental deformation (δl) to the initial length (l_0) is given by Equation (6.19). In fact, Equations (6.19)–(6.26) are still valid in the current approach.

Due to the presence of an electrical voltage gradient across the thickness t^* of the gel, the gel strip is bent into a curved strip by a nonuniform distribution of fixed and mobile ions in the gel. Note that an electric field gradient may be imposed across the thickness of the gel by charged molecules.

Several simplifying assumptions are made to mathematically model the nonhomogeneous deformation or bending forces at work in an ionic polyelectrolyte gel strip. The first assumption is that the polymer segments carrying fixed charges are cylindrically distributed along a given polymer chain and independent of the cylindrical angle θ (Figure 6.20).

This assumption is not essential but greatly simplifies the analysis. Consider the field of attraction and repulsion among neighboring rows of fixed or mobile charges in an ionic gel. Let r_i and Z_i be the cylindrical polar coordinates with the ith row as an axis such that the origin is at a given polymer segment. Let the spacing in the ith row be b_i, and let the forces exerted by the atoms be central and of the form cr^{-s} such that the particular cases of $s = 2$ $s = 7$, and $s = 10$ represent, respectively, the Coulomb, van der Waals and repulsive forces. Then it can be shown that the component of the field per unit charge at the point (r_i, Z_i) perpendicular to the row is represented in a series such that:

$$R(r_i, Z_i) = \sum_{n=-\infty}^{\infty} \left\{ \frac{r_i}{\left[\left(Z_i - nb_i \right)^2 + r_i^2 \right]^{(1/2)(s+1)}} \right\} \tag{6.90}$$

This function is periodic in Z_i, of period b_i and is symmetrical about the origin. It may, therefore, be represented in a Fourier series in the form:

$$R(r_i, Z_i) = \left(\frac{1}{2} \right) C_{0i} + \sum_{m=1}^{\infty} C_{mi} \cos\left(\frac{2\pi m Z_i}{b_i} \right) \tag{6.91}$$

where

$$C_{mi} = \left(\frac{2}{b_i} \right) \sum_{n=-\infty}^{\infty} \int_0^{\alpha i} \left\{ \frac{r_i \cos\left(2smZ_i/b_i \right)}{\left[\left(Z_i - nb_i \right)^2 + r_i^2 \right]^{(1/2)(s+1)}} \right\} dZ_i \tag{6.92}$$

Evaluating the coefficients C_{mi}, $m = 0, 1, 2, \ldots \infty$, in Equation (6.92), it is found that, for $m = 0$,

$$C_{0i} = \frac{r_i \Gamma(1/2) \Gamma(1)}{b_i \Gamma(3/2)} \tag{6.93}$$

and for $m > 0$,

$$C_{mi} = \left(\frac{4r_i}{b_i} \right) \left(\frac{\pi m_i}{b_i r_i} \right)^{(1/2)_s} \left[\frac{\Gamma(1/2)}{\Gamma(1/2)(s+1)} \right] K_{(1/2)^s}$$
$$\left(2 \frac{\pi m_i r_i}{b_i} \right) \cos\left(2 \frac{\pi m_i Z_i}{b_i} \right) \tag{6.94}$$

where Γ and K are modified Bessel functions. In the remainder of this section, only the Coulomb types of attraction and repulsion forces will be considered to simplify these expressions. With this assumption, the expression for $R(r_i, Z_i)$ becomes:

$$R(r_i, Z_i) = \left(\frac{2r_i}{b_i} \right) \left[r_i^{-2} + 4 \sum_{m=1}^{\infty} m K1 \left(\frac{2\pi m r_i}{b_i} \right) \cos\left(2 \frac{\pi m Z_i}{b_i} \right) \right] \tag{6.95}$$

Now recall that many molecular strands are occasionally cross-linked, oriented and entangled in an ionic polymer network. Assuming that, in the presence of an imposed voltage gradient across the thickness of the gel strip, the rows of fixed and mobile ions line up as

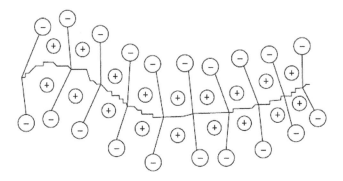

FIGURE 6.21   Spatial geometry of a local polymer segment with fixed charges.

shown in Figure 6.21, the mean-field can be obtained by superimposing the field corresponding to positive charges, namely:

$$R_+\left(\frac{2r_i}{b_i}\right)\left[r_i^{-2}+4\sum_{m=1}^{\infty}mK_1\left(\frac{2\pi mr_i}{b_i}\right)\cos\left(2\frac{\pi mZ_i}{b_i}\right)\right],\tag{6.96}$$

and corresponding to negative charges, namely:

$$R_-\left(r_i,Z_i\right)=\left(\frac{2r_i}{b_i}\right)\left\{r_i^{-2}+4\sum_{m=1}^{\infty}mK_1\left(\frac{2\pi mr_i}{b_i}\right)\cos\left[\left(2\frac{\pi mZ_i}{b_i}\right)\left(Z_i+\left(\frac{1}{2}\right)b_i\right)\right]\right\}\tag{6.97}$$

The resulting field for a pair of rows is:

$$R\left(r_i,Z_i\right)=\left(16\frac{\pi}{b_i^2}\right)\sum_{m=1,3,5,u}^{\infty}mK_1\left(\frac{2\pi mr_i}{b_i}\right)\cos\left[\left(2\frac{\pi mZ_i}{b_i}\right)\right],\tag{6.98}$$

The total field due to the presence of N strands is then given by

$$R^*\left(\eta,Z\right)=\sum_{i=1}^{N}16\pi b_i^{-2}\sum_{m=1,3,5,s}^{\infty}mK_1\left(\frac{2\pi mr_i}{b_i}\right)\cos\left[\left(2\frac{\pi mZ_i}{b_i}\right)\right],\tag{6.99}$$

where η is the cross section variable defined before.

For simplicity, let us assume that all α_i, Z_i and r are equal to α, Z, and η, respectively, where α is defined as the mean interior ionic distance in the gel. With this assumption, Equation (6.99) reduces to

$$R^*\left(\eta;Z\right)=16dNb^{-2}\sum_{m=1,3,5,s}^{\infty}mK_1\left(\frac{2\pi mn}{b}\right)\cos\left[\left(2\frac{\pi mZ}{b}\right)\right]\tag{6.100}$$

The mean Coulomb's attraction or repulsion force associated with the mean-field $R^*(\eta, Z)$ is then given by $F(\eta, Z)$ such that

$$F\left(\eta,Z\right)=Q^2R^*\left(\eta,Z\right)\tag{6.101}$$

where Q corresponds to the total charge between a pair of adjacent ionic surfaces in the gel strip, according to whether like or unlike charges lie in the adjacent rows of charges, this force is repulsive

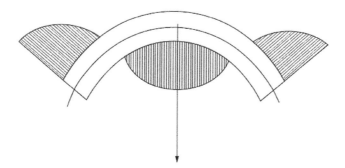

FIGURE 6.22 Nonuniform distribution of Coulomb's forces along the gel axis.

(positive) or attractive (negative) according to whether like or unlike charges lie in the adjacent rows of charges.

Experimental observations (Shahinpoor and Kim, 2001g) on bending of ionic gels in the presence of a voltage gradient generally indicated no gross motion in the direction of the field, suggesting that the force field $F(\eta, Z)$ along the long axis of the gel may be nonuniformly distributed. These facts are also suggested in Figures 6.22–6.24. In fact, the nature of $F(\eta, Z)$, namely:

$$F(\eta, Z) = 16 N \pi Q^2 b^{-2} \sum_{m=1,3,5,\ldots}^{\infty} m K_1 \left(\frac{2\pi m\eta}{b} \right) \cos \left(2 \frac{\pi m Z}{b} \right). \tag{6.102}$$

suggests that even one term series solution gives rise to a possible solution, namely:

$$F_1(\eta, Z) = \left[16 N \pi Q^2 b^{-2} \right] K_1 \left(\frac{2\pi\eta}{b} \right) \cos \left(2 \frac{\pi Z}{b} \right) \right] \tag{6.103}$$

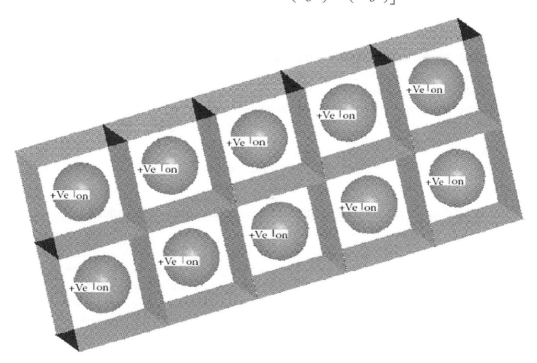

FIGURE 6.23 The simplest solution for a configuration with a few cations before activation.

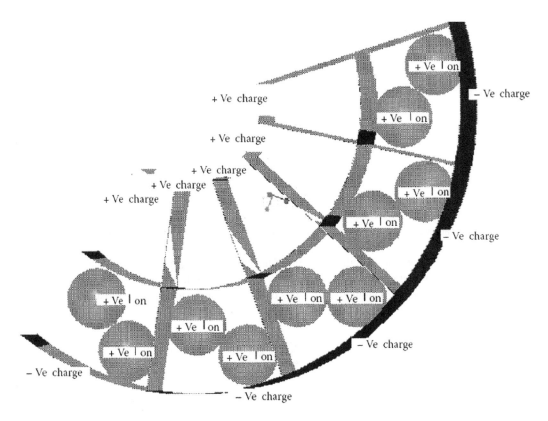

FIGURE 6.24 The simplest solution for a configuration with a few cations after activation.

Equation (6.103) implies that one possible force configuration is when a few fixed ions are located at the ends of the strip with the mobile ions located in the middle, as shown in Figure 6.23 (before actuation) and Figure 6.24 (after actuation).

Clearly, the situation shown in Figure 6.21 can be generalized to produce local bending toward, say, the anode, as shown in Figure 6.22. Thus, the solution for the force field given by Equation (6.102) is quite capable of creating a bending or a series of bendings in the gel in the presence of an electric field gradient. Furthermore, since the net force along the length of the gel strip is a distributed force and should give rise to a net Coulomb force of zero, the following equation holds:

$$\int_{-L/2}^{L_o/2} F(\eta,Z)\,dZ = 0, \tag{6.104}$$

This implies that the spacing between the fixed charges (i.e., b) obeys the equation

$$b = m^{-1}L_o \tag{6.105}$$

If Equation (6.104) does not hold for a given strip of ionic gel, then some overall movement of the gel strip occurs in addition to the bending. In fact, recent experimental results (Shahinpoor and Kim, 2001a, 2001b, 2001c, 2001d, 2001e, 2001f, 2001g) confirm this observation. Note that, from Equation (6.103), average stress over the cross section of the gel strip (width = w_g; thickness = t_g) may be defined as

$$\sigma(\eta,Z) = \left[\frac{-16N\pi Q^2 b^{-2}}{w_g t_g}\right]K_1\left(\frac{2\pi\eta}{b}\right)\cos\left(\frac{2\pi Z}{b}\right) \tag{6.106}$$

where the negative sign ensures that the stress is negative when the gel is in a contracted state.

To find the curvature κ_E in Equation (6.46), the difference $\lambda_+ - \lambda_-$ should be calculated. Assuming a neo-Hookean type of constitutive equation for polymer elasticity, it is concluded that

$$\sigma_+ = -\left(\frac{E_+}{3}\right)\left(\lambda_+ - \lambda_+^{-2}\right) \tag{6.107}$$

$$\sigma_- = -\left(\frac{E_-}{3}\right)\left(\lambda_- - \lambda_-^{-2}\right) \tag{6.108}$$

where E_+ and E_- are elastic moduli and functions of the pH or local ionic concentration and temperature, and λ_+ and λ_- refer to the stretches in the most remote fibers in the gel under bending. From Equation (6.106), it is calculated that

$$\sigma_+ = \sigma(C,Z) = \left[\frac{-16N\pi Q^2 b^{-2}}{w_g t_g}\right] K_1\left(\frac{2\pi C}{b}\right)\cos\left(2\frac{\pi Z}{b}\right) \tag{6.109}$$

$$\sigma_- = \sigma(-C,Z) = \left[\frac{-16N\pi Q^2 b^{-2}}{w_g t_g}\right] K_1\left(\frac{2\pi C}{b}\right)\cos\left(2\frac{\pi Z}{b}\right) \tag{6.110}$$

Thus, Equations (6.107)–(6.110) give rise to the following set of cubic equations for $\lambda_+ - \lambda_-$

$$\lambda_+^3 + \left(\left(\frac{3\sigma_+}{E_+}\right)\right)\lambda_+^2 - 1 = 0 \tag{6.111}$$

$$\lambda_-^3 + \left(\left(\frac{3\sigma_-}{E_-}\right)\right)\lambda_-^2 - 1 = 0 \tag{6.112}$$

The possible curvature with a real value is then calculated as

$$\kappa_E = \left(\frac{(\lambda_+ - \lambda_-)}{2C^*}\right), \tag{6.113}$$

Now, from Equations (6.101)–(6.113), it is clear that the nonhomogeneous force field and the curvature in bending of ionic gels can be electrically controlled utilizing an imposed voltage V across the gel. In this context, the difference $\lambda_+ - \lambda_-$ is related through Equations (6.107) and (6.108) to the stresses $\sigma_+ - \sigma_-$. These stresses are in turn related to the charge Q and other physical parameters by Equations (6.109) and (6.110). The total charge Q is then related to the voltage V and other electrical parameters C_g and R_g by Equation (6.26) as described before. This then allows the designer to control such deformations in ionic gels robotically employing a voltage controller.

Thus, it turns out that the effect of a coulombic type of ionic interaction tends to be the opposite of the cationic electroosmotic drag. This means that the migration of hydrated cations toward the cathode tends to swell the cathode side of the IPMNC strip and increase the osmotic pressure while, at the same time, decreasing the osmotic pressure on the anode side and thus giving rise to bending toward the anode side for a cationic IPMNC.

On the other hand, the redistribution of cations causes more negatively charged SO_3^- pendant groups and branches in the network on the anode side to repel each other more strongly and give rise to induced tensile strain on the anode side; the abundance of cations on the cathode side tends to cause the stronger attraction between them and the SO_3^- Groups and clusters near the cathode side and cause induced compressive strain near the cathode side. The overall effect is then bending toward the cathode side, which is the opposite of migrating hydrated cations. However, experimental observations

(Shahinpoor and Kim, 2000c, 2001b, 2001g, 2002h, 2002j; Kim and Shahinpoor, 2003b) consistently establish that the migration of hydrated cations and the induced swelling on the cathode side have a predominant effect. Thus, it will be illuminating to present another discussion on coulombic types of long-range collective ionic interactions connected with the bending of IPMNC strips in a transverse electric field. However, it must be mentioned that the electric field-induced bending of IPMNC strips has another complication due to the presence of non-hydrated (loose) water in the network.

Nonhydrated water tends to be dragged toward the cathode by the hydrated cations moving in a D.C. (step voltage) electric field and thus initially quickly creates more swelling on the cathode side and more deswelling on the anode side. However, once the strip has come to a stop under the step voltage, the back diffusion of non-hydrated water causes some small relaxation of bending in the opposite direction. If the IPMNC strip is partially dry in the sense of non-hydrated water, then no back relaxation is observed.

6.6   TIME-DEPENDENT PHENOMENOLOGICAL MODEL

6.6.1   Two-Component Transport Model

The most plausible mode of actuation and sensing is expected to be the following: Under the field $\underline{E}$, the cations drift, and they carry with them a certain number of water molecules depending on the hydration number, n, of the cation $M^{(+)}$. When these hydrated water molecules carried by cations pile up near the cathode, they create a local overpressure (see Figure 6.24), which tends to deform the material. A thin membrane, which was originally flat, then tends to acquire a certain spontaneous curvature $C(\underline{E}) = 1/\rho_c$, where ρ_c is the radius of curvature.

The hydrated water migration role becomes clear if we compare the response of an IPMNC strip with $Na^{(+)}$ and $Li^{(+)}$, as well as other cations. It turns out that, for example, for similar IPMNC strips under the same voltage, the deformation and maximum generated force almost double (Shahinpoor and Kim, 2000c) with $Li^{(+)}$ compared to $K^{(+)}$ because the hydration number of $Li^{(+)}$ is about 6 while that of $K^{(+)}$ is about 3. Note that the response would have been very similar from the point of view of ionic redistribution and in connection with osmotic pressures, but it is not, as shown in Figures 6.25 and 6.26.

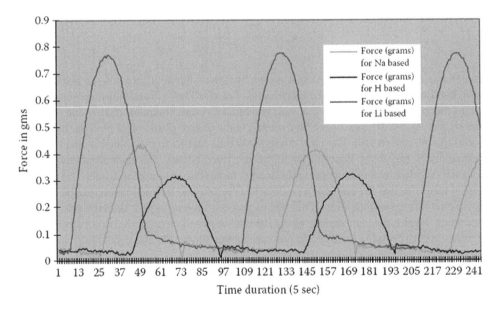

FIGURE 6.25   Force improvement by chemical tweaking showing the effect of changing cations from H^+ to Na^+ to Li^+.

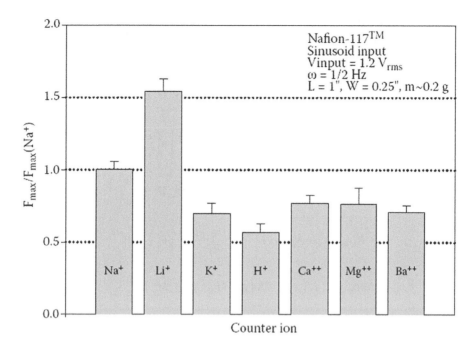

FIGURE 6.26 Experimental evidence for the effect of different ions and their hydration numbers on the tip force and deformation of an IPMNC strip.

These observations clearly establish that the main mechanism of actuation is water migration in the form of cationic hydrated water and loose water dragged by the migration of hydrated cations; the coulombic type of ionic redistribution and forces may play a minor role.

Application of an electric field creates a potential according to Poisson's equation, such that

$$\nabla^2 \phi(x,y,z,t) = \frac{\rho^*(x,y,z,t)}{\varepsilon} \tag{6.114}$$

where $\phi(x, y, z, t)$ is an electric potential, $\rho^*(x, y, z, t)$ is an electric charge density and ε is the composite's dielectric constant. $\rho^*(x, y, z, t)$ is determined by the equivalent weight of the precursor ion-containing polymer and, in particular, the molar concentration of cations (lithium) and the charged groups in the polymer (sulfonic or carboxylic), such that

$$\rho^*(x,y,z,t) = \left(\rho_{M+}(x,y,z,t) - \rho_{so3}(x,y,z,t)\right)Ne \tag{6.115}$$

where $\rho_{M+}(x, y, z, t)$ and $\rho_{so3}(x, y, z, t)$ are, respectively, the molal density of cations and sulfons; N is the Avogadro's number (6.023×10^{26} molecules per kilogram-mole in meter-kilogram-second units) and e is the elementary charge of an electron (-1.602×10^{-19} C).

The electric field within the ionic polymeric structure is

$$\underline{E}(x,y,z,t) = -\nabla \phi(x,y,z,t) \tag{6.116}$$

Balance of forces on individual cations hydrated with n molecules of water inside the molecular network, clusters and channels based on the diffusion-drift model of ionic media due to Nernst and Plank (Nernst-Plank equation) can be stated as:

$$Ne\rho_{M+}E_x(x,y,z,t) = N(\rho_{M+}M_{M+} + n\rho_w M_w)\left(\frac{dv_x}{dt}\right)$$
$$+ N\rho_{M+}\eta v_x + N\rho_{M+}kT\left(\frac{\partial \ln(\rho_{M+}n\rho_w)}{\partial x}\right) + \left(\frac{\partial P}{\partial x}\right)$$

(6.117)

$$Ne\rho_{M+}E_y(x,y,z,t) = N(\rho_{M+}M_{M+} + n\rho_w M_w)\left(\frac{dv_y}{dt}\right)$$
$$+ N\rho_{M+}\eta v_y + N\rho_{M+}kT\left(\frac{\partial \ln(\rho_{M+}n\rho_w)}{\partial y}\right) + \left(\frac{\partial P}{\partial y}\right)$$

(6.118)

$$Ne\rho_{M+}E_z(x,y,z,t) = N(\rho_{M+}M_{M+} + n\rho_w M_w)\left(\frac{dv_z}{dt}\right)$$
$$+ N\rho_{M+}\eta v_z + N\rho_{M+}kT\left(\frac{\partial \ln(\rho_{M+}n\rho_w)}{\partial z}\right) + \left(\frac{\partial P}{\partial z}\right)$$

(6.119)

where M_{M+} and M_w are the molecular weight of cations and water, respectively; P is the local osmotic fluid pressure, P_f, minus the local swelling pressure or stress, σ^*, such that $P = P_f - \sigma^*$,

$$e^E(x,y,z,t) = \left[eE(x,y,z,t)_x, eE(x,y,z,t)_y, eE(x,y,z,t)_z\right]^T$$

(6.120)

is the force vector on an individual cation due to the electroosmotic motion of an ion in an electric field, k is the Boltzmann's constant,

$$v(x,y,z,t) = \left[v_x(x,y,z,t), v_y(x,y,z,t), v_z(x,y,z,t)\right]^T$$

(6.121)

is the velocity vector of the hydrated cations,

$$\eta^v(x,y,z,t) = \left[\eta v_x(x,y,z,t), \eta v_y(x,y,z,t), \eta v_z(x,y,z,t)\right]^T$$

(6.122)

is the force vector of the viscous resistance to the motion of individual hydrated cations in the presence of a viscous fluid medium with a viscosity of η, and

$$kT^\nabla\left[\ln(\rho_{M+}(x,y,z,t) + n\rho_w(x,y,z,t)\right]$$

(6.123)

is the force vector due to diffusion of individual cations and accompanying molecules of hydrated water in the polymer network with the following x, y, and z components, respectively:

$$kT\left(\frac{\partial \ln[\rho_{M+}(x,y,z,t) + n\rho_w(x,y,z,t)]}{\partial x}\right)$$

(6.124)

$$kT\left(\frac{\partial \ln[\rho_{M+}(x,y,z,t) + n\rho_w(x,y,z,t)]}{\partial y}\right)$$

(6.125)

$$kT\left(\frac{\partial \ln\left[\rho_{M+}(x,y,z,t)+n\rho_w(x,y,z,t)\right]}{\partial z}\right)\tag{6.126}$$

The force vector due to inertial effects on an individual hydrated cation is:

$$\left(M_{M+}+nM_w\right)\left(\frac{\mathrm{d}v}{\mathrm{d}t}\right)\tag{6.127}$$

such that, in a compact vector form, the force balance equation reads:

$$Ne\rho_{M+}\underset{\sim}{E}=N\left(\rho_{M+}\mathrm{M}_{M+}+n\rho_w\mathrm{M}_w\right)\left(\frac{\mathrm{d}\underset{\sim}{v}}{\mathrm{d}t}\right)+N\rho_{M+}\eta\underset{\sim}{v}$$

$$+N\rho_{M+}kT\underset{\sim}{\nabla}\ell n\left(\rho_{M+}+n\rho_w\right)+\underset{\sim}{\nabla}P_f-\underset{\sim}{\nabla}\cdot\underset{\approx}{\sigma}^*\tag{6.128}$$

where the stress tensor $\underset{\approx}{\sigma}^*$ can be expressed in terms of the deformation gradients in a nonlinear manner such as in neo-Hookean or Mooney-Rivlin types of constitutive equations as suggested by Segalman and coworkers (1991, 1992a, 1992b, 1992c, 1993). The flux of hydrated cations is given by

$$\underset{\sim}{Q}=\left[\rho_{M+}(x,y,z,t)+n\rho_w(x,y,z,t)\underset{\sim}{v}(x,y,z,t)\right]\tag{6.129}$$

Thus, the equation of continuity becomes

$$\left(\frac{\partial\left[\rho_{M+}(x,y,z,t)+n\rho_w(x,y,z,t)\right]}{\partial t}\right)=-\underset{\sim}{\nabla}\cdot\underset{\sim}{Q}\tag{6.130}$$

Equations 6.114–6.130 are the governing equations for the dynamics of IPMCs. Clearly, they are highly nonlinear and require detailed numerical simulations currently underway and will be reported later.

Next, a linear steady-state version of the formulation is presented to obtain some preliminary understanding of the complex ionic diffusion and drift in these electronic materials.

6.6.2 Linear Irreversible Thermodynamic Modeling

6.6.2.1 Introduction

Figure 6.27 depicts the general structure of the IPMNCs after chemical plating and composite manufacturing. The structure bends toward the anode. The nature of water and hydrated ions transport within the IPMNC can affect the moduli at different frequencies.

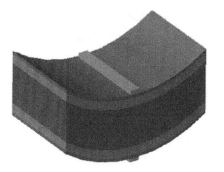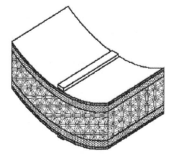

FIGURE 6.27 General structures of an IPMNC or IPCNC film with near-boundary functionally graded electrodes and surface electrodes.

6.6.2.2 Steady-State Solutions

Let us now summarize the underlying principle of the IPMNCs' actuation and sensing capabilities, which can be described by the standard Onsager formulation using linear irreversible thermodynamics. When *static conditions* are imposed, a simple description of *mechanoelectric effect* is possible based upon two forms of transport: *ion transport* (with a current density, J, normal to the material) and *solvent transport* (with a flux, Q, we can assume that this term is water flux). The conjugate forces include the electric field, E, and the pressure gradient, $-\nabla p$. The resulting equation has the concise form of

$$J(x,y,z,t) = \sigma E(x,y,z,t) - L_{12} \nabla p(x,y,z,t) \tag{6.131}$$

$$Q(x,y,zt) = L_{21} E(x,y,z,t) - K \nabla p(x,y,z,t) \tag{6.132}$$

where o and K are the material electric conductance and the Darcy permeability, respectively. A cross-coefficient is usually $L = L_{12} = L_{21}$. The simplicity of the preceding equations provides a close view of the underlying principles of actuation, transduction and sensing of the IPMNCs, as shown in Figure 6.28.

When we measure the *direct* effect (actuation mode, Figure 6.29), we work (ideally) with electrodes impervious to water, and thus we have $Q = 0$. This gives:

$$\nabla p(x,y,z,t) = \frac{L}{K} E(x,y,z,t) \tag{6.133}$$

This $\nabla p(x, y, z, t)$ will, in turn, induce a curvature κ proportional to $\nabla p(x, y, z, t)$. The relationships between the curvature κ and pressure gradient $\nabla p(x, y, z, t)$ are fully derived and described in de Gennes et al. (2000). Let us just mention that $(1/\rho_c) = M(E)/YI$, where $M(E)$ is the locally induced bending moment and is a function of the imposed electric field E, Y is Young's modulus

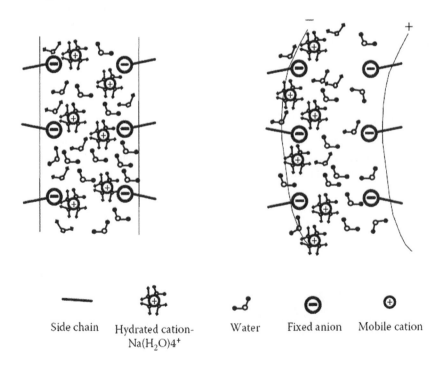

Side chain   Hydrated cation-   Water   Fixed anion   Mobile cation
              Na(H$_2$O)4$^+$

FIGURE 6.28 Schematics of the electroosmotic migration of hydrated counterions within the IPMNC network.

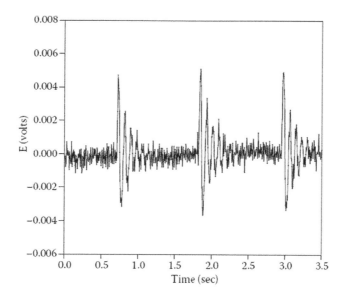

FIGURE 6.29 Dynamic sensing of the ionic polymer due to imposed deformation.

(elastic stiffness) of the strip, which is a function of the hydration H of the IPMNC, and I is the moment of inertia of the strip. Note that, locally, M(E) is related to the pressure gradient such that, in a simplified scalar format:

$$\nabla_p(x,y,z,t) = \left(\frac{2p}{t^*}\right) = \left(\frac{M}{1}\right) = \frac{Y}{\rho_c} = Y\underset{\sim}{\kappa} \tag{6.134}$$

Now, from Equation (6.134), it is clear that the vector form of curvature $\underset{\sim}{\kappa}_E$ is related to the imposed electric field **E** by

$$\underset{\sim}{\kappa}E = \left(\frac{L}{KY}\right)_{\sim}^{E} \tag{6.135}$$

Based on this simplified model, the tip bending deflection δ_{max} of an IPMNC strip of length l_g should be almost linearly related to the imposed electric field because

$$\underset{\sim}{\kappa}E \cong \left[\frac{2\delta_{max}}{\left(l_g^2 + \delta^2 max\right)}\right] \cong \frac{2\delta_{max}}{l_g^2} \cong \left(\frac{L}{KY}\right)_{\sim}^{E} \tag{6.136}$$

The experimental deformation characteristics depicted in Figure 6.30 are clearly consistent with the preceding predictions obtained by the previous linear irreversible thermodynamics formulation. This is also consistent with Equations (6.133) and (6.134) in the steady-state conditions and has been used to estimate the value of the Onsager coefficient, L, to be of the order of 10^{-8} m^2/(V s). Here, we have used a low-frequency electric field to minimize the effect of loose water back diffusion under a step voltage or a D.C. electric field.

Other parameters have been experimentally measured to be K ~ 10^{-18} m^2/CP, σ ~ 1 A/mV, or S/m. Figure 6.31 depicts a more detailed set of data pertaining to Onsager coefficient L.

The role of loose water (nonhydrated on the cations) is also of interest to our experimental and theoretical investigations. In the presence of loose water in the network and within and in the vicinity of ionic clusters under a step-voltage activation, there is a clear final deformation of tip deflection

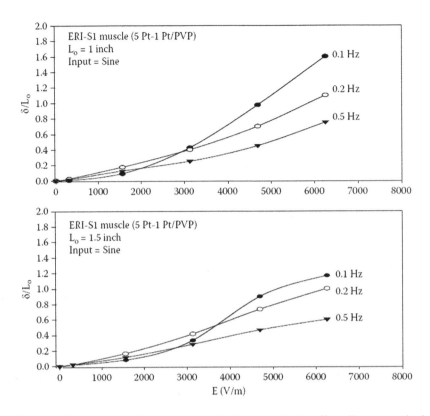

FIGURE 6.30 Actuation under a low-frequency electric field to minimize the effect of loose water back diffusion.

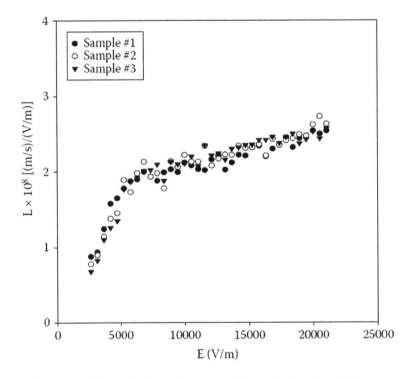

FIGURE 6.31 Experimental determination of Onsager coefficient L using three different samples.

accompanied by a small relaxation due to back diffusion of such loose water back toward the anode. In other words, the loose water is dragged by the hydrated cations, similar to the added mass effect in the fluid mechanics of moving objects in a viscous fluid. Thus, once the step voltage causes all cations to move toward the cathode and accumulate there and finally come to osmotic equilibrium, the loose water simply diffuses back due to the local pressure gradient. This causes a slight back relaxation of the bending deformation of the IPMNC strip (see Figure 6.32(a)). If the strip is dry enough to be just devoid of loose water, no such back relaxation occurs (see Figure 6.32(b)). To achieve this, the IPMC strip needs to be dried partially to remove any loose water in the network. This objective has been successfully achieved using a controlled-humidity environmental chamber.

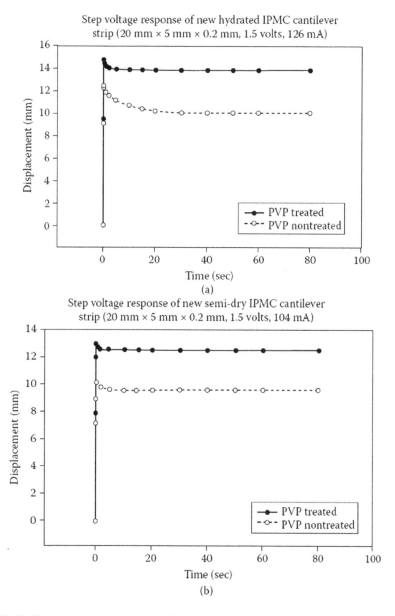

FIGURE 6.32 Deflection under a step voltage for fully hydrated and semidry samples. Note a small back relaxation due to the presence of loose water (a) and virtually no back relaxation due to the absence of loose water (b).

Figure 6.35 depicts a typical spectacular steady-state deformation mode of a strip of such ionic polymers under a step voltage.

When we study the *inverse effect* (transduction mode, Figure 6.28), we apply a bending moment *M* to the cantilever membrane, and we impose two conditions:

No current is produced ($J = 0$).
The strip stays bent with a curvature κ_E

Then, as fully derived and described in de Gennes et al. (2000), the water pressure gradient $\nabla(x, y, z, t)$ turns out to be proportional to *M*.

The condition $J = 0$ gives, from Equation (6.131),

$$E(x,y,z,t) = \frac{L}{\sigma} \nabla p(x,y,z,t) \tag{6.137}$$

Since

$$\nabla p(x,y,z,t) = \left(\frac{M}{I}\right) \tag{6.138}$$

the electric field $E(x, y, z, t)$ generated is thus proportional to *M* according to

$$E(x,y,z,t) = M\left(\frac{L}{\sigma I}\right) = \kappa E\left(\frac{LY}{\sigma}\right) \tag{6.139}$$

It clearly establishes that, for any imposed curvature, κ_e, by applying a bending moment, *M*, an electric field $E(x, y, z, t)$ is generated.

It must, however, be noted that the value of L/σ is considerably smaller, by almost two orders of magnitude than the corresponding one in the actuation mode in the presence of an imposed electric field. This is because, in actuation mode, the imposition of an external electric field changes the values of L and σ. Thus, imposing the same deformation creates an electric field almost two orders of magnitude smaller than the electric field necessary to generate the same deformation. The reader is referred to Shahinpoor and Kim (2000c, 2001b, 2001g, 2002h, 2002j) and Kim and Shahinpoor (2003b) to discuss these issues further.

This completes our coverage of mechanoelectrical phenomena in ionic polymers. More extensive modeling is underway and will be reported later. For other formulations of micromechanics of IPMC actuation and sensing, the reader is referred to Asaka et al. (2000), Asaka and Oguro (2000), Nemat-Nasser and Li (2000), and Tadokoro (2000).

6.6.3 Expanded Ion Transport Modeling for Complex Multicomponent and Multicationic Systems and Ionic Networks

This section introduces the reader to ionic polymer networks that are copolymers and contain multiple components and multiple cationic and anionic clusters and systems within the copolymerized charged network. In these charged networks, only water and hydrated cations are transported, and each of these two species is assumed to be incompressible and governed by the mass conservation principle:

$$\nabla \cdot N_w = 0, \tag{6.140}$$

$$\nabla \cdot N_c^i = 0 \tag{6.141}$$

where N_w and N_c^i are the molar flux of water and the *i*th cation, respectively.

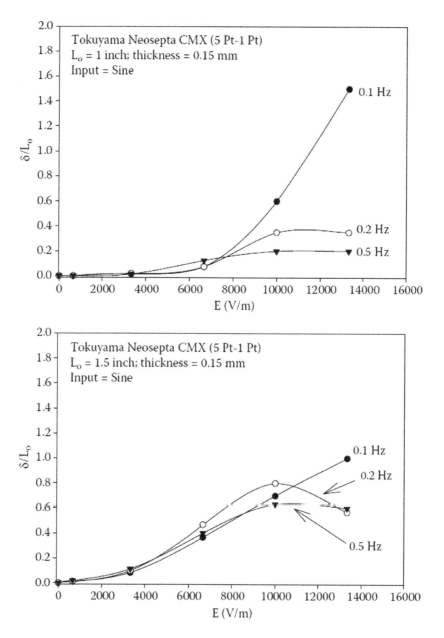

FIGURE 6.33 Displacement characteristics of an IPMNC, Tokuyama Neosepta CMX (sty- rene/divinylbenzene-based polymer). δ: arc length; L_o: effective beam length; $L_o = 1.0$ in. (top) and 1.5 in. (bottom).

The molar flux of water in the membrane is determined by three processes – diffusion, pressure gradient and electroosmotic drag – that lead to:

$$N_w = -D_w \nabla C_w - C_w \varepsilon_w^m \left(\frac{k_h}{\mu_i} \right) \nabla p_l + \left(\frac{n_d^i}{F} \right), \tag{6.142}$$

where

C_w = molar concentration of water, which is the variable of our interest

D_w = diffusion coefficient

ε_w^m = volume fraction of the water in the membrane

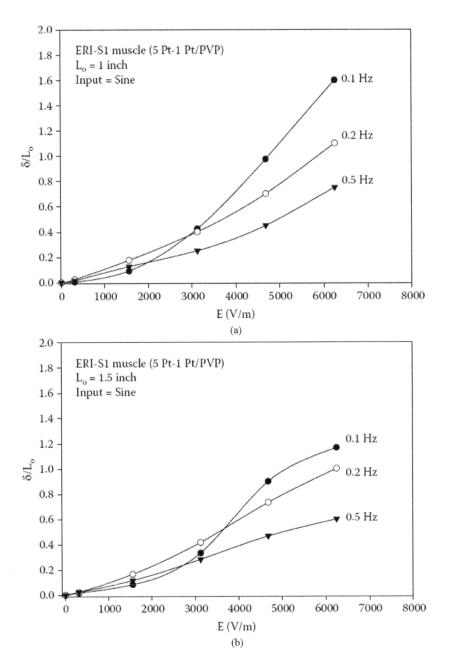

FIGURE 6.34 Displacement characteristics of an IPMNC, ERI-S1. δ: arc length; L_o: effective beam length; (a) $L_o = 1.0$ in. and (b) 1.5 in.

k_h = permeability of the membrane
μ_l = viscosity
p_l = pressure of liquid water
n_d = electroosmotic drag coefficient, which will be discussed later
$\underset{\sim}{i}$ = current density flux.

Taking into account the current conservation law:

$$\nabla \cdot \underset{\sim}{i} = 0, \tag{6.143}$$

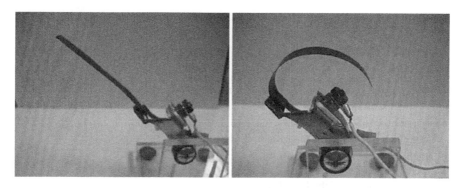

FIGURE 6.35 Typical deformation of strips ($10 \times 80 \times 0.34$ mm^3) of ionic polymers under a step voltage of 4 V.

Equations (6.140)–(6.142) now result in:

$$-D_w \nabla^2 C_w = C_w \varepsilon_w^m \left(\frac{k_h}{\mu_l} \right) \nabla^2 p_l - \varepsilon_w^m \left(\frac{k_h}{\mu_l} \right) \left(\nabla p_l \cdot \nabla C_w \right) + \left(\frac{\nabla n_d \cdot i}{F} \right) = 0, \tag{6.144}$$

The determination of the electroosmotic drag coefficient is based on the work of Springer and colleagues (1991, 1993), who propose a functional relationship between this coefficient and the membrane water content:

$$n_d = \left(\frac{2.5}{22} \right) \lambda, \tag{6.145}$$

where λ is the hydration index (defined as the number of moles of water per equivalent sulfonic or carboxylic acid groups SO_3^- or COOH– in the membrane) and the numerical values of 2.5 and 22 correspond to the number of water molecules dragged per migration of cations (Li$^+$, Na$^+$, H$^+$, etc.), ion, and the possible maximum hydration index, respectively. Springer et al. also present an empirical formula relating C_w to λ as follows:

$$C_w = \frac{e\lambda}{f\lambda + 1}, \tag{6.146}$$

where f is an experimentally determined swelling coefficient for the membrane, and e is expressed as

$$e = \frac{\rho_m^{\text{dry}}}{E_m} \tag{6.147}$$

with ρ_m^{dry} corresponding to the density of the dry membrane, and E_m the equivalent weight of the membrane. Rearranging Equations (6.144)–(6.147) yields:

$$-D_w \nabla^2 C_w - C_w \varepsilon_w^m \left(\frac{k_h}{\mu_l} \right) \nabla^2 p_l - \varepsilon_w^m \left(\frac{kh}{\mu_l} \right) \left(\nabla p_l \cdot \nabla C_w \right) + \frac{5e}{44(e - fC_w)^2} \left(\frac{\nabla C_w \cdot i}{F} \right) = 0 \tag{6.148}$$

Given distributions of pressure p_l and current density i, we can solve Equation (6.148) to obtain C_w. To investigate a given case with a specific partial hydration level, C_w can be prescribed at the anode-side membrane border while prescribing the water content at the cathode–membrane interface such that water balance conditions described by Equations (6.140) and (6.141) are satisfied on a point-by-point basis along the length of the membrane.

The Nernst–Planck equation describes the flux of protons through the membrane:

$$N_c^i = -Z_c^i \left(\frac{F}{RT} \right) D_c^i C_c^i \, \nabla\phi - D_c^i \, \nabla C_c^i + C_c^i \underset{\sim}{U}_w, \tag{6.149}$$

where the terms on the right-hand side represent migration due to the electric field, diffusion, and convection of the dissolved protons;

Z_c^i = charge number of the ith cation
D_c^i = diffusion coefficient pertaining to the ith cation
C_c^i = molar concentration of the ith cation
ϕ = electrical potential
$\underset{\sim}{U}_w$ = convective velocity of the liquid water.

The velocity of the liquid water in the membrane is modeled by the Schlögl equation (Verbrugge and Hill, 1993; Singh et al., 1999), which states that the convection is caused by the electric potential and the pressure gradient:

$$U_w = \varepsilon_w^m \left[\left(\frac{k_\phi}{\mu_w} \right) Z_f C_f F \nabla\phi - \left(\frac{k_c^i}{\mu_w} \right) \nabla p_w \right], \tag{6.150}$$

Where

k_ϕ = electric permeability
k_c^i = hydraulic permeability of the ith cation
Z_f = charge number of the fixed charges
C_f = fixed-charge concentration.

The flow of charged species is related to the current density by

$$\underset{\sim}{i} = F \sum_i Z_c^i \underset{\sim}{N}_c^i, \tag{6.151}$$

The membrane conductivity is defined as

$$\kappa = \left(\frac{F^2}{RT} \right) \sum_i \left(Z_c^i \right)^2 D_c^i C_c^i, \tag{6.152}$$

Also, the electroneutrality holds

$$Z_f + C_f + \sum_i Z_c^i C_c^i = 0, \tag{6.153}$$

Here, we note that the only mobile ions in the membrane are the monovalent cations, and for them, we have $Z_c^i = 1$.

Equations (6.148)–(6.153) and Equation (6.140) can be recombined, yielding

$$-\Delta\phi = -\kappa^{-1} \left[\left(\nabla \ell n C_c^i \right) \cdot \underset{\sim}{i} \right] + \left(\frac{RT}{F} \right) \underset{\sim}{\nabla} \cdot \left[\underset{\sim}{\nabla} \left(\ell n C_c^i \right) \right] \tag{6.154}$$

Since the membrane swells due to internal hydration, $p\,c$ is a hydration function rather than a constant. Similarly to Equation (6.146), the proton concentration can be described by:

$$C_c^i = \frac{e}{f\lambda + 1} \tag{6.155}$$

Combining Equations (6.146) and (6.155) yields

$$C_c^i = e - fC_w \tag{6.156}$$

Equations (6.145)–(6.156) should be substituted in Equation (6.144) to obtain the ion transport and deformation dynamics governing equations. However, such sophisticated detail will not be expanded in this chapter and will be left for future work.

6.6.4 EQUIVALENT CIRCUIT MODELING

Newbury and Leo (2002) present a gray box model in the form of an equivalent circuit. The model is based on the assumption of linear electromechanical coupling among electric field, charge strain, and applied stress. Under this assumption, an impedance model of a cantilever transducer is derived and applied to transducer performance analysis. This model enables simultaneous modeling of sensing and actuation in the material. Furthermore, the model demonstrates the existence of reciprocity between sensing and actuation.

The model is validated experimentally for changes in transducer length and width. Experimental results demonstrate the improved tracking capability through the use of model-based feedback. Recent publications by Newbury (2002) and Newbury and Leo (2002) have produced a model of electromechanical coupling in ionic polymer materials. The model's fundamental assumption is that the electromechanical coupling is a linear relationship among strain, stress, electric field and charge. Thus, they assume:

$$S = \frac{T}{Y} + dE$$

$$D = dT + \varepsilon E \tag{6.157}$$

where
S = strain in the material
T = stress (Pa)
E = electric field (V/m)
D = charge density (C/m?).

The three material properties of the model are the elastic modulus Y (Pa), the electric permittivity (F/m) and the linear coupling coefficient (m/V or C/N).

For a cantilevered sample of ionic polymer material, Equation (6.157) can be integrated over the transducer volume to produce a relationship among tip displacement, tip force, voltage, and charge (Newbury, 2002). The only limiting assumption in this derivation is that the stress acts at the bender element's surface. This assumption is consistent with other actuation models in ionic polymers (Nemat-Nasser and Li, 2000). Transforming the integrated equations into the frequency domain produces an impedance relationship among voltage, force, velocity and current of the form:

$$\begin{pmatrix} v \\ f \end{pmatrix} = \begin{bmatrix} Z_{11}(j\omega) & Z_{12}(j\omega) \\ Z_{12}(j\omega) & Z_{22}(j\omega) \end{bmatrix} \begin{pmatrix} i \\ \dot{u} \end{pmatrix} \tag{6.158}$$

The impedance terms are determined from the curve fit of experimental data. As discussed in Newbury (2002), these parameters can be determined from a set of three measurements of the polymer response. The measurements incorporate step response and frequency-response data to produce an accurate model over a broad frequency range. The frequency ranges tested in Newbury (2002) are 0–20 Hz. Unlike an ideal transformer, this model incorporates a frequency-dependent transformer coefficient that produces a frequency-dependent coupling parameter. The terms in the equivalent circuit are derived in Newbury (2002).

The impedance model is shown in Equation (6.158) was utilized as a basis for modeling the coupled system. The model was rewritten with voltage and velocity as the inputs and force and current as outputs:

$$
\begin{pmatrix} i \\ f_a \end{pmatrix} = \begin{bmatrix} g_{11} & g_{12} \\ g_{21} & g_{22} \end{bmatrix} \begin{pmatrix} v \\ \dot{u} \end{pmatrix}
\tag{6.159}
$$

The force at the actuator tip was then written as a combination of the motor's inertial force and the resistance force of the sensor.

Several other models of electromechanical coupling have been presented in the literature. Models based on measured data have been presented by Kanno et al. (1994, 1996). Models developed from first principles have been proposed by Nemat-Nasser and Li (2000), Nemat-Nasser (2002), Asaka et al. (1995), Asaka and Oguro (2000), de Gennes et al. (2000), and Tadokoro (2000). Surveying these models, we see several explanations for the electromechanical coupling in ionic polymer materials. Most of the debate centers around the relative importance of electrostatic effects and hydraulic effects within the material.

6.7 CONCLUSIONS

This chapter presented a detailed description of various modeling and simulation techniques and the associated experimental results connected with ionic polymer-metal composites (IPMNCs) as soft biomimetic sensors, actuators, transducers and artificial muscles. These techniques included the continuum electrodynamics of ionic polymeric gels swelling and deswelling, continuum-diffusion electromechanical model for asymmetric bending of ionic polymeric gels, continuum microelectromechanical models, microelectromechanical modeling of asymmetric deformation of ionic gels, time-dependent phenomenological modeling, steady-state solutions based on linear irreversible thermodynamics, expanded ion transport modeling, and, finally, equivalent circuit modeling. An exact expression for the curvature and maximum tip deflection of an IPMNC strip in an imposed electric field was also derived.

7 Sensing, Transduction, Feedback Control and Robotic Applications of Polymeric Artificial Muscles

7.1 INTRODUCTION

This chapter covers ionic polymer metal composites (IPMCs) sensing, transduction, feedback control, robotic actuation capabilities, energy harvesting, tactile sensing, haptic feedback and related issues. Ionic polymer–metal nanocomposites (IPMNCs) and ionic polymer conductor nanocomposites (IPCNCs) are unique tools for soft robotic actuation and built-in sensing and transduction in a distributed manner. One can even see the distributed biomimetic, noiseless nanosensing, nanotransduction and nanoactuation capabilities of IPMCs.

7.2 SENSING CAPABILITIES OF IPMCs

This section presents a brief description and testing results of ionic polymer–metal composites (IPMNCs) as dynamic sensors. As previously noted, a strip of IPMNC can exhibit large dynamic deformation if placed in a time-varying electric field. Conversely, the dynamic deformation of such ionic polymers produces dynamic electric fields. The underlying principle of such a mechanoelectric effect in IPMNCs can be explained by the linear irreversible thermodynamics in which ion and solvent transport are the fluxes. The electric field and solvent pressure gradient are the forces, as described in Chapter 6. Important parameters include material capacitance, conductance and stiffness, which are related to material permeability.

The dynamic sensing response of a strip of IPMNC under an impact type of loading is also discussed. A damped electric response is observed that is highly repeatable with a broad bandwidth to megahertz frequencies. Such direct mechanoelectric behaviors are related to endo-ionic mobility due to imposed stresses. This mobility means that if one imposes a finite solvent flux without allowing a current flux, the material creates a particular conjugate electric field that can be dynamically monitored. IPMCs are observed to be highly capacitive at low frequencies and highly resistive under high-frequency excitations. Current efforts are to study the low- and high-frequency responses and sensitivity of IPMNCs that might conceivably replace piezoresistive and piezoelectric sensors with just one sensor for broad frequency range sensing and transduction capabilities.

7.2.1 BASICS OF SENSING AND TRANSDUCTION OF IPMCs

It is so far established that ionic polymers (such as a perfluorinated sulfonic acid polymer) in a composite form with a conductive metallic medium can exhibit large dynamic deformation if placed in a time-varying electric field (see Figure 7.1). Conversely, dynamic deformation of such ionic polymers produces dynamic electric fields (see Figure 7.2). A recently presented model by de Gennes et al. (2000) presents a plausible description of the underlying principle of electro-thermodynamics in ionic polymers based on internal ion and solvent transport and electrophoresis. IPMCs show great

FIGURE 7.1 Successive photographs of an IPMC strip that shows substantial deformation (up to 4 cm) in the presence of low voltage. The sample is 1-cm wide, 4-cm long and 0.2-mm thick. The time interval is 1 s. The actuation voltage is 2-V D.C.

potential as dynamic sensors, soft robotic actuators and artificial muscles in a broad range of nano- to micro- to macro-scale sizes.

A recent study by de Gennes and coworkers (2000) has presented the standard Onsager formulation on the underlining principle of IPMC actuation/sensing phenomena using linear irreversible thermodynamics: When static conditions are imposed, a simple description of *mechanoelectric effect* is possible based upon two forms of transport: *ion transport* (with a current density, J, across the thickness of the material) and *electrophoretic solvent transport.* (with a flux Q, one can assume that this term is the water flux.) The conjugate forces include the electric field vector **E** and the pressure gradient vector $-\nabla p$. The resulting ***linear irreversible thermodynamic equations*** have a concise form of

$$J = \sigma E - L_{12}\nabla p \tag{7.1}$$

$$Q = L_{21}E - K\nabla p \tag{7.2}$$

where σ and K are the membrane conductance and the Darcy permeability coefficient, respectively. A cross-coefficient is usually $L_{12} = L_{21} = L$, estimated to be on the order of 10^{-8} (ms^{-1})/(volt-m^{-1}) (Shahinpoor and Kim, 2000a, 2001g). The preceding equations' simplicity provides a close view of the underlining principles of actuation and sensing of IPMCs.

Figure 7.2(a) and (b) shows the dynamic actuation and sensing response of a strip of an IPMC (thickness of 0.2 mm) subject to a dynamic impact loading in a cantilever configuration. A damped electric response is observed that is highly repeatable with high bandwidth of up to 100 Hz. Such direct mechanoelectric behaviors are related to endo-ionic mobility due to imposed stresses. This mobility implies that, if we impose a finite solvent (= water) flux, $|Q|$ – not allowing a current flux, $J = 0$ – a particular conjugate electric field, $\vec{E}$, is produced that can be dynamically monitored.

From Equations (7.1) and (7.2), one imposes a finite solvent flux Q while having zero current ($J = 0$) and nonzero bending curvature. This situation certainly creates an intrinsic electric field $\vec{E}$, which has a form of

$$E = \frac{L}{\sigma}\nabla p \frac{12(1-v_p)}{(1-2v_p)}\left\{\frac{L}{\sigma h^3}\right\}\Gamma \tag{7.3}$$

Note that notations v_p, h and Γ are, respectively, the Poisson ratio, the strip thickness and an imposed torque at the built-in end produced by a force F applied to the free end multiplied by the free length of the strip l_g.

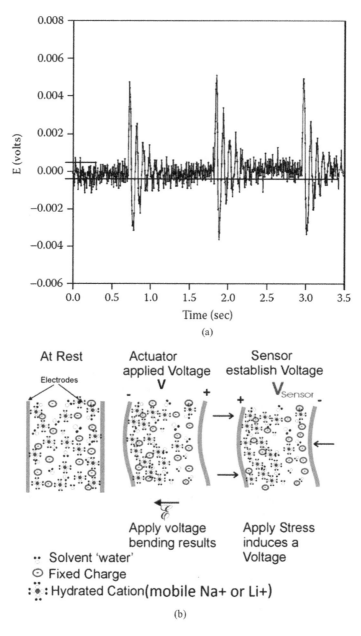

FIGURE 7.2 (a) A typical sensing response of an IPMNC. The IPMNC ($5 \times 20 \times 0.2$ mm³) in a cantilever mode, as depicted in Figure 7.1, is connected to an oscilloscope and is manually flipped to vibrate and come to rest by vibrational damping. (b) Actuation and sensing mechanisms in IPMCs.

7.2.2　Electrical Properties

To assess the electrical properties of the IPMNC, the standard A.C. impedance method that can reveal the equivalent electric circuit has been adopted. A typically measured impedance plot, provided in Figure 7.3, shows the frequency dependency of the impedance of the IPMNC.

Overall, it is interesting to note that the IPMC is nearly resistive (>50 Ω) in the high-frequency range and fairly capacitive (>100 μF) in the low-frequency range. IPMCs generally have a surface

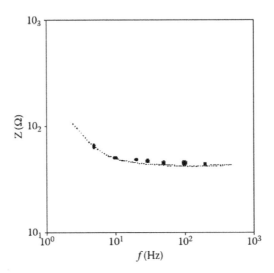

FIGURE 7.3 The measured A.C. impedance characteristics of an IPMC sample (dimension = 5-mm width, 20-mm length and 0.2-mm thickness).

resistance, R_{SS}, of about a few ohms per centimeter, near-boundary resistance, R_S, of a few tens of ohms per centimeter and cross-resistance, R_P, of a few hundreds of ohms per millimeter; typical cross capacitance, C_g, is a few hundreds of microfarads per millimeter.

Based on these findings, we consider a simplified equivalent electric circuit of the typical IPMNC, such as the one shown in Figure 7.4 (de Gennes et al., 2000). In this approach, each single unit circuit (i) is assumed to be connected in a series of arbitrary surface resistance (R_{ss}) on the surface.

This approach is based upon the considerable surface electrode resistance (see Figure 7.4). We assume that there are four components to each single-unit circuit: the surface electrode resistance (R_s), the polymer resistance (R_p), the capacitance related to the ionic polymer, the double layer at the surface–electrode/electrolyte interface (C_d) and an impedance (Z_w) due to a charge transfer resistance near the surface electrode. For the typical IPMNC, the importance of R_{ss} relative to R_s may be interpreted from $\sum R_{ss}/R_s \approx L/t \gg 1$, where notations L and t are the length and thickness of the electrode. Note that the problem now becomes a two-dimensional one; the fact that the typical value of t is ~1–10 μm makes the previous assumption.

Thus, a significant overpotential is required to maintain the effective voltage condition along the surface of the typical IPMNC. An effective technique to solve this problem is to overlay a thin layer of a highly conductive metal (such as gold) on top of the platinum surface electrode (de Gennes et al., 2000). Figure 7.4 depicts a digitized rendition of the equivalent circuit for IPMC, a continuous material. Figure 7.5 depicts the measured surface resistance, R_s, of a typical IPMNC strip as a function of platinum particle penetration depth. Note that SEM was used to estimate the penetration depth of platinum into the membrane.

The four-probe method was used to measure the surface resistance, R_s, of the IPMCs. Obviously, the deeper the penetration of metallic particles is, the lower the surface resistance is.

Figure 7.6 depicts measured chronoamperometry responses of a typical IPMC sample in which the current response is recorded after a step potential is applied.

It should be noted that the important physical phenomena occurring in the vicinity of the electrodes and the associated processes, particularly within a few micron depths, play an important role in the sensing characteristics of IPMCs.

Mass transfer of cations and their hydrated water molecules involved as they move into and out of the bulk material and through the porous electrodes is another important sensor feature. Diffusion

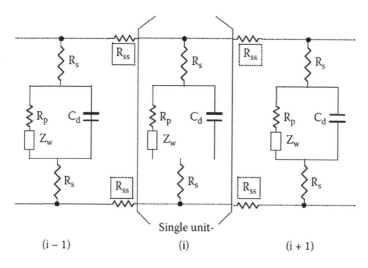

FIGURE 7.4   An equivalent electronic circuit for a typical IPMNC strip obtained by an impedance analyzer.

can be considered the sole transport process of electroactive species (cations and hydrated water). The plausible treatment of diffusion is to use the Cottrell equation having a form of

$$i(t) = \frac{nFAD^{1/2}C}{(\pi t)^{1/2}} = Kt^{1/2} \tag{7.4}$$

where $i(t)$, n, A, D, C and F are the current at time t and the number of electrons involved in the process, the surface area, diffusivity, concentration and Faraday constant, respectively. Equation (7.4) states that the current is inversely proportional to the square root of time.

Figure 7.6 shows the overall current versus the square root of time. Also, it states that the product $i(t) \times t$ should be a constant K for a diffusional electrode. Deviation from this constancy could

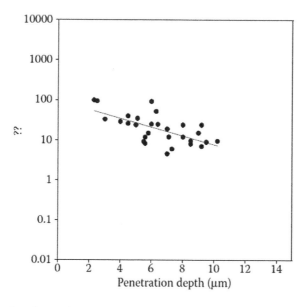

FIGURE 7.5   Measured surface resistance, R_s, of a typical IPMNC strip, as a function of platinum particle penetration depth.

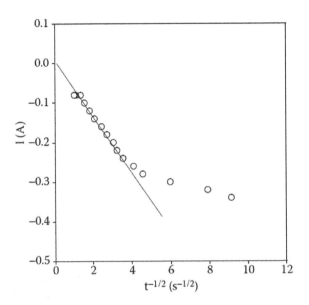

FIGURE 7.6   The current, $i(t)$, versus $t^{-1/2}$ (chronoamperometry data). Note that $A = 6.45$ cm^2, $E = -3$ V.

be caused by several factors, including slow capacitive charging of the electrode during the step-voltage input and coupled chemical reactions (hydrolysis). This figure is constructed under a step potential of -3 V for a typical IPMNC. As can be seen, the characteristics follow the simple Cottrell equation that confirms the fact that the electrochemical process is diffusion controlled.

A more recent equivalent circuit proposed by Paquette and Kim (2002) is shown in Figure 7.7. The two loops that include R_1, C_1, R_3 and C_2 represent the two composited effective electrodes of the IPMNC. R_2 represents the effective resistance of the polymer matrix. E's value is the electric field applied across the material for actuation: $E = V/h$, where V is the voltage applied, and h is the membrane thickness.

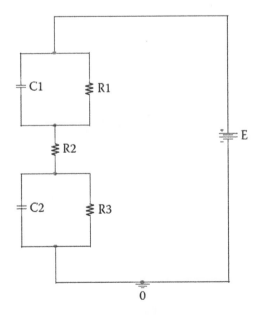

FIGURE 7.7   Equivalent circuit model representing an IPMC.

The effective capacitance values are $C_1 = C_2$ and $R_1 = R_3$. Upon examination of the proposed circuit, the current for the equivalent circuit is:

$$i(t) = V_A \cdot \left[\frac{1}{(R_2 + 2R_1)} + 2 \cdot \frac{R_1}{(R_2 + 2R_1)} \cdot \frac{\exp\left[-(R_2 + 2 \cdot R_1) \cdot (t/(R_2 \cdot R_1 \cdot C_1)) \right]}{R_2} \right] \tag{7.5}$$

where V_A is the applied step voltage (at time $t \geq 0$ s). The resulting voltage across either capacitive loop is

$$V_C(t) = \frac{V_A - V_A \cdot R_2 \cdot \left[\begin{array}{c} \left(1/(R_2 + 2 \cdot R_1)\right) + 2 \cdot \left(R_1/(R_2 + 2 \cdot R_1)\right) \cdot \\ \dfrac{\exp\left[-(R_2 + 2 \cdot R_1) \cdot (t/(R_2 \cdot R_1 \cdot C_1)) \right]}{R_2} \end{array} \right]}{2} \tag{7.6}$$

7.2.3 EXPERIMENT AND DISCUSSION

As discussed extensively before and to refresh the reader's memory, the manufacturing of an IPMNC starts with an anionic polymer subjected to a chemical transformation (REDOX), which creates a functionally graded composite ionic polymer with a conductive phase. *Ionic polymeric material* selectively passes through ions of a single charge (cations or anions). They are often manufactured from polymers that consist of fixed covalent, ionic groups. The currently available ionic polymers are:

1. Perfluorinated alkenes with short side chains terminated by ionic groups (typically sulfonate or carboxylate [SO_3^- or COO^-] for cation exchange)
2. Styrene/divinylbenzene-based polymers in which the ionic groups have been substituted from the phenyl rings where the nitrogen atoms are fixed to ionic groups

Figure 7.8 is an SEM micrograph showing the cross section of a typical IPMNC used in this study. In Figure 7.9, preliminary quasistatic D.C. sensing data is provided in terms of the voltage produced at different displacements. Note that the displacement is shown in terms of the deformed angle relative to the standing position in degree. The dimension of the IPMNC sample sensor is $5 \times 25 \times 0.12$ mm^3. Such a direct mechanoelectric effect is convenient because the produced voltage is large, and the applicable displacement is large.

Comparing such unique features of IPMCs as sensing devices relative to other current state-of-the-art sensing technologies such as piezoresistive or piezoelectric devices, one can find more flexibility in connection with IPMCs.

The slow current leakage due to the redistribution of ions is often observable. Additional investigations of the current leakage are necessary to stabilize the voltage output of the IPMNCs in a sensing mode.

7.3 EVALUATION OF IPMCS FOR USE AS NEAR-D.C. ELECTROMECHANICAL SENSORS

7.3.1 INTRODUCTION

Henderson and coworkers (2001) and Shahinpoor et al. (2001) offer information on using IPMNCs as near-D.C. mechanical sensors. IPMNC active elements enable near-D.C. acceleration measurement devices with modest power, volume, mass and complexity requirements, provided their unique

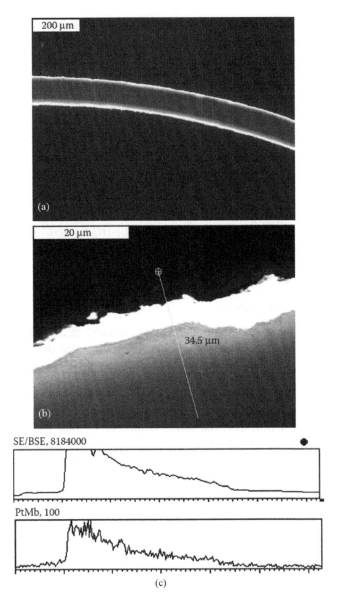

FIGURE 7.8   An SEM micrograph shows the cross-section of an IPMNC sensor. It depicts a (a) cross-section of an IPMNC strip, (b) its close-up and (c) the X-ray line scan.

properties are accounted for in the design. Advantages over conventional piezoelectric elements are documented for some applications.

Acceleration measurements are necessary for various dynamics experiments and often serve as sensing inputs in structural control systems. In conventional practice, a piezoelectric element, as part of a single-degree-of-freedom harmonic system, is often used to sense acceleration. Figure 7.10(a) and 7.10(b) are representations of two approaches. Figure 7.10(a) depicts the most basic embodiment, a piezoelectric ceramic element, such as lead zirconium titanate (PZT), in line with a mass. When the base is accelerated vertically, the inertia of the mass causes strain in the active element

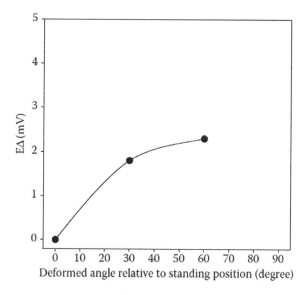

FIGURE 7.9 D.C. sensing data in terms of produced voltages, ΔE, versus displacement. Note that the displacement is shown in terms of the deformed angle relative to the standing position in degree. The dimension of the sample sensor is $5 \times 25 \times 0.12$ mm^3.

such that when electroded, it generates a charge proportional to the strain. The first-order system's damping may be tuned to provide a nearly flat mechanical response over a desirable range below the system resonance frequency. Since piezoceramic materials are brittle and cannot support significant tension loads, a mechanical or electrical preload must be applied in this embodiment to avoid such a situation.

Figure 7.10(b) shows a slightly more common approach for an acceleration sensor. A mass is suspended from three sides by piezoelectric elements affixed to the accelerometer enclosure boundary. In this case, as the enclosure is vibrated, the active elements undergo a shear strain, which also generates a proportional voltage. The piezoceramic material can support shear loads in either direction, so no vertical preload is necessary for this embodiment.

In either of the conventional embodiments with realistic active element sizes, low-frequency response (<1 Hz) is generally poor without additional electronics. Piezoelectric elements are

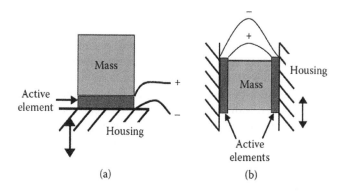

FIGURE 7.10 Accelerometer implementations using PZT in (a) 3–3 mode and (b) 3–1 mode.

largely capacitive, and when connected directly to an oscilloscope ($\approx$1-MΩ input impedance), the charge tends to bleed off quickly at low frequencies in the effective RC circuit. Voltage followers or charge amplifiers are usually included in the circuit to lower this frequency and eliminate measurement sensitivity to cable noise and environmental parameters. However, low-frequency performance comes at the expense of size, weight and complexity. Furthermore, functional performance at frequencies below 1 Hz is extremely difficult to achieve. For this reason, piezoelectric accelerometers make a poor choice for near-D.C. acceleration measurements. Here, we investigate using a fairly new material, IPMNC, for use as an active element in near-D.C. accelerometers.

7.3.2 Background of near DC sensing

Recently, IPMCs consisting of a thin Nafion-117 sheet plated with gold or platinum on both sides have received much attention for their possible applications to sensing and actuation (see Shahinpoor and Kim, 2000, for an extensive review). The materials work via internal ionic transport phenomena, requiring water internal to the system and coupling between the electrical field and bending deformation. Not only do such materials exhibit relatively large deformations with a mild voltage input, but they also have a significant electrical response to bending deformation, even at subhertz frequencies. The long response time of IPMNCs – on the order of 4–10 s – limits their use in applications requiring a short response time. Although this limitation may be mitigated to some extent through feedback control, it becomes an advantage when considering the material for low-frequency applications. The usefulness of IPMNC material in a near-D.C. accelerometer application and potential problems and design considerations is examined here.

7.3.3 Experiment Setup

The complex impedance of the IPMNC was measured by the well-known voltage divider method. The impedance was measured over a range from 0.05 to 5000 Hz, using a 15-kΩ resistor in series and a Siglab data acquisition system to record the voltages' magnitude and phase.

For the dynamic experiments, a Ling 5-lb shaker was mounted on an optical table in the vertical orientation with the stinger pointing down. A machined aluminum block was mounted on the end of the stinger to provide a flat movable surface. An eddy current probe was mounted such that the head was held beneath the block, near one edge, with a 0.25-mm standoff distance. The probe was calibrated to provide a 1.2-mm/V displacement response near an aluminum surface. For the control case, a piezoelectric patch (PZT-5A with nickel electrodes, $11 \times 29 \times 0.26$ mm^3, 0.646-g mass, and 20-nF capacitance) was used as an active element. In the experiment case, an IPMNC patch (with gold electrodes, 0.237-g mass) was used as the active element and cut to the same planar dimensions as the piezoelectric patch but with 0.32-mm thickness.

For the control case, the mounting apparatus consisted of an aluminum-base block with a piezoelectric patch (PZT) sandwiched between the base block and the shaker block. The shaker and block were lowered onto the optical table mounting post to generate an initial compressive preload in the patch and to ensure that it remained in compression over the entire range of motion applied to the shaker. Both blocks were covered with Kapton tape, where they contacted the piezoelectric element to prevent current bleed-off into the optical table ground. Copper shims were cut and affixed to the top and bottom of the piezoelectric patch with conductive grease to provide external leads. When the shaker was actuated, the piezoelectric element was compressed in the 3–3 direction, generating a voltage between the copper leads. The leads were connected via a BNC cable to an HP digital signal analyzer (DSA 35665A) with 1-MΩ input impedance. No conditioning circuitry was applied. The DSA source channel was used to actuate the shaker with a 5-V peak-to-peak sine sweep input over a frequency range from 0.015 to 30 Hz. The frequency response function from eddy current displacement to piezoelectric element voltage output was measured over this frequency range and converted to a volts/strain spectrum.

The IPMNC mounting apparatus consisted of a Plexiglas cantilevered clamping device with conductive contact points and wire leads. Again, the leads were connected by BNC cable to the DSA input, without any conditioning circuitry. The apparatus was positioned under the shaker block with an initial static displacement such that the tip of the cantilevered IPMNC patch was deflected by the actuation of the shaker over its entire range of motion. The block was covered with Kapton tape at the contact point to avoid bleed-off of the current into the optical table during the experiment. The response of the IPMNC in cantilevered mode to a sine sweep of the shaker was measured over a frequency range of 0.015–50 Hz. The frequency response spectrum was calculated in terms of voltage output versus end deflection.

7.3.4 EXPERIMENT RESULTS

Any conditioning circuitry design for an accelerometer will require an accurate understanding of the active element's complex impedance over the application's frequency range. Hence, this spectrum was measured for the IPMC.

Figures 7.11 and 7.12 represent the complex impedance of the sample. This IPMC patch had been left to dry in the open atmosphere for approximately one month before these data were taken. Note that the element is fairly capacitive at low frequencies but is almost entirely resistive above 100 Hz. Despite the curve's simple appearance, this impedance response cannot be modeled simply as a three-element resistor and capacitor circuit but rather requires a series solution.

Figures 7.13 and 7.14 depict the impedance curves for similarly sized IPMNC samples immediately after being removed from their water-filled storage bags. Note that the magnitudes between wet and dry samples differ by almost three orders of magnitude. Moreover, Shahinpoor and Kim (2001g) measured an impedance magnitude spectrum (not shown) between these two extremes.

This surprising variability underscores the importance of accurately knowing or controlling the active element's moisture state before IPMC materials may be used effectively in practice.

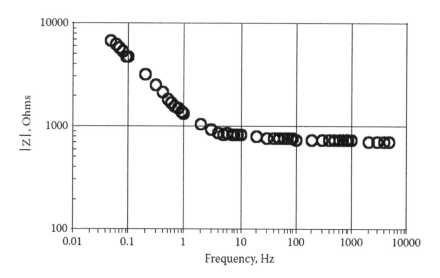

FIGURE 7.11 Dry IPMNC impedance magnitude.

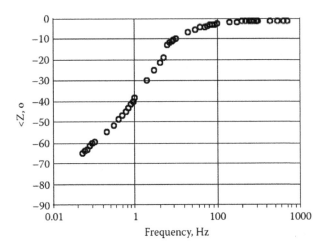

FIGURE 7.12 Dry IPMNC impedance phase.

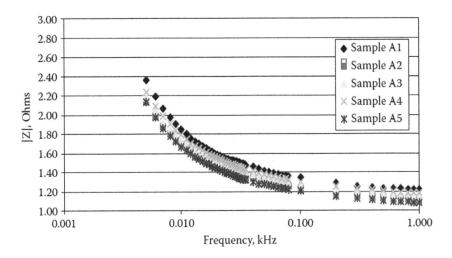

FIGURE 7.13 Wet IPMNC impedance magnitude.

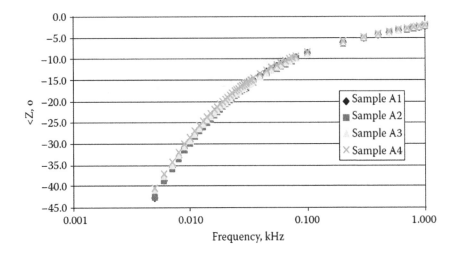

FIGURE 7.14 Wet IPMNC impedance phase.

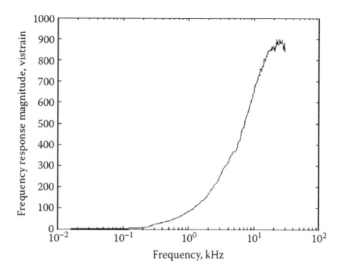

FIGURE 7.15 PZT frequency response magnitude.

Figure 7.15 is the frequency response magnitude of the piezoelectric patch in 3–3 compression. Note that the break frequency (the point at which the response is 0.707 of the maximum value) occurs at approximately 10 Hz. This corresponds closely to the theoretical value of 8 Hz for a pure RC circuit ($\omega = 1/RC$), based on the measured capacitance of the element (20 nF) and the input impedance of the DSA (1 MΩ).

In conventional practice, the useful range of a piezoelectric element for an accelerometer is considered to include only the region where the response deviates from the norm by no more than 5%. This further limits the usefulness of the PZT for low-frequency accelerometer applications. A voltage follower or charge amplifier circuit would lower this break frequency in a real implementation. However, this requires increased mass, volume and complexity, and there are practical limits to how low this frequency may be set. In general, subhertz conditioning with a piezoelectric element is rare and difficult to achieve.

Figure 7.16 is the frequency response magnitude of the IPMNC patch under cantilever excitation.

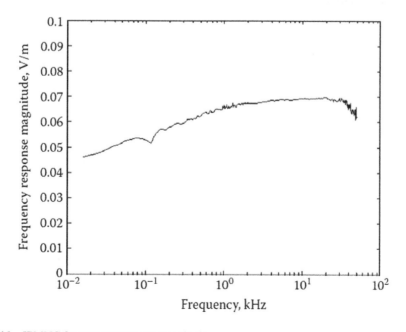

FIGURE 7.16 IPMNC frequency response magnitude.

The voltage magnitude of the IPMNC output is much lower than that of the piezoelectric element. However, a fair comparison cannot be made from these data since they were actuated in different modes. However, note that the break frequency for the IPMNC cantilevered beam occurs at approximately 0.03 Hz. This frequency is significantly better than the 10-Hz break frequency for the comparably sized piezoelectric patch for low-frequency accelerometer applications. Recall that this low break frequency was achieved without any conditioning electronics. However, the IPMNC response begins to roll off above 40 Hz. This high-frequency limit, which depends on the patch's geometry, must be taken into consideration when designing IPMNC accelerometers for specific applications.

7.3.5 Discussion and Conclusions

Based on the frequency response spectrum of an IPMNC patch under cantilever excitation, it seems likely that this material would be useful in low-frequency accelerometer applications. Unlike piezoelectric elements, the IPMNC specimen tested produced a useful output in the sub-hertz frequency range without conditioning electronics. Care must be taken when designing circuits utilizing the material as an active element due to their extreme sensitivity to moisture. In response to this concern, Shahinpoor has developed a polymer-coated IPMNC patch that severely retards moisture evaporation, allowing for consistently performing devices. However, such specimens' moisture-retaining performance must be quantified in detail before device shelf-life estimations may be made.

The nature of the material requires that accelerometers utilizing IPMNC active elements must include a mechanism to induce bending in the material under acceleration. This mechanism may take the form of a cantilever in the manner of the experiments described here, a drumhead type membrane with a mass in the center such as that discussed by Sadeghipour et al. (1992) or even a low-profile flat spring.

Overall, IPMNC active elements enable near-D.C. acceleration measurement devices with modest power, volume, mass and complexity requirements, provided their unique properties are accounted for in the design. The advantages of IPMNC over conventional piezoelectric approaches open up the possibility of thin-film or extremely lightweight accelerometers or, if integrated, devices with rough position-sensing capability.

7.3.6 Advances in Sensing and Transduction

A recent paper by Shahinpoor (2004d) presents a review of IPCNCs' sensing and transduction properties. In 1995 and 1996, Shahinpoor reported that by themselves and not in hydrogen pressure electrochemical cells as reported by Sadeghipour and coworkers (1992), IPMNCs can generate electrical power like an electromechanical battery if flexed, bent or squeezed. He also reported the discovery of a new effect in ionic polymeric gels – namely, the *ionic flexoelectric* effect in which flexing, compression or loading of IPMNC strips in the air created an output voltage like a dynamic sensor or a transducer converting mechanical energy to electrical energy. Keshavarzi and colleagues (1999) applied the transduction capability of IPMNCs to measuring blood pressure, pulse rate and rhythm measurement using thin sheets of IPMCs.

Motivated by the idea of measuring pressure in the human spine, Ferrara et al. (1999) applied pressure across the thickness of an IPMNC strip while measuring the output voltage. Typically, flexing such material in a cantilever form sets them into a damped vibration mode that can generate a similar damped signal in electrical power (voltage or current), as shown in Figure 7.3. The experimental results for mechanoelectrical voltage generation of IPMNCs in a flexing mode are shown in Figure 7.17(a) and (b). Figure 7.17(a) also depicts the current output for a sample of thin sheets of IPMCs. Figure 7.16(b) depicts the power output corresponding to the data presented in Figure 7.17(a).

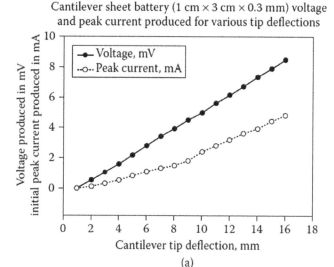

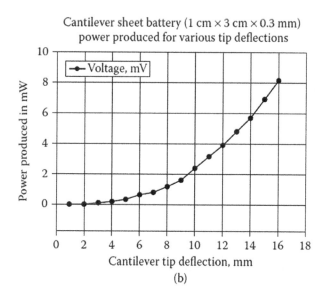

FIGURE 7.17 Typical voltage/current output (a) and power output (b) of IPMNC samples.

The experimental results showed that almost a linear relationship exists between the voltage output and the imposed displacement of the IMPC sensor (Figure 7.17). IPMC sheets can also generate power under normal pressure. Thin sheets of IPMC were stacked and subjected to normal pressure and normal impacts and were observed to generate a large output voltage. Endo-ionic motion within IPMC thin-sheet batteries produced an induced voltage across these sheets' thickness when a normal or shear load was applied.

A material testing system (MTS) was used to apply consistent, pure compressive loads of 200 and 350 N across the surface of an IPMNC 2×2-cm^2 sheet. The output pressure response for the 200-N load (73 psi) was 80 mV in amplitude; for the 350-N load (127 psi), it was 108 mV. This type of power generation may be useful in the heels of boots and shoes or places where there is a lot of foot or car traffic. Figure 7.18 depicts the output voltage of the thin sheet IPMNC batteries under a 200-N normal load. The output voltage is generally about 2-mV/cm length of the IPMNC sheet.

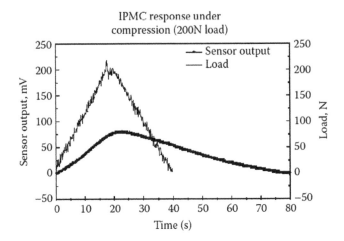

FIGURE 7.18 Outvoltage due to the normal impact of 200-N load on a 2-cm × 2-cm × 0.2-mm IPMNC sample.

7.4 SIMULATION AND CONTROL OF IONOELASTIC BEAM DYNAMIC DEFLECTION MODEL

An effort to model the dynamic motion of an IPMNC elastic beam was undertaken. The development of the model's static portion was begun by assuming the beam behaves following the nonlinear equation used to describe the large-angle deflection of elastic cantilever beams. A Simulink simulation was developed to estimate the final deflection of a beam due to a constant moment. The model's dynamic portion was developed by assuming that each beam segment could be represented as a simple second-order system.

7.4.1 INTRODUCTION TO MODELING OF ELASTIC BEAMS

This section presents a summary of the effort to model the deflection dynamics of ionoelastic beams made of IPMCs. Strips of these composites can undergo large bending and flapping displacement if an electric field is imposed across their thickness. IPMNC beams show large deformation in the presence of low applied voltage and exhibit low impedance. They have been modeled as capacitive and resistive element actuators that behave like biological muscles and provide an attractive actuation as artificial muscles for biomechanics and biomimetic applications. Essentially, the polyelectrolyte membrane inside the composite possesses ionizable groups on its molecular backbone. These groups have the property of disassociating and attaining a net charge in a variety of solvent media. In particular, if the interstitial space of a polyelectrolyte network is filled with liquid containing ions, then the electrophoretic migration of those ions inside the structure due to an imposed electric field can cause the macromolecular network to deform accordingly.

7.4.2 STATIC DEFLECTION

The ionoelastic beam is depicted as an elastic cantilever beam, shown in Figure 7.19. The nonlinear equation for large-angle deflections in elastic cantilever beams is:

$$\frac{(\partial^2/\partial x^2)(v)}{\left(1+(\partial v/\partial x)^2\right)^{3/2}} = -\frac{M}{EI} \tag{7.7}$$

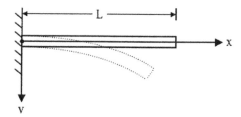

FIGURE 7.19   IPMNC cantilever beam.

Equation (7.7) can be rearranged algebraically to:

$$\frac{\partial^2}{\partial x^2}(v) = -\frac{M}{EI}\left(1+\left(\frac{\partial v}{\partial x}\right)^2\right)^{3/2} \tag{7.8}$$

Solving Equation (7.7) for a constant moment, M, will produce a function, $v(x)$, that is the beam deflection of each point on the beam as a function of the wall's distance. Determining the solution for Equation (7.7) subject to a constant moment, M, can be accomplished in five steps:

1. Change the independent variable (temporarily) from x to t:

$$\frac{\partial^2}{\partial t^2}(v) = -\frac{M}{EI}\left(1+\left(\frac{\partial v}{\partial t}\right)^2\right)^{3/2} \tag{7.9}$$

   This can be rewritten as

$$v'' = -\left(\frac{M}{EI}\right)\left[1+\left(v'\right)^2\right]^{3/2} \tag{7.10}$$

2. Change the second order differential Equation (7.10) into a set of first-order differential equations:

$$\text{set } x_1 = v \tag{7.11}$$

$$\text{then } x_1' = v' \tag{7.12}$$

$$\text{set } x_2 = x_1' = v' \tag{7.13}$$

then

$$x_2' = x_1'' = v'' = -\left(\frac{M}{EI}\right)\left[1+\left(v'\right)^2\right]^{3/2} \tag{7.14}$$

or

$$v'' = -\left(\frac{M}{EI}\right)\left[1+\left(x_2\right)^2\right]^{3/2} \tag{7.15}$$

3. Model the set of first-order differential equations using Matlab's Simulink tool, as shown in Figure 7.20.
4. Integrate the Simulink model. Simulink defaults to a fourth-order Runge-Kutta integration method, which was used for this effort. The time step used during integration was

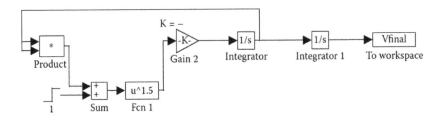

FIGURE 7.20   Simulink model.

constrained to be constant since this would correspond (after the transformation described in step 5) to a fixed-length fraction of the beam.

5. Change the independent variable back to x from t:

$$v(x) = v(t) \tag{7.16}$$

A hypothetical deflection case, subject to the following boundary conditions, was run through the five-step modeling process and resulted in the beam deflection shown in Figure 7.21:

$$M = 0.5 \; v(t_0) = 0 \tag{7.17a}$$

$$E = 1 \; v'(t_0) = 0 \tag{7.17b}$$

$$I = 1 \; v''(t_0) = 0 \tag{7.17c}$$

$$L = 1 \; dx = 0.01 \tag{7.17d}$$

One of the implications of solving Equation (7.7) is the "stretching" of the beam. As depicted in Figure 7.22, the beam tip will deflect a distance, v, and the resulting curved beam will be longer by a small distance than it originally was (though no extension force has been placed on the beam).

To rectify this discrepancy, the straight-line displacement of each portion of the beam is summed together until the curved beam's length equals that of the original beam, as shown in Figure 7.23. Once that length is reached, the rest of the beam is not included in the deflection plot. As a result of

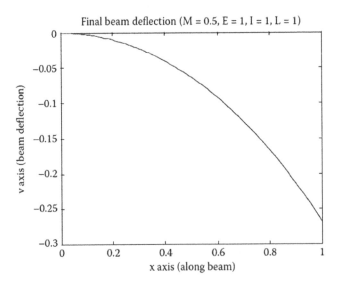

FIGURE 7.21   Hypothetical deflection plot.

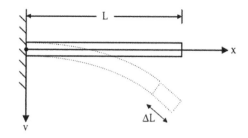

FIGURE 7.22   Beam, "lengthening".

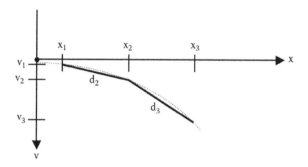

FIGURE 7.23   Segmented beam.

this beam "reshortening", the deflection of the hypothetical beam shown in Figure 7.23 will eventually look as shown in Figure 7.24.

For the static case, there is a distinct final deflection solution for each moment value. The plot in Figure 7.25 shows the distinct solutions for various moment values.

7.4.3   DYNAMIC CASE

The previous section describes the initial and final positions of the beam but does not describe the dynamics it undergoes to reach the final position. This section will describe the effort to build a dynamic model.

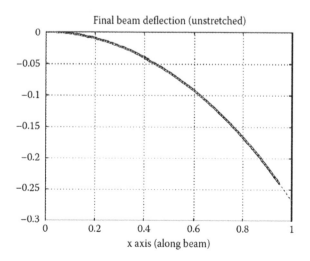

FIGURE 7.24   Unstretched hypothetical deflection plot.

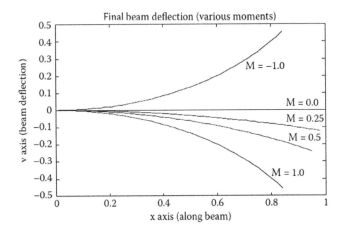

FIGURE 7.25 Deflections due to various moments.

For a step input of 2 V on a 1 × 0.25 in.²-strip (0.2-mm thick), the moment will be constant; such a constant moment will produce a distinct final deflection. The step response of the tip to such a deflection command is shown in Shahinpoor (2003). From that step response, it would appear that the beam could be modeled as a simple second-order transfer function.

The step response of a second-order transfer function of the following form will approximate the step response if $\omega = 0.364$ Hz and $\zeta = 0.3$:

$$G(s) = \frac{\omega^2}{s^2} + 2\zeta\omega s + \omega^2 \tag{7.18}$$

The step response of this model is shown in Figure 7.26.

The approximate transfer function's step response does correspond to that presented in Shahinpoor and Alvarez (2002). The only significant discrepancy is in the maximum value during the initial portion of the step response. Shahinpoor and Alvarez (2002) report the maximum value at 0.170; the approximate solution reports the maximum value at 0.146.

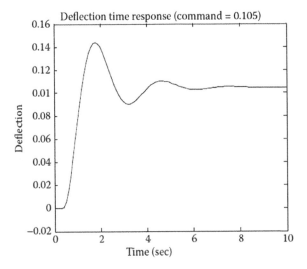

FIGURE 7.26 Beam tip step response.

The frequency-domain parameters that characterize the model's open-loop transfer function should be assumed to apply to only the 1×0.25-in.^{2} beam. The natural frequency and damping coefficient of beams with other lengths, widths and thickness will undoubtedly be very different. The function that relates those frequency domain parameters to beam dimensions was not researched during this modeling effort; an effort to determine this function must be undertaken before model development can be considered complete.

Once the shape of the tip's open-loop transfer function was determined, a rather significant assumption was made. For model development purposes, it was assumed that each point on the beam responded like the tip; that is, the beam was a set of second-order transfer functions responding independently to their final deflection commands. Given that assumption, the process to simulate the entire beam's deflection dynamically is detailed graphically in Figure 7.27.

In this graphic, the modeling input is the moment value (assumed constant for this portion of the effort). The moment is used in the nonlinear large-angle deflection equation; the final deflection position is the solution to that equation and is in the form of a single vector (with individual elements for each portion of the curved beam). Each of the final deflection vector elements is used as a step command input into its second-order open-loop transfer function. The independent transfer functions' outputs are the individual deflection time responses (one for each portion of the curved beam). Those outputs are collected into a deflection time response matrix; each column of that matrix represents the beam deflection at a certain simulation time.

Figure 7.28 depicts the beam deflection as a function of time for the hypothetical case initially described in support of Figure 7.21. The figure's darker line represents the final deflection vector command (solution to the nonlinear large-angle deflection equation).

7.4.4 VARIABLE MOMENTS

The previous sections have all constrained the input moment to a constant value. The model developed, however, will work equally well under variable moment conditions. To model the beam deflection properly under variable moment conditions, the moment at each time step is used to create a final deflection command vector each time step. That variable command vector replaces the constant step command used as an input to the open-loop transfer functions.

To develop the variable moment deflection plot shown in Figure 7.29, a square wave – switching from $+0.5$ to -0.5 and back every second for 10 s – was used as the moment input. The resultant motion is the expected "flapping" motion seen during hardware tests.

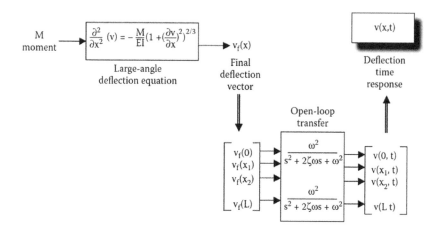

FIGURE 7.27 Beam modeling process.

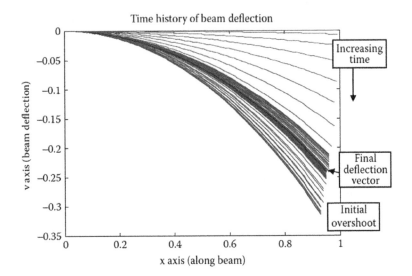

FIGURE 7.28   Time history of beam deflection.

The beam model presented in the last sections is only an initial model. To upgrade this model's fidelity, additional work in three areas must be performed as outlined next.

7.4.4.1   Moment Modification

The simplifying assumption inherent to implementing the large-angle deflection Equation (7.5) is that the moment, which produces the deflection, is simply the moment at the wall. Though this assumption is reasonable for a first-order model, it is easy to appreciate the errors induced by making such an assumption. Shahinpoor (2000, 2002e, 2002f) has shown that the electric field produces a unique moment upon each beam segment.

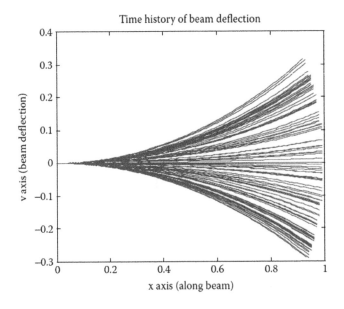

FIGURE 7.29   "Flapping" beam.

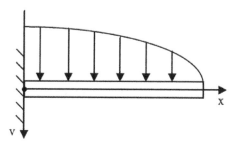

FIGURE 7.30 Diagram of the moment induced by the electric field.

The moment produced by an electric field can be approximated using a parabola with a maximum value near the wall and a minimum value (zero) at the beam tip. Figure 7.30 presents a hypothetical moment.

The current model implementation reduces the complex moment model to a single composite moment at a given distance from the wall. The equivalence of those representations is presented in Figure 7.31. The moments at the wall for both cases in Figure 7.31 are equal, so both beams' deflection would be identical. Yet, even a cursory analysis of each of those cases' steady-state deflection would show that the beams would indeed deflect very differently.

The future work that must be undertaken in this area involves dividing the beam into numerous small segments, calculating each segment's deflection due to the moment applied there and integrating the individual segment dynamics to produce a composite beam deflection picture.

7.4.4.2 Extension

The dynamics portion of the first-order model is currently tied to the step response of the 1×0.25-in.^{2} beam. To make the model usable for a larger swath of the beam population requires that three specific tasks be performed:

1. Data describing the step responses of a representative number of different length and width beams must be collected and archived.
2. The frequency and damping coefficient parameters that would produce similar step responses for each of the beams must be selected.
3. The simple functions that relate the model's parameters of all the various beams (dependent only upon each beam's dimension) must be derived.

7.4.4.3 Validation

Finally, and most importantly, the upgraded first-order model's performance must be validated against actual beam performance. Several beams' dynamic responses must be run; conditions must be inputted into the model to drive equivalent simulation runs. The data collected and archived during the muscle and simulation tests must then be compared toward modifying model parameters or

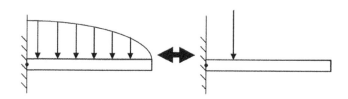

FIGURE 7.31 Hypothetical moment simplification.

algorithms until the simulation runs match the beam responses throughout the population of potential beam dimensions and forcing functions.

7.4.5 SUMMARY

An effort to model the dynamic motion of an IPMNC elastic beam was undertaken. The development of the model's static portion was begun by assuming the beam behaves following the nonlinear equation used to describe the large-angle deflection of elastic cantilever beams.

A Simulink simulation was developed to estimate the final deflection of a beam due to a constant moment. The model's dynamic portion was developed by assuming that each beam segment could be represented as a simple second-order system. Future efforts to model beam motion more accurately by modifying the method used to model the forcing moment, by expanding the model to predict the performance of beams of all dimensions and by validating model performance against actual beam motion were recommended

7.4.6 FEEDBACK CONTROL IN BENDING RESPONSE OF IPMNC ACTUATORS

One of the disadvantages of ionic polymer materials is that their rather slow time constant limits the actuation bandwidth. A feedback controller has been developed by Mallavarapu and colleagues (Mallavarapu and Leo, 2001) to reduce the open-loop response time of cantilevered actuators from 4 to 10 s to closed-loop response time from 0.1 to 1.5 s using linear quadratic regulator (LQR) control.

Figure 7.32 shows an open-loop response for a $40 \times 10 \times 0.2$-mm^3 IPMNC actuator for an unsupported length of 30 mm. The inset in Figure 7.29 shows the resonant modes on closer observation of the open-loop step response. Figure 7.33 shows the control of a $10 \times 20 \times 0.2$-mm^3 IPMNC actuator polymer.

This section demonstrates the use of feedback control to overcome these resonance modes. An empirical control model was developed after measuring the open-loop step response of a $40 \times 10 \times 0.2$-mm^3 IPMNC actuator in a cantilever configuration. A compensator was designed using a linear observer-estimator in state space. The design objectives were to constrain the control voltage to less than 2 V and minimize the settling time using feedback control. The controller

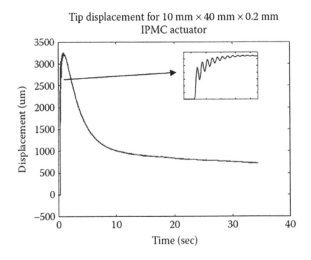

FIGURE 7.32 Tip displacement of IPMNC actuator for 1 V.

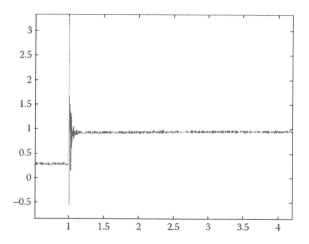

FIGURE 7.33 Experimental closed-loop tip response for $10 \times 20\text{-mm}^2$ IPMNC actuator for a step voltage of 1 V.

was designed using LQR techniques, which reduced the number of design parameters to one variable. This LQR parameter was varied, and simulations were performed, which reduced the closed-loop settling time. The controller was later used in experimentation to check simulations. Results obtained were consistent to a high degree. The electromechanical impedance of five sample actuators was measured.

The test's purpose was to determine the voltage-to-current relationship in the actuator over the frequency range from 5 to 1000 Hz. Multiple actuators were tested to determine the uniformity of the impedance properties over the surface of the sheet sent to us.

7.4.7 RESULTS

Parts a and b in Figure 7.34 are pictures of the test setup used by Mallavarapu and colleagues (Mallavarapu and Leo, 2001). The setup consisted of a small fixture to hold the IPMNC samples. The fixture was electroded and connected to a BNC jack to facilitate actuator testing. An HP 4192A LF (for low frequency) impedance analyzer was used to collect the impedance data.

Five samples from the Artificial Muscle Research Institute were cut from the material sent to Mallavarapu and coworkers at Virginia Tech. The samples were labeled A1 through A5 and placed

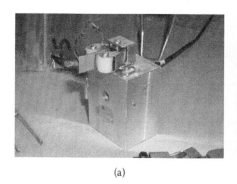

(a)

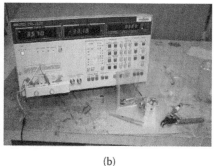

(b)

FIGURE 7.34 (a) Fixture for impedance test; (b) impedance analyzer.

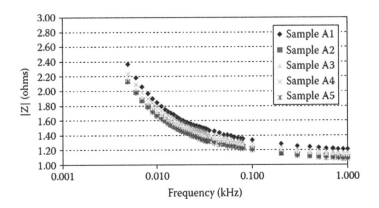

FIGURE 7.35 Measured impedance at an input voltage of 0.5 V.

in water-filled baggies to keep them hydrated. Each sample was tested in the impedance analyzer by manually sweeping through a range of frequencies and measuring the magnitude and phase of the impedance at each frequency.

The impedance of the samples was on the order of 2–5 Ω. The test fixture impedance was on the order of 1 Ω; therefore, the measured data were corrected to account for the fixture's impedance. Assuming that the sample was in series with the fixture, the data were corrected by first transforming the measured data into real and imaginary components through the expressions:

$$\mathrm{Re}\left(Z_{\mathrm{meas}}\right) = \left|Z_{\mathrm{meas}}\right|\cos\left(\angle Z_{\mathrm{meas}}\right)$$
$$\mathrm{Im}\left(Z_{\mathrm{meas}}\right) = \left|Z_{\mathrm{meas}}\right|\sin\left(\angle Z_{\mathrm{meas}}\right) \tag{7.19}$$

The data were corrected by subtracting the fixture's impedance from the actuator's measured impedance and transforming it back into magnitude and phase.

$$\mathrm{Re}\left(Z_{\mathrm{corr}}\right) = \mathrm{Re}\left|Z_{\mathrm{meas}}\right| - \mathrm{Re}\left(Z_{\mathrm{fixture}}\right)$$
$$\mathrm{Im}\left(Z_{\mathrm{corr}}\right) = \mathrm{Im}\left|Z_{\mathrm{meas}}\right| - \mathrm{Im}\left(Z_{\mathrm{fixture}}\right) \tag{7.20}$$

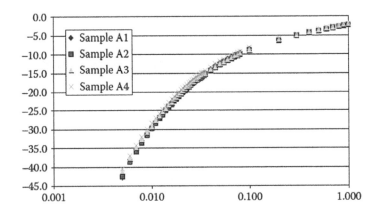

FIGURE 7.36 Phase of the electromechanical impedance at 0.5 V.

Figures 7.35 and 7.36 are the magnitude and phase plots for the corrected data. The impedance magnitude varies between 1.1 and 2.4 Ω over the frequency range of 5 Hz–1 kHz. The reactive component of the impedance is approximately equal to the active component below approximately 10 Hz.

7.5 CONCLUSIONS

1. The electromechanical impedance of the ionic polymer material was consistent over the five samples tested.
2. The samples' impedance had a significant reactive component in the range of 5–10 Hz but became primarily real (resistive) above approximately 100 Hz. The resistance of the samples was on the order of 1.2 Ω in the range of 100–1000 Hz.
3. The samples were highly capacitive at low frequencies. This attribute could complicate the development of actuator power electronics.

8 Conductive or Ion-Conjugated Polymers as Artificial Muscles

8.1 INTRODUCTION

This chapter offers a brief presentation on the impact of conductive or conjugated polymers on the general field of artificial and synthetic muscles. Certainly, the pioneering work and discoveries of the three Nobel laureates in chemistry in 2000 –Alan J. Heeger (Noble Prize Lecture, 2001), Alan MacDiarmid (Noble Prize Lecture, 2001) and Hideki Shirakawa (Noble Prize Lecture, 2001) – in the field of conductive polymers and synthetic metals paved the way to current knowledge and discoveries on conductive polymers, as can also be evidenced in the early papers of Shirakawa et al. (1977), Chiang et al. (1977, 1978) and McGehee and coworkers in *Twenty Years of Synthetic Metals* in 1999.

Following Shahinpoor (who, as early as 1991, presented biomimetic robotic fish equipped with ionic polymers as undulating fin and artificial muscles), Otero and colleagues (1992a, 1992b) were the first to discuss the properties of polypyrrole as a conductive polymer actuator that mimicked natural muscles and was named an artificial muscle. They also discussed the electrochemomechanical phenomena involved in such electrochemical reactions. *Conjugated polymers*, also known as *conducting polymers*, are distinguished by alternating single and double bonds between carbon atoms on the polymer backbone. The conjugated polymer with the simplest chemical structure is polyacetylene, shown in Figure 8.1. Figure 8.2 depicts the molecular structure of a more popular conductive polymer polypyrrole.

Note that conjugated polymers are organic semiconductors and have a bandgap. They can emit light, the color of which can be tailored through the chemical structure. They can generate a current upon absorbing light and thus can be used in photovoltaic devices as well. The conductivity of conductive or conjugated polymers depends on the doping level. They are usually p-doped. The dopant concentration is about 25–30% for polypyrrole. The doping level depends on the oxidation state of the polymer, which can be electrochemically controlled. For actuator applications, one has to change the oxidation level by the application of a potential. Thus, many materials change properties, including their volume, color, mechanical properties and hydrophobicity.

Note that conjugated conductive polymers (CCPs) can act biomimetic and like artificial muscles due to the requisite requirements pertaining to the collapse of the internal network structures due to electronic jump between macromolecular chains. Therefore, conducting conjugated polymers formed by polymers' principal families, such as polyacetylenes, polypyrroles (Ppy), polythiophenes, and polyanilines (PANi), has been the focus of some recent research and development on artificial muscles. The electronic conductivity in the ion-conjugated polymers is essentially due to electrons' and ions' ability to jump between polymeric molecular chains. The presence of dopant agents, which modify electrons' local density on the electronic valence bands, causes such electronic jumps. The dopants known as type "p" remove electrons from a valence band, leaving the molecule positively charged or oxidized. The "n" dopants add electrons to the electronic valence band, so the molecule's net charge will be negative, or the polymer will be in a reduced state.

Thus, similar to ionic polymer conductor nanocomposites, such deformations in conductive polymers are governed by oxidation-reduction (REDOX) processes. Thus, the conjugated polymer can be oxidized (p-doped) or reduced (n-doped) by introducing positive or negative ions or photons. These changes are all electrochemically controlled. The neutral state, the reduced state, the oxidized state or any other intermediate state of the polymer can be reached to apply the appropriate electric potential.

DOI: 10.1201/9781003015239-8

FIGURE 8.1 Simple structures of polyacetylene alternating single and double bonds between carbon atoms.

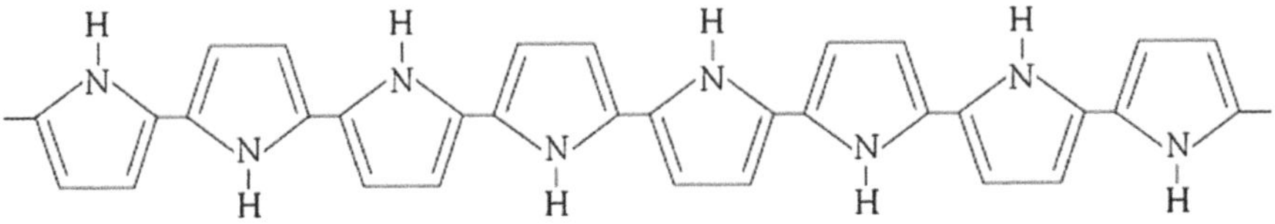

FIGURE 8.2 Molecular structure of a simple polypyrrole conductive polymer.

8.2 DEFORMATION OF CONDUCTING OR CONJUGATED POLYMERS

There has been a tremendous number of pioneering works on this subject by Baughman and coworkers from 1990 through 1996 and Otero and coworkers from 1990 through 1997 on conducting polymers as artificial muscles; by Smela, Pei, Inganäs and Lundström on microactuators in the form of bending bilayer strips built from polyaniline for artificial electrochemical muscles; and De Rossi, Della-Santa and Mazzoldi and coworkers from 1990 through 1998 on characterization and modeling of conducting polymers for muscle-like linear actuator applications.

It should be noted that Burgmayer and Murray first reported the deformational change in the CCP in 1982. Using PPy membranes, they showed that certain ions' membrane permeability could be changed by two orders of magnitude under polarization at different potentials. The volume changes in the CCP are due to the ionic movement produced during an electrochemical reaction. For example, if a p-doped polymer such as PPy is oxidized, there are two possibilities to maintain electroneutrality. If the polymer is doped with a mobile and small anion in an electrolytic solution with both mobile cations and anions, the insertion and desertion of the anions during the REDOX reaction will maintain the electroneutrality (Equations 8.1 and 8.2 and Figure 8.3) (see Baughman et al., 1991; Burgmayer et al., 1982; Matencio et al., 1995; Otero, 1998):

$$PPy^+(CLO_4^-) + e^- \underset{oxidation}{\overset{reduction}{\rightleftharpoons}} PPy^0 + CLO_{4(aq)}^- \tag{8.1}$$

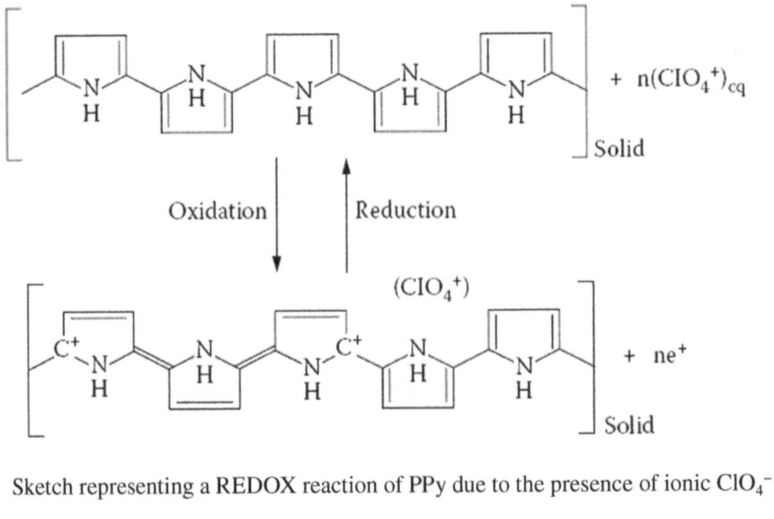

FIGURE 8.3 Sketch representing a REDOX reaction of PPy due to the presence of ionic CLO_4^- anions.

Suppose the polymer is doped with a nonmobile and big anion as dodecyl benzene sulfonate anion (DBS−) into an electrolyte containing mobile and small cations X+. In that case, cations will be inserted and de-inserted during the REDOX reaction to maintain the electro neutrality:

$$PPy^+(DBS^-) + X^+_{aq} \xrightleftharpoons[\text{oxidation}]{\text{reduction}} PPy^0(DBS^-X^+) \qquad (8.2)$$

Note that chemical Equation (8.1) shows that in the first case, the CCP is expanded during the oxidation phase due to the insertion of ClO_4^- anions. In the second case (chemical reaction 8.2), the polymer is expanded during the reduction by the insertion of X^+ cations.

These investigations showed that the ionic flow in the CCP depends on cation and anion size, the thickness of the CP film, applied voltage and the time scale.

Upon application of a voltage or imposition of an electric field, a volume change is induced because the oxidation state of CCP is modified. During the volume increase of CCP, the inserted ions and their hydrated shells occupy any available free space. When the inserted anions or cations get out of the polymer, the network shrinks and expands when they get in. According to the properties of volume change observed in the CCP, Baughman et al. (1991) reported on the possibility of using this volume change due to charge insertion for mechanical actuation.

The behavior of polypyrrole is dramatically altered with chemical doping. The p-doped polypyrrole conductive conjugated polymers have several applications in electrochromic devices, rechargeable batteries, capacitors, ionic membranes, charge dissipation and electromagnetic shielding. Generally, polypyrrole (PPy) is partially oxidized to produce p-doped materials, as shown next.

Otero et al. (1992b) and Otero (1998) reported how a bilayer of $PPy(ClO_4^-)$ with another neutral, flexible polymer substrate could bend in a cantilever fashion in $LiClO_4$ electrolyte solution. Otero and colleagues further studied this bilayer's behavior in solutions of acetonitrile/$LiClO_4$, propylene carbonate/$LiClO_4$ and water/$LiClO_4$. As one of the pioneers in this field, Otero and coworkers (1992a, 1992b, 1993, 1994, 1995, 1996a, 1996b, 1997, 1998) have developed a model explaining the volume change in the PC, taking into account the electrostatic repulsions between charged polymeric chains (Otero et al.,1992a, 1992b, 1993, 1994). According to this model, when a CCP like polypyrrole (PPy) is subjected to an oxidation reaction, positive charges are generated along the polymeric chains. These positive charges produce electrostatic repulsions between them. Due to these repulsions, some conformational changes are generated in the polymeric structure.

Tourillon and Garnier (1984) also demonstrated that the anions' expulsion accompanies these polymer network reconfigurations into the electrolytic medium by XPS measurements. Pei and Inganäs (1993) have studied the behavior of PPy(DBS−) films. In this case, they observed an increase in mass when hydrated cations were inserted into the CCP network and decreased mass when they were taken out. Thus, the CCP became swollen during the reduction and shrunk during the oxidation. Smela and Gadegaard (2001) have studied the *in situ* volume change in these PPy films by atomic force microscopy (AFM).

They found that the film thickness increased between 30 and 40% in the reduced state compared to the oxidized state (Smela and Gadegaard, 2001).

Polyaniline (PANI) is another commonly used CCP for robotic actuator applications. Okabayashi et al. (1987) determined that a volume variation in PANi according to its oxidation state can occur. They observed the weight change of PANi in propylene carbonate/$LiClO_4$ during a REDOX reaction by an electrogravimetric method. The polymer's weight increased during the oxidized doped (with ClO_4^- anions solvated) up to eight times. The emeraldine form of PANi also can be electrochemically oxidized or reduced in an aqueous or acidic environment, resulting in pernigraniline (PS) and leucoemeraldine (LS) salts, respectively, as is shown in Figure 8.4. The REDOX reaction occurs with protons and electrons' motion in strong acid (pH <3).

The addition of protons and electrons in nitrogen is observed during the reduction; this leads to the phenyl ring changing to quinonoid structures and vice versa during oxidation and reduction,

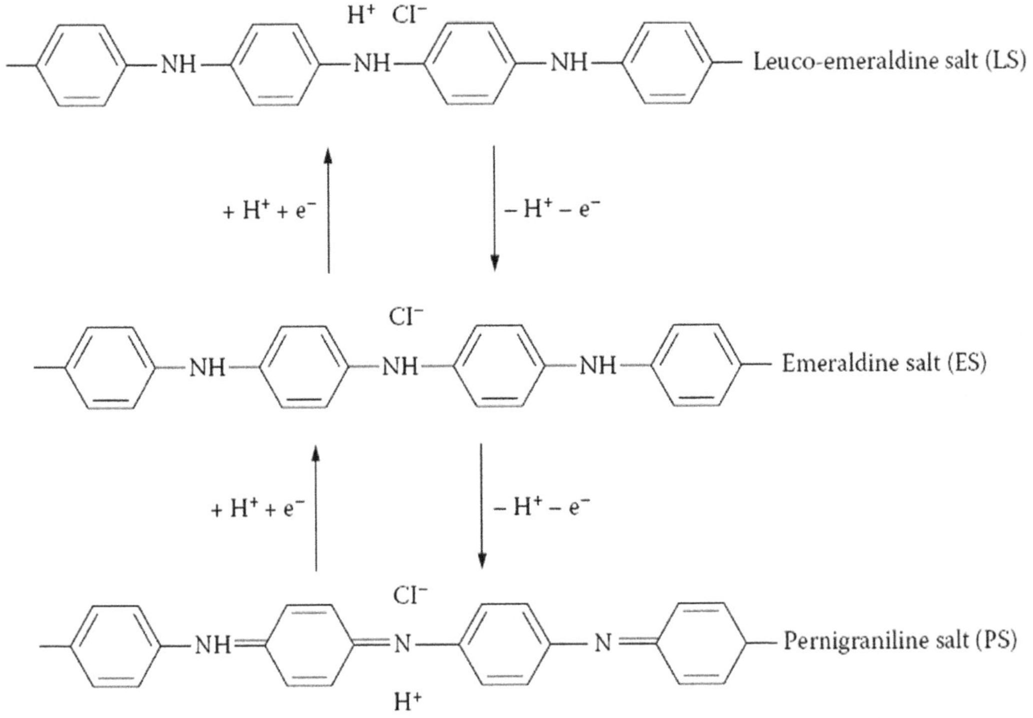

$$\text{H}^+ \quad \text{Cl}^-$$

—NH—⟨⟩—NH—⟨⟩—NH—⟨⟩— Leuco-emeraldine salt (LS)

$+ \text{H}^+ + e^-$ ↑ $- \text{H}^+ - e^-$ ↓

$$\text{Cl}^-$$

—NH—⟨⟩—NH—⟨⟩—NH—⟨⟩— Emeraldine salt (ES)

$+ \text{H}^+ + e^-$ ↑ $- \text{H}^+ - e^-$ ↓

$$\text{Cl}^-$$

—NH=⟨⟩=N—⟨⟩—N=⟨⟩— Pernigraniline salt (PS)

$$\text{H}^+$$

FIGURE 8.4 REDOX cycle of PANi in HCl aqueous solution. The emeraldine salt is oxidized into perni-graniline (PS) salt or reduced into leucoemeraldine (LS) salt. (From Tourillon, G. and F. Garnier. 1984. *J. Electroanal. Chem.* 161:51.).

respectively (Figure 8.5). In this case, the structural changes (phenyl to quinonoid to phenyl) lead to deformation and strain in PANi. Kaneto et al. (1995) have shown that PANi is more compact in the reduced state than in the oxidized state.

Polythiophene (PT) and its derivatives are not as well studied as PPy and PANi, but they are also a subject of research as actuators at present. Most actuators based on PTs have been fabricated from monomers synthesized for particular actuation purposes. Tourillon and Garnier (1984) studied the behavior of substituted PTs in acetonitrile. The swelling of the polymer was observed when this was oxidized. They have demonstrated, for example, that the thickness varies from 160 nm in the undoped state to 200 nm in the oxidized doped state. The doping level was determined by elemental microanalysis to about 25% (one positive charge developed in every four monomeric units).

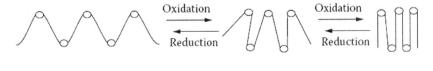

   ○ Hinge molecules (for example, calix [4] arene)

   ╲ Rigid rods: Thiophene chains

FIGURE 8.5 Picture showing the actuation mechanism of a molecular actuator based on PT chains intercon-nected with a highly versatile molecule as a calixarene. (From Anquetil, P. A. et al. 2002. In *Proceedings of SPIE smart structures and materials symposium*, 4695).

The dopants do not modify the electrochemical properties of the PTs, contrary to the case of polypyrrole. This difference has been attributed to different morphology types; polythiophene derivatives show a fibrillar structure, whereas polypyrrole is much more compact. Besides, the doping process becomes more difficult by the steric hindrance increase, leading to decreased conductivity. Unlike PPy and PANi actuators, in PT actuators, the actuation mechanism is not simply due to ion intercalation in the polymer chains. In this class of materials, the actuation results from stacking of thiophene oligomers upon oxidation, producing a reversible molecular displacement, promoting large strains. One possible molecular rearrangement is the formation of dimers by the tendency of orbitals to align due to Pauli's exclusion principle during the material's oxidation.

Figure 8.5 shows the actuation mechanism of a material composed of hinge molecules (as a calixarene) interconnected with seven rigid thiophene chains. The thiophene chains attract to each other in the oxidized state, contracting the material. This strain is reversible during the reduction of the polymer (Anquetil et al., 2002). Thus, these novel polymers hold the promise of improving the speed limits of the PPy and PANi actuators caused by ionic diffusion rates.

9 Engineering, Industrial and Medical Applications of Ionic Polymer–Metal Nanocomposites

9.1 INTRODUCTION

There are numerous potential engineering, scientific and biomedical applications that could benefit from ionic polymer–metal nanocomposites (IPMNCs) as actuators, sensors, energy harvesters, artificial muscles and transducers. In this chapter, some critical applications using IPMCs are presented. Industrial, biomedical and aerospace applications are identified and discussed, along with brief illustrations in the following sections.

It is certainly clear that the extent of applications of ionic polymeric conductor nanocomposites (IPCNCs) and IPMNCs goes beyond the scope of or space allocated to this chapter. However, it will present the breadth and the depth of all such applications of IPMCs, IPCNCs and IPMNCs as biomimetic, robotic, distributed nanosensors, nanoactuators, nanotransducers and artificial/synthetic muscles.

9.2 ENGINEERING AND INDUSTRIAL APPLICATIONS

9.2.1 Mechanical Grippers

IPMCs can be fabricated to act as micro- or macrogrippers (e.g., tweezers when two membranes are wired and sandwiched in a way such that they bend in opposing directions). Figure 9.1 is a perspective view of the mechanical gripper concept showing two treated IPMC actuators packaged as an electrically controlled gripper. The two IPMC actuators are placed parallel to each other, with top surfaces facing each other. The terminals are attached to the top and bottom surfaces of each actuator. Terminals are connected to one pole of the power supply by an electrical wire. The wire length depends on the required gap between the two IPMC actuators depending on the application.

The most important advantage of such IPMC microgrippers originates from their intrinsic material softness relative to conventional actuators. They are considered soft robotic devices.

As seen in Figure 9.1, the fingers are shown as vertical gray bars; the electrical wiring, where the films are connected back to back, can be seen in the middle portion of the figure. Upon electrical activation, this wiring configuration allows the fingers to bend inward or outward, similar to a hand's operation, and thus close or open the gripper fingers as desired. The hooks at the end of the fingers represent the nails' concept and secure the gripped object encircled by the fingers.

To date, multi-finger grippers that consist of two, four and eight fingers have been produced, where the four-finger gripper shown in Figure 9.1 was able to lift a 10.3-g mass. This gripper prototype was mounted on a 5-mm diameter graphite/epoxy composite rod to emulate a lightweight robotic arm.

DOI: 10.1201/9781003015239-9

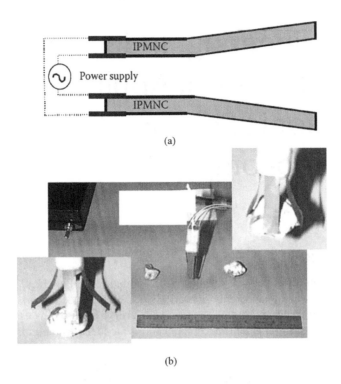

FIGURE 9.1 The IPMC (a) gripper concept and (b) a four-finger gripper.

This gripper was driven by a 5-V square wave signal at a frequency of 0.1 Hz to allow sufficient time to perform a desirable demonstration of the capability of the gripper (opening the gripper fingers, bringing the gripper near the collected object, closing the fingers and lifting an object with the arm). The demonstration of this gripper's capability to lift a rock was intended to pave the way for a future potential application of the gripper to planetary sample collection tasks using ultra-dexterous and versatile end-effectors or to handle soft biological objectives. Interestingly, the work at NASA/JPL (Shahinpoor et al., 1998) reported that the actuation properties of IPMNCs' muscles in a harsh space environment, such as 1 torr of pressure and −140°C temperature, are noticeable for space applications.

9.2.2 THREE-DIMENSIONAL ACTUATOR

Figure 9.2 shows an illustrative view of a three-dimensional IPMC actuator C packaged in three-dimensional form for use with a three-phase generator box M. IPMNC actuator C is a hollow triangular tube configuration consisting of three independent membrane actuators, A, attached and electrically insulated along the long edges and having three outer faces, B. The tube is fixed to the generator box M. One pair of terminals, D, is located on each of the three actuators C for connection to electrodes E incorporated in generator box M.

The IPMNC actuator C is designed to produce a three-dimensional movement by positioning each actuator to be stimulated at a phase angle apart from the adjacent actuator by a low-amplitude alternating signal inducing wobble-like motion around the long imaginary axis of the combined actuator tube in the null position. Each IPMNC actuator has terminal connections to each phase of a typical three-phase power generator (M) or a multiphase power supply (programmable function generators/power supplies exist that have phase-separated outputs). Figure 9.2 details this arrangement. Each of these IPMC actuators has its external and internal faces similar to the gripper's top and bottom faces shown in Figure 9.1.

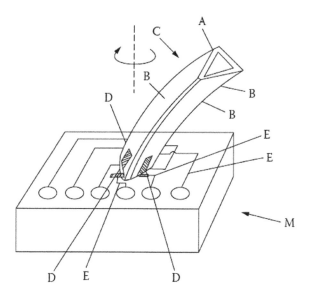

FIGURE 9.2 The three-dimensional IPMNC actuator concept.

Other configurations of three-dimensional motion actuators, such as the four-sided square rod shown in Figure 9.3 and undulating and morphing actuators shown in Figure 9.4, are also possible. The motion produced by any such device would be used to power soft mixers, production line feeders and other task-specific equipment for many industrial usages. Also, in sensing modes, they can be used as joysticks or X-Y locators.

9.2.3 Robotic Swimming Structure

Figure 9.5 shows one embodiment of a robotic swimming structure made by cutting and packaging strips of IPMNCs *A* to desired size and shape and consequently placing an alternating low voltage (a few volts' peak per strip) across the muscle assembly *E*. In this figure, muscle assembly *E* is formed of IPMNC strips *B* encapsulated into an elastic membrane *C* with electrodes *D* imprinted on each

FIGURE 9.3 The fabricated IPMNC in a square rod form.

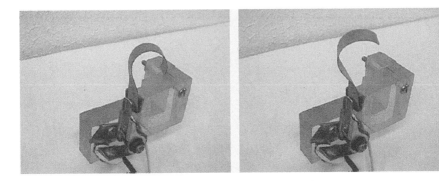

FIGURE 9.4 The undulating and morphing actuator made with an IPMNC.

strip there and with a first end and a second end. Second end *F* is attached to an appropriate electronics and wiring structure *G* for providing guidance and control to actuate the muscle assembly
E. Structure *G* as shown comprises a sealed housing module *H* containing a means for generating a
signal and a means for generating power *J*.

The tail assembly consists of electrically actuated artificial muscles such as IPMNCs cut in tiny
fibers or strips. The tail is then encapsulated in an elastic membrane. The ends of fibers closer to the
head assembly *H* are wired to a miniature printed circuit board (PCB) or similar assembly to a signal
generator assembly consisting of an oscillator circuit and batteries or other power sources. The head

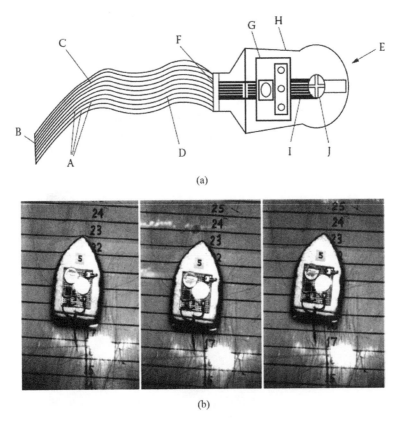

FIGURE 9.5 (a) Robotic swimming structure and (b) swimmer with muscle undulation frequency of 3 Hz
(bottom). The scale shown is in centimeters.

assembly is preferably sealed to protect the circuitry and electronics from the elements. By varying the frequency of the applied voltage to the membrane muscle, the speed of muscle-bending oscillation of muscle assembly E and, therefore, propulsion of the swimming structure can be modulated.

In this manner, robotic swimming fish and submarine structures containing a sealed signal and power-generating module (preferably in the head assembly) can be made to swim at various depths by varying the buoyancy of the structure by conventional means. Remote commands via radio signals can then be sent to modulate propulsion speed and buoyancy. Based on such dynamic deformation design and observed characteristics, a noiseless swimming robotic structure, as shown in Figure 9.5, was constructed and also tested for collective vibrational dynamics.

9.2.4 Biomimetic Noiseless Swimming Robotic Fish

Figure 9.6 presents another arrangement of the IPMNC actuator showing an elastic construction with imprinted electrodes for use as a robotic swimming structure – more specifically, a robotic fish. The figure shows a robotic swimming structure made by cutting and packaging strips of IPMNCs A in two rows of desired size and shape and imprinted with electrodes B spaced throughout and in a single structure. In this figure, muscle assembly structure F is formed of polymer gel strips A and is encapsulated into an elastic membrane C with multiple electrodes B imprinted there, with a first end D (tail) and a second end E (head). Head assembly E contains appropriate electronics and wiring structure H for providing power, guidance and controls to the muscle assembly F. Structure H is contained in a sealed housing module I containing a means for generating a signal M and a means for generating power.

The power source J at end E places an alternating low voltage (a few volts' peak per strip) across the muscle assembly F as shown. Power source J includes an erasable, programmable chip K and batteries. Note the two rows of small actuators in parallel. Each has two terminals connected individually to the multiphase signal generator M located in the head assembly E. Batteries (or other power sources) are also housed in this section for required voltage input. By energizing one pair (across) of actuators at a time and then the consequent pairs downstream, one can produce a propagating or traveling wave downstream on each side of the fish. This will produce a sting-ray type of motion, which propels the swimming structure forward. The middle terminals or spines act as conductors that connect the signal generator outputs in the head assembly to each actuator in the tail or wing assembly.

By varying the frequency of the applied voltage, the speed of muscle-bending oscillation of the membranes A and, therefore, propulsion of the swimming structure F can be modulated. In this manner, robotic swimming fishes and submarine structures containing a sealed signal and power-generating module in the head assembly can be made to swim at various depths by varying the buoyancy of the structure by conventional means. Remote commands via radio signals can then be sent to modulate propulsion speed and buoyancy.

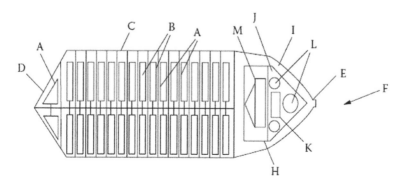

FIGURE 9.6 An illustrative design of robotic fish.

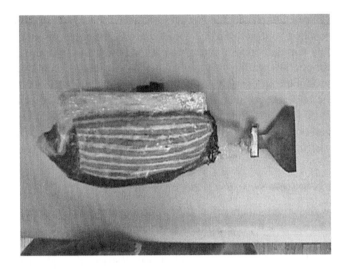

FIGURE 9.7 A robotic fish equipped with a single IPMNC tail fin.

In Figure 9.7, another robotic fish design is presented. This robotic fish, equipped with a tail fin made from a single piece of IPMNC material, has demonstrated that such a structure is feasible for mimicking biological fish locomotion. Furthermore, the noiseless propulsion is attractive in nature. A maximum speed of approximately 2 m/min was achieved under an applied voltage of 2 V.

Actual electrically controllable caudal actuator fins (propulsion, and gross turning and maneuvering) and pectoral actuator fins (fine turning and maneuvering) and remotely controllable stealthy, noiseless, biomimetic swimming robotic fish made with IPMNCs were designed, manufactured and tested. Some of these are shown next. It is also important to consider the optimal design of fish fin actuators such as the ones shown in Figures 9.8 and 9.9. The strategy here is to design different kinds of fins for noiseless fish propulsion. Note that there are five different kinds of fins:

Caudal or tail fin, which is primarily used for propulsion
Dorsal or backfin used for sudden turns and stability
Pectoral fins (paired) on the sides of a fish, primarily used for turning and stability

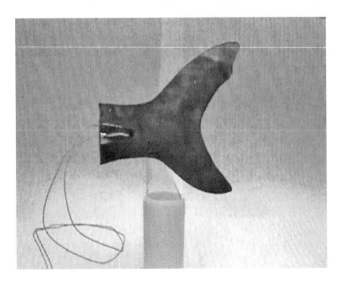

FIGURE 9.8 A designed and fabricated undulating shark caudal fin actuator.

FIGURE 9.9   ERI's biomimetic fish with an emarginated type of caudal fin design.

Pelvic fins (paired) on the sides, primarily used for braking or slowing down the propulsion
Anal fin under the fish, near the belly and the tail, to add stability

Figure 9.10 depicts some natural designs for caudal fins. However, the design of caudal fins may take many shapes, as depicted in Figure 9.11.

Figures 9.12–9.14 depict several designed biomimetic robotic fish that are remotely controllable and some unique caudal fin designs. The fish is designed so that its caudal fin can be replaced and different caudal fins can be tested.

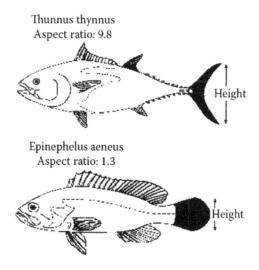

FIGURE 9.10   Some typical naturally evolved designs for caudal fins.

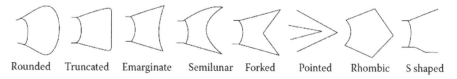

FIGURE 9.11   Some natural designs for caudal fins.

FIGURE 9.12   Another ERI's biomimetic fish with an emarginated type of caudal fin design.

FIGURE 9.13   ERI's biomimetic fish with a shark type of caudal fin design.

FIGURE 9.14   An assortment of ERI's biomimetic robotic fish equipped with IPMCs.

The main applications of such efforts are noiseless propulsion, undulating fins and smart sonar evading skins made with IPMNCs to be used in a noiseless biomimetic swimming robotic fish for naval applications. The requirements include:

- The IPMNC fins must be able to survive in water and must sustain the harsh ocean environment while performing sensing and actuation for propulsion.
- The IPMNC fins must have good force density for propulsion. That is, for a typical caudal fin of 20-cm^2 surface area, an undulating force of 1 N or about 100 gf (gram forces) will be required.
- The IPMNC undulating fin must have a good bandwidth to undulating frequencies of at least 10 Hz.

There are basically no competing technologies with these specifications. Shape memory alloy actuators may come close but are still deficient in bandwidth for undulations and frequency of actuation.

Figure 9.15 depicts the newest generation of completely watertight and impermeable underwater biomimetic robotic fish equipped with IPMNCs.

9.2.5 LINEAR ACTUATORS

Linear actuators can be made to produce a variety of robotic manipulators, including platform type or parallel platform IPMNC actuators, as shown in Figure 9.16. Also, multiple degrees of freedom of motion can be obtained by controlling each IPMNC with a robotic controller. Since polyelectrolytes are for the most part three-dimensional networks of macromolecules often cross-linked nonuniformly, the concentration of certain ionic charge groups is also nonuniform within the polymer matrix. Based on dynamic deformation characteristics, linear- and platform-type actuators can be designed and made dynamically operational.

Other variations in design are also possible. The assortment of linear actuators and the bistrip type linear actuator shown in Figure 9.17 are simple versions of such linear actuators. A film pair weighing 0.2 g was configured as a linear actuator and, when 5 V and 20 mW were used, successfully induced more than 11% contraction displacement. Also, the film pair displayed a significant expansion capability, where a stack of two film pairs 0.2-cm thick expanded to about 2.5-cm wide (see Figure 9.17).

FIGURE 9.15 The newest generation of completely watertight and impermeable underwater biomimetic robotic fish equipped with IPMNCs.

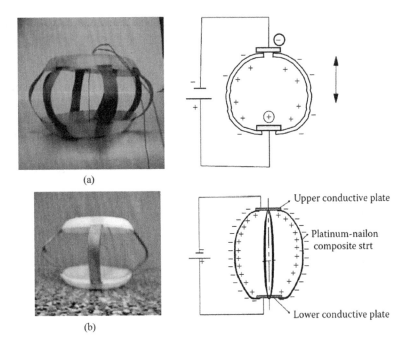

FIGURE 9.16 (a) A photograph of a platform actuator driven by eight IPMNCs. This design can feature two-dimensional motion of the platform. (b) The operating principle is illustrated.

Another possibility is to create long linear actuators by proper placement of electrodes on a cylindrical body of an IPMNC such as the one shown in Figure 9.18(a). Additional design configurations and linear actuators made with IPMNCs are depicted in Figures 9.18(b)–(i).

9.2.6 IPMNC CONTRACTILE SERPENTINE AND SLITHERING CONFIGURATIONS

Some efforts were directed toward creating certain contractile serpentine and slithering artificial muscle configurations for the strips of IPMNC by placing alternating electrodes on the surfaces of IPMNC strips as shown in Figures 9.19 and 9.20.

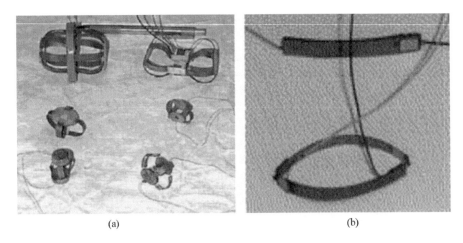

FIGURE 9.17 Assortment of bilinear and linear IPMNC actuators: (a) a reference pair and (b) an activated pair.

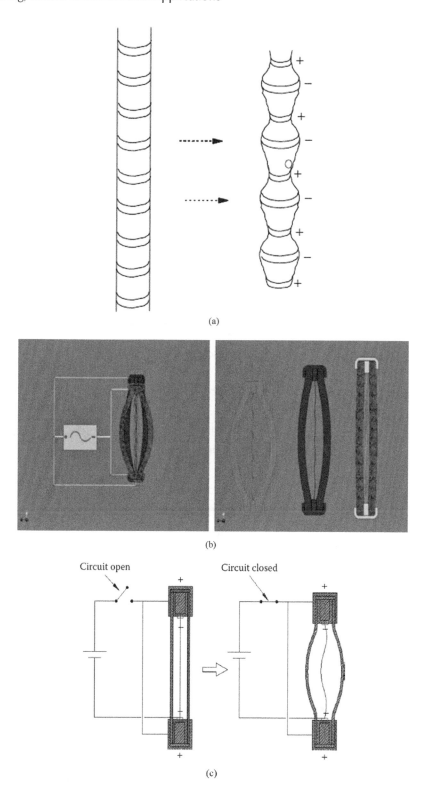

FIGURE 9.18 (a) Schematic representation of an IPMNC cylindrical linear actuator with discretely arranged ring electrodes. (b) Basic design configurations to make IPMNC linear actuators. (c) Basic operational configurations of the IPMNC-based linear actuators. *(Continued)*

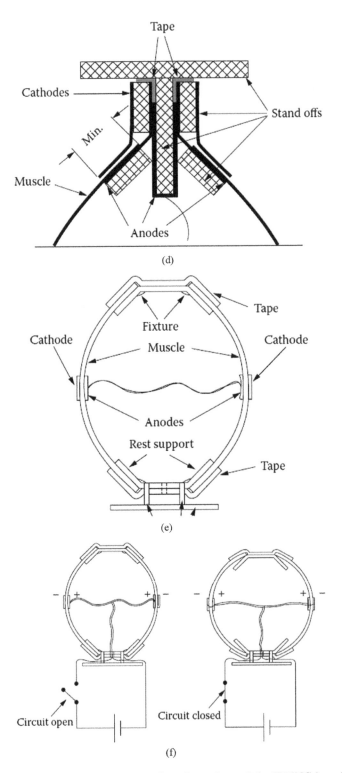

FIGURE 9.18 (d) Basic electrode placement and configurations of the IPMNC-based linear actuators. (e) Alternative design configurations of the IPMNC-based linear actuators. (f) Operational configurations of the alternative design of IPMNC-based linear actuators. *(Continued)*

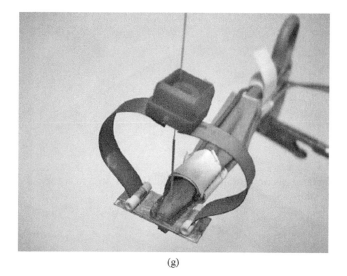

(g)

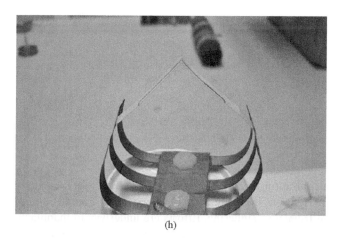

(h)

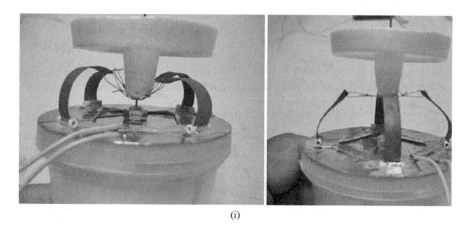

(i)

FIGURE 9.18 *(Continued)* (g) Laboratory prototype configurations of the IPMNC-based linear actuators. (h) Another laboratory prototype of the IPMNC-based linear actuators. (i) IPMNC-based linear actuator producing over 30% linear actuation.

FIGURE 9.19   Interdigitated electrode arrangement on an IPMNC strip to create a serpentine-like contractile and slithering artificial muscle.

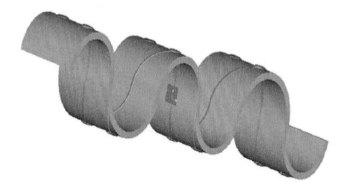

FIGURE 9.20   Another interdigitated electrode arrangement on a slithering IPMNC strip to create a serpentine-like contractile and slithering artificial muscle.

Some actual configurations of slithering IPMNC strips were constructed, as shown in Figure 9.21(a) and (b). However, the results are not yet very encouraging because the stiffness of the strips prevented them from easy slithering. Efforts are underway to manufacture thinner IPMNC strips and to repeat such experiments to observe more profound serpentine-like or snake-like slithering and maneuvering motions of IPMNC strips.

9.2.7   Metering Valves

Metering valves can be manufactured from IPMNCs. By applying a calibrated amount of direct voltage/current to the IPMNC metering valve attached to any tubes and, consequently, varying

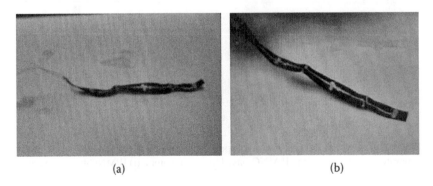

(a)   (b)

FIGURE 9.21   Actual interdigitated electrode arrangement on a slithering IPMNC strip to create a serpentine-like contractile and slithering artificial muscle.

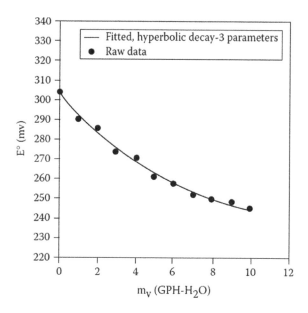

FIGURE 9.22 Metering valve data obtained by a strip of IPMNC undulating in a pipe flow.

the degree of bending displacement of the IPMNC, the control of aqueous fluid flow can be attained. Figure 9.22 depicts a set of data obtained in using an IPMNC strip in a fluttering mode in a pipe flow.

9.2.8 DIAPHRAGM PUMPS USING FLEXING IPMNC STRIPS AND DIAPHRAGMS

Bellows pumps can be made by attaching two planar sections of slightly different sizes of IPMNC sections and properly placing electrodes on the resulting cavity. This permits modulation of the volume trapped between the IPMNCs. The applied voltage amplitude and frequency can be adjusted to control the flow and volume of fluid being pumped.

IPMNC diaphragm pumps can also be made in various ways. Single or multiple IPMNCs can function as the diaphragms that create positive volume displacement. In Figure 9.23, we present a

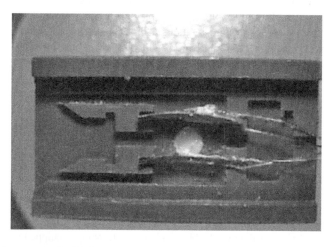

FIGURE 9.23 A photograph of a fabricated double-diaphragm pump. The size of the IPMNC is 1-mm width × 5-mm length × 0.2-mm thickness.

FIGURE 9.24   Perspective view of two (rectangular and circular chamber) double-diaphragm minipumps equipped with IPMNC muscles and an inductive receiving coil.

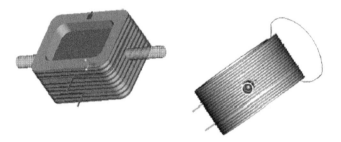

FIGURE 9.25   Side view of the two double-diaphragm minipumps equipped with synthetic muscles and an inductive receiving coil.

miniaturized double diaphragm pump constructed of an IPMNC. Such a pump produces no noise and has a controllable flow rate in the range of a few microliters per minute.

9.2.8.1   Diaphragm Pump Designs

Each of these pump systems includes a pumping chamber with an anterior end attached to an implantable influent conduit. In the case of an ocular pressure control device, the influent conduit is inserted into the anterior chamber of the eye. A flexing ionic polymer conductor nanocomposite (IPCNC) synthetic muscle, which is a type of IPMNC synthetic muscle, functions as the primary actuator. The posterior end of the pumping chamber is connected to an effluent or drainage conduit, which may drain bodily fluids or dispense drugs to an area of the body. Figures 9.24–9.27 depict various configurations of such mini diaphragm pumps with rectangular and circular chambers.

An alternative external power system includes a biocompatible induction coil with gold wire armature that can be transcutaneously activated, adjusted, computer interrogated and controlled

FIGURE 9.26   Exploded view of the two double-diaphragm minipumps equipped with synthetic muscles and an inductive receiving coil.

FIGURE 9.27 Cutaway view of the double-diaphragm minipumps equipped with synthetic muscles and an inductive receiving coil.

by a surgeon. The device of the invention is further equipped with a pair of adjustable variable flow valves placed at the juncture of the inlet and effluent conduits with the pumping chamber. The valves are used to regulate fluid flow through the pumping chamber. A pressure-regulating system, including a pressure sensor and pump-controlling microprocessor, may also be used with the inventive system.

Based on the preceding designs, a number of minipumps with rectangular and circular chamber configurations were built with flexing IPMNC diaphragms. The IPMNC diaphragms were sandwiched between two 24-K gold-plated ring electrodes, which were circular or rectangular. The ring electrodes were cut from circular copper tubing or rectangular copper channel tubing and then gold plated with 24-K gold.

The chambers were also equipped with gold-plated armature windings to act as receiving inductive coils to energize the minipump, in case it was implanted in a patient's body or a remote location not easily accessible to a direct source of electricity.

Figure 9.28 depicts a double diaphragm minipump for which a series of experiments for pumping characteristics were conducted.

Test results were obtained for this first generation of medipumps. The housing of the pump was PMMA (polymethyl methacrylate, i.e., Plexiglas) and the synthetic muscle (diaphragm) was 30-μm thick fluorinated Teflon (polytetrafluoroethylene, PTFE) coated with platinum. The ring electrodes were copper coated with 24-K gold, the tubings were Teflon and the inductive coil was enameled gold armature wire with the stripped ends gold plated with 24-K gold. The pump had to be sterilized before any medical implanting surgery. The results of the flow rate measurements under a head

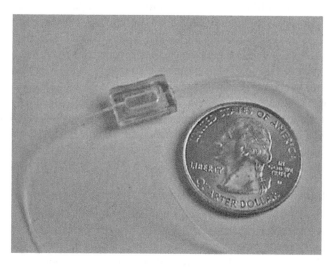

FIGURE 9.28 A fabricated double-diaphragm minipump equipped with IPMNC diaphragms.

pressure of 20, 40 and 60 mmHg with or without the inductive generator powering the pumps are discussed next.

At 20-mmHg input pressure:

Average flow rate of water at room temperature (~70°F) without the inductive generator was 0.6 μL/min.

Average flow rate of water at room temperature (~70°F) with the inductive generator was 0.8 μL/min.

It is estimated that these flow rates in the eye and with aqueous humor replacing water will drop to almost 0 μL/min without inductive generator activation and 0.1 μL/m with inductive generator activation.

At 40-mmHg input pressure:

Average flow rate of water at room temperature (~70°F) without the inductive generator was 1.1 μL/min.

Average flow rate of water at room temperature (~70°F) with the inductive generator was 1.36 μL/min.

It is estimated that these flow rates in the eye and with aqueous humor replacing water will drop to almost 0.8 μL/min without the inductive generator activation and 1.11 μL/min with the inductive generator activation

At 60-mmHg input pressure:

Average flow rate of water at room temperature (~70°F) without the inductive generator was 1.7 μL/min.

Average flow rate of water at room temperature (~70°F) with inductive generator was 1.9 μL/min.

It is estimated that these flow rates in the eye and with aqueous humor replacing water will drop to almost 1.4 μL/min without the inductive generator activation and 1.7 μL/min with the inductive generator activation.

In conclusion, it should be emphasized that implantable, pressure-adjustable, diaphragm pump systems can be fabricated with IPMNCs. Furthermore, these minipumps are scalable and are characterized by a common type of actuating mechanism in the form of synthetic muscles made with ionic polymeric conductor composites (IPCNCs). The pumps may be inductively and transcutaneously powered via adjacent, mutually inductive electromagnetic coils. Alternatively, the pumps may be effectively "self"-powered using a synthetic muscle attached to a local bending or twisting force.

A key feature of the pump is the self- or secondary power generation system in the form of a much larger piece of IPCNC synthetic muscle, which, in the case of glaucoma-prevention systems, may be placed on the globe surface (sclera) of the eye and attached to and secured by the extraocular muscles of the eye. An alternative external power system includes a biocompatible induction coil with gold wire armature that can be transcutaneously activated, adjusted, computer interrogated and controlled by a surgeon.

9.2.8.2 Exoskeleton Human Joint Power Augmentation (ESHPA)

IPMNC artificial muscles can be used in certain attire to augment human joint power. Human skeletons have on the average 98 skeletal joints. Some of these joints, such as the jaw's temporo-mandibular joint, hand's radio carpal (wrist) joint, fingers' interphalangeal (IP) joints or thumb's carpometacarpal (CM) joint, are highly active. Others, such as the foot's subtalar joint or transverse tarsal joint, are less active. Yet other joints are rather integrated joints, such as the spine

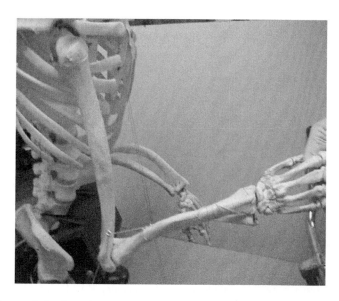

FIGURE 9.29 Human skeletal arm joints.

cervical, thoracic or lumbar vertebrate joints. The human skeletal joints are exoskeletally powered by elaborate systems of skeletal muscles – some 4,000 of them – mostly operating in an antagonist configuration in which families of pairs of contractile muscles perform articulated joint motions (Figures 9.29 and 9.30).

The powering sequence of skeletal muscles starts with an initial electrical polarization wave signal from the brain through the human spine and nervous system to cause an ATP–ADP release of chemical energy to power the muscles. Therefore, in order to fabricate the proposed family of ESHPA systems equipped with solid-state polymeric sensors and actuators, the full integration of triggering signals, energy sources, power converters, sensors and actuators into a complete exoskeleton system will not be discussed here because it is beyond the scope of the book.

FIGURE 9.30 Human skeletal arm muscles.

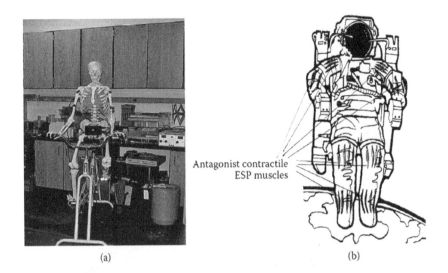

Antagonist contractile
ESP muscles

(a)   (b)

FIGURE 9.31   (a) "Myster bony", a plastic human skeleton equipped with electrically contractile muscle fabrics, riding an exercycle. (b) Schematic representation of an astronaut in a pressurized space suit equipped with joint power augmentation artificial muscles.

Some preliminary results obtained in our research have clearly established that such integrated anthropomorphic systems can be designed and made operational as exoskeleton power augmentation systems on human skeletons.

Figure 9.31(a) depicts "Myster Bony" of our Artificial Muscle Research Institute riding an exercycle while equipped with a system of polymeric contractile muscles. Figure 9.31(b) depicts a schematic representation of an astronaut in his or her pressurized space suit equipped with joint power augmentation artificial muscles. These systems are intended to improve the quality of an individual and can be extended to power augmentation of pressurized space suits for astronauts (Figure 9.31(b)), empowering paraplegics, quadriplegics and disabled and elderly people, as well as a variety of other robotic and medical applications.

9.2.9   MICROELECTROMECHANICAL SYSTEMS

Microelectromechanical systems (MEMS), microrobots made with electroactive polymers, and, in particular, IPMNCs represent an enabling technology for manufacturing sensor and actuator microarrays, disposable microbiosensors for real-time medical applications and a variety of microfabrication processes requiring the manipulation of small objects. The IPMNC actuator microarrays will have immediate applications in micromirror-based photonic optical fiber switches. IPMNC microgrippers are actuated with low voltages (less than 0.5 V), are fast (minimum of 50-Hz bandwidth), and can be cut arbitrarily small (see Figure 9.32) from sheets of the IPMNC material (a typical thickness of 30 μm; see Figure 9.32).

As MEMS technology develops, the most obvious problem is how to build small devices. It is equally important to develop techniques to manipulate and assemble the MEMS components into systems. Historically, grasping and manipulating objects of any size has been a challenge. As components become smaller, the problem becomes even more pronounced. For the most part, there are no suitable actuators for the range of around 10–100 μm. Electroceramic materials (piezoelectric and electrostrictive) offer effective, compact actuation materials to replace electromagnetic motors. A wide variety of electroactive ceramic (EAC) materials are incorporated into motors, translators and manipulators and devices such as ultrasonic motors and inchworms.

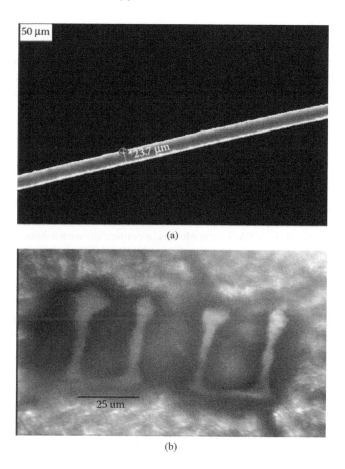

FIGURE 9.32 (a) A photograph of a manufactured, micron-scale IPMNC that can be used for MEMS applications. (b) Assembly of microstrips of IPPC cut in a laser-microscope workstation for microsensing and microactuation.

In contrast to electroceramics, IPMNCs are emerging as new actuation materials with displacement capabilities that cannot be matched by the striction-limited and rigid ceramics. Table 9.1 shows a comparison between the capability of IPMNC materials and electroceramics and shape memory alloys (SMAs). As shown in the table, IPMNC materials are lighter and their potential striction capability can be as high as two orders of magnitude more than that of EAC materials. Further, their

TABLE 9.1

Comparison of the Properties of IPMNCs, SMAs and EACs

Property	IPMNC	SMA	EAC
Actuation displacement	>8%	<6% short fatigue life	0.1–0.3%
Force (MPa)	10–30	About 700	30–40
Reaction speed	μs to s	s to min	μs to s
Density	1–2.5 g/cc	5–6 g/cc	6–8 g/cc
Drive voltage	0.1–7 V	NA	50–800 V
Fracture toughness	Resilient, elastic	Elastic	Fragile

Abbreviations: IPMNC, ionic polymer–metal nanocomposite; SMA, shape memory alloys.

response time is significantly higher than that of SMAs. The current study is directed toward taking advantage of these polymers' resilience and the ability to engineer their properties to meet robotic microarticulation and MEMS requirements. The mass produceability of polymers and the fact that electroactive polymer materials do not require poling (in contrast to piezoelectric materials) help to reduce cost. IPMNC materials can be easily formed in any desired shape and can be used to build MEMS-type mechanisms (actuators and sensors). They can be designed to emulate the operation of biological muscles and they have unique characteristics of low density as well as high toughness, large actuation strain constant and inherent vibration damping.

When electroactive ceramics or SMAs are applied to micromanipulation, a variety of creative approaches have been taken to compensate for each actuator's limitations. For example, many creative systems have been proposed, including nonlinear, high-ratio transmission systems made with a piezoelectric actuator and micromanipulation using SMAs and the use of temperature change to modify the pressure inside microholes on the surface of the end-effector.

The current state of the art in MEMS technologies in connection with robotic micromanipulation and assembly, as well as sensing and actuation, is that small micron-size components can be made by traditional micromachining in the semiconductor industry. Sensors, valves, pumps, manipulators, filters, probes and connectors are just a few examples of MEMS-based devices. Fabrication processes involve silicon surface micromachining, silicon bulk micromachining and wafer bonding, LIGA, EDM (electrodischarge machining) and single-point diamond machining. MEMS are the integration of mechanical elements, sensors, actuators and electronics on a common silicon substrate through the utilization of the preceding microfabrication technology. Since MEMS devices are manufactured using batch fabrication techniques, similar to ICs, unprecedented levels of functionality, reliability and sophistication can be placed on a small silicon chip at a relatively low cost.

IPMNC sensors and actuators can be naturally integrated with the current MEMS technology because they can be easily batch processed and manufactured and they can be made as small as desired and in any desired geometry, as we have proven. IPMNC-MEMS technology will definitely become an enabling new technology to help in biotechnology as well. Technologies such as the polymerase chain reaction (PCR), microsystems for DNA amplification and identification, the micromachined scanning tunneling microscopes (STMs), biochips for detection of hazardous chemical and biological agents and microsystems for high-throughput drug screening and selection will particularly benefit from IPMNC-MEMS integration. IPMNC-MEMS can also easily integrate into high-output dynamic sensing systems such as accelerometers and dynamic motion and force sensors as well.

Although MEMS devices are extremely small, MEMS technology is not about size. Furthermore, MEMS is not about making things out of silicon but is a manufacturing technology: a new way of making complex electromechanical systems using batch fabrication techniques similar to the way in which integrated circuits are made and making these electromechanical elements along with electronics. It is in this spirit that IPMNC sensors and actuators can easily be integrated into MEMS technologies and manufacturing techniques.

These new manufacturing technologies will have several distinct advantages. First, MEMS is an extremely diverse technology that potentially could have a significant impact on every category of commercial and military products. MEMSs are currently used for everything from indwelling blood pressure monitoring to active suspension systems for automobiles to airbag accelerometers. Historically, sensors and actuators are the most costly and unreliable parts of a macroscale sensory-actuator electronics system. In comparison, MEMS technology allows these complex electromechanical systems to be manufactured using batch fabrication methods. In this context, the use of IPMNCs to make large MEMS-based microarrays of sensors and actuators for distributed types of applications is quite promising. Examples of these applications are distributed microactuator arrays for photonic optical fiber switching and tactile biosensing. These new applications will allow the cost and reliability of the sensors and actuators to be put into parity with those of integrated circuits.

IPMNC-based MEMS switches have the potential to form low-cost, high-performance, ultra-broadband, quasi-optical control elements for advanced defense and commercial applications. IPMNC-based MEMS quasi-optical switches offer numerous advantages over conventional switches. Another potential application will be in military and commercial microwave systems requiring monolithic solutions for the realization of low-cost, compact systems. The IPMNC-actuated micromachined switch has great potential for microwave applications due to its extremely high power-handling capability and compatibility with other state-of-the-art fabrication technologies for higher level integrated circuits or systems.

The developments in state-of-the-art MEMS technology have made possible the design and fabrication of micromachined control devices suitable for switching microwave signals. IPMNC-MEMS switches will have low parasitics at microwave frequencies (due to their small size) and will be amenable to achieving low resistive switching or high capacitive (on-capacitance) switching. Also, in MEMS technologies, micromanipulation has always been the most difficult problem.

The first and most obvious way to make a microgripper is to miniaturize an industrial size gripper. Unfortunately, this does not take into account the physics of changing the scale of the problem. Normally, gravity is the predominant force, and when a gripper opens (and sometimes sooner), the carried object falls to the floor. In the microworld, gravity is no longer the predominant force. Adhesive forces, such as electrostatic, van der Waals and surface tension forces, dominate in the small scale. It has been shown that, at a 10-μm object radius, the attractive forces between a sphere and a plane show 10^{-10}, 10^{-10}, 10^{-8} and 10^{-5} N for gravity, electrostatic, van der Waals and surface tension forces, respectively.

Given the challenges to micromanipulation, it is not surprising to find a wide variety of approaches to the problem. Recent approaches to fabricate microgrippers have attempted to use:

- Piezoresistive strain gauges (Hexsil process) for tactile feedback or the assembly of precision optical and magnetic components
- A vacuum system with Lithographic, Galvanoformung, Abformung (LIGA) fabrication temperature change causing a change in pressure
- SMAs (i.e., rotary microjoint)
- Laser trapping
- Dielectrophoresis effects

The technologies that have been applied to micromanipulation do not satisfy all of the requirements necessary for an economically viable approach. One would expect a material that is flexible rather than brittle has long life rather than short life, reacts quickly rather than slowly and is simple rather than complex. IPMNCs are believed to satisfy such requirements of MEMS microactuation technologies. Since these muscles can be cut as small as desired, they present a tremendous potential to MEMS sensing and actuation applications. Figure 9.32 displays a micron-sized array of IPMNC muscles cut in a laser microscope workstation.

A variety of MEMSs can be made by packaging and fabricating IPMNCs in small, miniature and micro sizes. Some examples include micropropulsion engines for material transport in liquid media and biomedical applications such as active microsurgical tools. Other applications involve micropumps, microvalves and microactuators. Flagella- and cilia-type IPMNC actuators fall under this category. Figure 9.32 shows a manufactured IPMNC in a thickness of 25 μm. Note that an effective way of manufacturing such micro-sized IPMNCs is to incorporate solution recasting techniques.

As noted, IPMNCs have shown remarkable displacement under a relatively low voltage drive, using a very low power. However, these ionomers have demonstrated a relatively low force actuation capability. Since the IPMNCs are made of a relatively strong material with a large displacement capability, we investigated their application to emulate fingers. As seen in Figure 9.33, a gripper is shown that uses IPMNC fingers in the form of an end-effector of a miniature low-mass robotic arm. The fingers are shown as vertical gray bars. Upon electrical activation, this wiring configuration

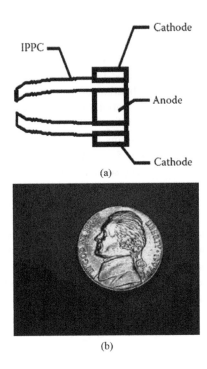

(a)

(b)

FIGURE 9.33   Microgripper (a) and fabricated (b) designs.

allows the fingers to bend inward or outward, similarly to the operation of a hand, and thus close or open the gripper fingers as desired. The hooks at the ends of the fingers represent the concept of nails and allow securing the gripped object encircled by the fingers.

A two-dimensional schematic of the microgripper is provided in Figure 9.33. The gripper would normally be attached to a gross manipulation device (e.g., a small robot) and the artificial muscles are actuated under voltage control. When actuated, the muscles will move together and grip the object in a compliant manner. By increasing the control voltage, the amount of gripping force is increased, and a firmer grasp is achieved. Since the artificial muscle also can act as a sensor, gluing muscles together provides an interesting mechanism to explore how closed-loop controlled micro-gripping is best achieved. It is envisaged using the sensing capabilities of the muscle to provide feedback for gripper closure. A variety of experiments must be performed to determine the best shape for the muscles to achieve grasping variously sized objects from about 10–100 μm. One can vary the number of fingers on the microgripper as well as the artificial muscles sensors attached to the mechanism as shown in Figure 9.34.

9.2.10   Electromechanical Relay Switches

Nonmagnetic, self-contained, electromechanical relay switches can be made from IPMNCs by utilizing their good conductivity and bending characteristics in small applied voltages to close a circuit. In this manner, several of these IPMNC actuators can be arranged to make a multipole-multithrow relay switch.

9.2.11   Continuous Variable Aperture Mirrors and Antenna Dishes

Continuous variable aperture mirrors and antenna dishes can be made by cutting circular sections of the IPMNC and placing electrodes at strategic locations. The focal point of the

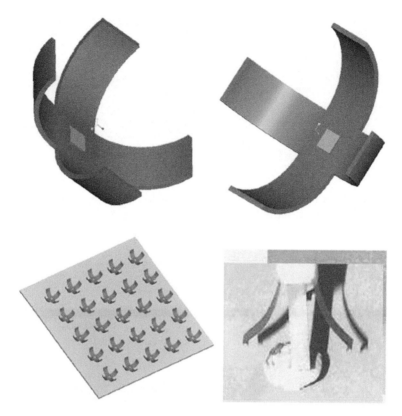

FIGURE 9.34　Array of micro- and/or nanogrippers.

resulting parabolic dish can be controlled by varying the amplitude of the applied voltage to selected electrodes.

9.2.12　SLITHERING DEVICE

Snake-like locomotion can be accomplished by arranging proper segments of the IPMNC in series and controlling each segment's bending by applying sequential input power to each segment in a cascade mode.

9.2.13　PARTS ORIENTATION/FEEDING

Soft-part orientors or feeders for delicate handling of parts in a manufacturing assembly line can be made from flaps made out of IPMNC membrane (Figure 9.35).

9.2.14　MUSICAL INSTRUMENTS

Because mechanical flexing of IPMNC materials generates a voltage and if these materials are already stretched, they create different frequency output signals, one can use them as a musical instrument. Figure 9.36 depicts a one-string musical instrument that operates like a cello or a counterbase and generates very low-frequency musical tones.

9.2.15　FLAT KEYBOARDS, DATA ATTIRE AND BRAILLE ALPHABET

The flat keyboard, data attire, boots and gloves and, particularly, the Braille alphabet applications of IPMNC are rather straightforward in the sense that the material is active everywhere and can be

An eight finger synthetic muscle.
It has a thickness of approximately
2 mm.

A rod shape synthetic muscle.
It has a rectangular cross-section of
approximately 8 mm × 8 mm.

A coil type synthetic muscle.

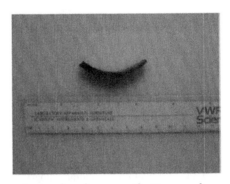

A circular shape synthetic muscle.

FIGURE 9.35 Various shapes of IPMNCs with three-dimensional shapes.

used as a flat keyboard, data attire and gloves and boots (Figure 9.37) with continuity of movement for Braille (dot) alphabet applications by blind people. For Braille alphabet applications, a common electrode is placed on top of the alphabet reading surface. An interdigitated electrode network in the back of the reading surface crates a combination of dots in the form of miniflexing of the IPMNC sheet to enable a blind person to read in a dynamic fashion.

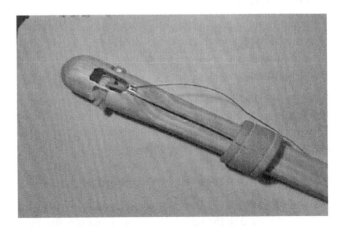

FIGURE 9.36 One-string musical instrument.

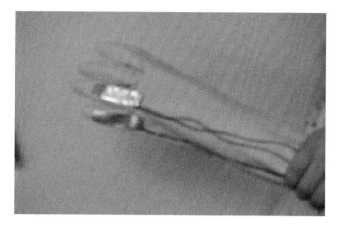

FIGURE 9.37   Data gloves with embedded IPMNC sensor elements.

9.3   BIOMEDICAL APPLICATIONS

The softness and flexibility of polyelectrolytes are definite advantages that can be used in biomedical applications. Here, we identify a number of potential biomedical applications that have been or are currently being developed.

9.3.1   ARTIFICIAL VENTRICULAR OR CARDIAC-ASSIST MUSCLES

Artificial ventricular assist types of muscles can be made for patients with heart abnormalities associated with cardiac muscle functions. We present the broad category of heart compression and assist and arrhythmia control devices – in particular, IPMNC biomimetic sensors, actuators and artificial muscles integrated as a heart compression device that can be implanted external to the patient's heart and partly sutured to the heart without contacting or interfering with the internal blood circulation. Thus, the potential IPMNC device can avoid thrombosis and similar complications common to current artificial heart or heart-assist devices, which may arise when the blood flow makes repeated contacts with nonbiological or nonself-surfaces.

In compressing a heart ventricle, the device must be soft and electronically robust in order not to damage the ventricle. This means that the device should contain control means such as bradycardic (pacing) and tachyarrhythmic (cardioverting/defibrillating) to facilitate device operation in synchronism with the left ventricular contraction and should be capable of transcutaneous recharging of the implanted batteries. The general idea is presented in Figure 9.10. Note that the device is implanted essentially in the ribcage of the patient but is supported on a slender flexible stem that extends to the abdomen. The stem allows the systolic and diastolic cycles of the heart to continue and yet allows the body of the heart to make swinging motions to one side or the other without unnecessary restriction. It is also possible to place the supporting structure of the heart compression device on the diaphragm muscle. These details can be worked out during the clinical testing and operation of such devices.

Note in Figure 9.38 that *42* is the patient body, *44* is the abdomen area, *46* is the ribcage, *5* is the heart, *3* is the polymeric compression finger made with IPMNCs, *30* is the base of the compression device, *10* is a slender conduit carrying the electronic wires to the muscle and acting as a flexible support column and *12* is the power/microprocessor housing placed in the abdomen.

Figure 9.39 depicts a more detailed drawing of the compression device. Again *3* denotes the compression fingers made with IPMNCs, *5* is the heart, *4* depicts an encapsulated enclosure filled with water to create a soft cushion for the compression fingers, *4ds* are IPMNC-based sensor cilia to continuously monitor the compression forces applied to the heart and *3e* and *3f* are the associated wiring and electronics, respectively.

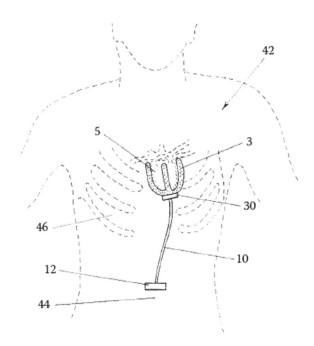

FIGURE 9.38   General configuration for the proposed heart compression device.

As designed, this device produces assisting or soft compression of the left ventricle of a weak heart to produce more internal pressure and to pump more blood from one or more sides in synchrony with the natural systolic contraction of the ventricle. Additionally, the system can provide arrhythmia control of the beating heart. The soft fingers incorporate suitably located electrodes for monitoring the ventricular stroke volume and pressure. A simpler design configuration uses a compression band to assist the heart in its systolic and diastolic cycles of compression-decompression as shown in Figure 9.40.

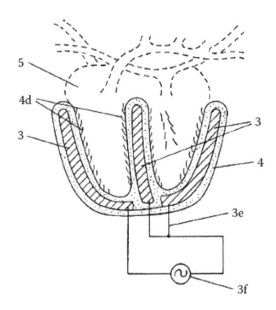

FIGURE 9.39   Heart compression device equipped with IPMNC fingers.

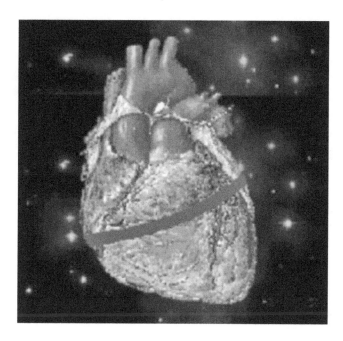

FIGURE 9.40 A heart with an IPMNC compression band.

The compression band can be designed so that it can encircle the heart, as shown in Figure 9.41. Other configurations are depicted in Figures 9.42 and 9.43. Compression devices shown in the latter two figures were designed and fabricated from thick (2 mm) IPMCs and were subsequently 24-K gold plated. These devices performed quite remarkably and showed that enough compression force can be generated with IPMNCs of reasonable thickness for heart compression applications, as will be discussed next.

9.3.1.1 Electroactive Polymer-Powered Miniature Heart Compression Experiment

Here some preliminary data concerning a miniature heart compression device equipped with IPMNCs are presented. First, the force generated by each strip at 5 V and 0.5 Å is measured and then the pressure generated when squeezing a small balloon or the rat's heart is measured. (These results were obtained by heart surgeon Dr. PierGiorgio Tozzi of Lausanne, Switzerland; see Figure 9.44.) Figure 9.45(a) and (b) depict the variation of pressure generated in millimeters of mercury with the voltage applied.

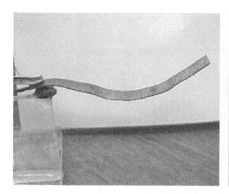

FIGURE 9.41 IPMNC compression bands in open and closed configurations.

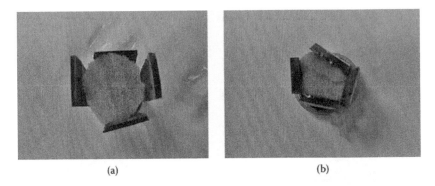

(a)   (b)

FIGURE 9.42 Four-fingered heart compression device equipped with thick IPMNCs: (a) before compression and (b) after compression.

FIGURE 9.43 The upright configuration of the heart compression device.

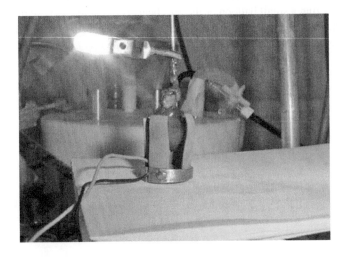

FIGURE 9.44 Miniature heart compression device equipped with IPMNC muscles.

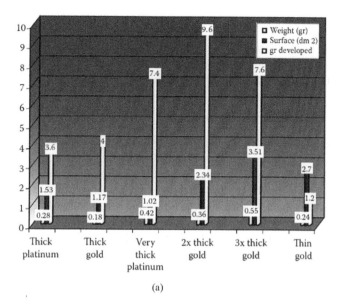

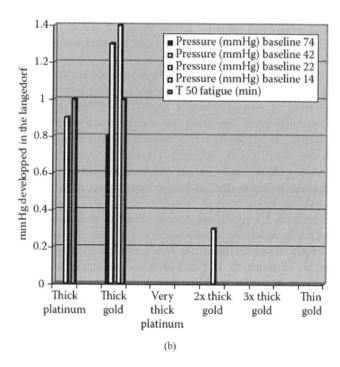

FIGURE 9.45 (a) Pressure generation versus electrode thickness. (b) Pressure generation versus electrode thickness.

9.3.2 Surgical Tool

The IPMNC actuator can be adopted for use as a guide wire or a microcatheter in biomedical applications for intracavity endoscopic surgery and diagnostics. Small internal cavities in the body can be navigated by using small strip- or fiber-like IPMNC actuators.

9.3.3 Peristaltic Pumps

Peristaltic pumps can be made from tubular sections of the membrane of an IPMNC and placement of the electrodes in appropriate locations. Modulating the volume trapped in the tube is possible by applying appropriate input voltage at the proper frequency.

9.3.4 Artificial Smooth Muscle Actuators

Artificial smooth muscle actuators similar to biological smooth muscles can be made by attaching several segments of tubular sections of IPMNC and employing a simple control scheme to sequentially activate each segment to produce a traveling wave of volume change in the combined tube sections. This motion can be used to transport material or liquid contained in tube volume. The activation of each segment is similar to the peristaltic pump described earlier. Artificial veins, arteries and intestines made with the IPMNCs can be fabricated and packaged in a variety of sizes depending on the application. Figure 9.46 shows an artificial smooth muscle actuator that mimics a human hand. It is made with an IPMNC.

Another method of using IPMNC actuators is to package them as human skeletal joint mobility and power augmentation systems in the form of wearable, electrically self-powered, exoskeletal prostheses, orthoses and integrated muscle fabric system components such as jackets, trousers, gloves and boots. These features are intended to improve the quality of a human system and can be extended to power augmentation of attire for advanced soldier and astronaut systems and as prosthetic devices to empower paraplegics, quadriplegics and disabled and elderly people, as well as a variety of other robotic and medical applications.

The essence of the operation of such prostheses, orthoses and wearable attire (smart muscle fabric) is that, for example, a skeletal joint such as the elbow will be equipped with a flexible strip-like bending muscle made from a family of IPMNCs. As noted, IPMNCs have the ability to sense any dynamic motion imparted to them by generating tens of millivolts of electricity (for a $10 \times 40 \times 2$ mm synthetic muscle bent by 1 cm in a cantilever configuration) and the same muscle can generate a torque of about 20 gf-cm, with 9 V and 100 mA, to augment the bending power of a skeletal joint.

Thus, such prostheses, orthoses and smooth muscle fabric systems can be integrated devices equipped with sensing and actuation that can be used for positive feedback robotic control for the mobility of any joint such as the knee, elbow, shoulder, neck, hip or fingers.

Human skeletons normally have 98 skeletal joints. Some of these joints, such as the jaw's temporomandibular joint, hand's radio carpal (wrist) joint, fingers' interphalangeal (IP) joints or thumb's CM joint, are highly active. Others, such as the foot's subtalar joint or transverse tarsal joint, are less active. Yet other joints are rather integrated joints, such as the spine cervical, thoracic or lumbar vertebrate joints. The human skeletal joints are exoskeletally powered by elaborate systems of skeletal muscles – some 4,000 of them – mostly operating in an antagonist configuration in which

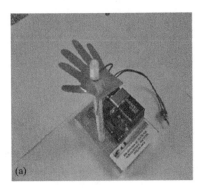 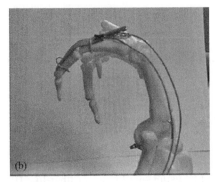

FIGURE 9.46 An artificial smooth muscle actuator that mimics (a) a human hand and (b) a fabricated human joint mobility and power augmentation system equipped with IPMNCs.

families of pairs of contractile muscles perform articulated joint motions. The integrated smooth muscle systems shown in Figure 9.46 as integrated joint power augmentation muscle systems will eventually allow robots to be anthropomorphic and thus capable of carrying distributed loads.

9.3.5 Artificial Sphincter and Ocular Muscles

Artificial sphincter and ocular muscles can also be made from the IPMNC by incorporating thin strips of the actuators in a bundle form similar to the parallel actuator configuration.

9.3.6 Incontinence Assist Devices

Various configurations of IPMNCs may be used in medical applications involving incontinence. In these systems, a patient can activate the muscles by means of a push-button switch or the like, which is preferably battery operated, to prevent leakage and control discharge.

9.3.7 Correction of Refractive Errors of the Human Eyes and Bionic Eyes and Vision

Various configurations of IPMNC may be used in medical applications involving dynamic or static surgical corrections of the refractive errors of the mammalian eyes. In these systems, a patient can activate the muscles by means of a push-button switch or the like, which is preferably battery operated, to prevent leakage and control discharge.

Described here are an apparatus and method to create an automatic or on-demand correction of refractive errors in the eye by the use of an active and smart (computer-controllable) scleral band equipped with composite IPMNC or IPCNC artificial muscles. The scleral band is an encircling band around the middle of the eye's globe to provide relief of intraretinal tractional forces, in cases of retinal detachment or buckle surgery, by indentation of the sclera as well as reposition of the retina and choroids. It can also induce myopia, depending on how much tension is placed on the buckle, by increasing the length of the eye globe in the direction of the optical axis and changing the corneal curvature.

By using the same kind of encircling scleral band, even in the absence of retinal detachment, one can actively change the axial length of the scleral globe and the corneal curvature in order to induce refractive error correction. Figure 9.47 depicts the proposed surgical correction of refractive errors

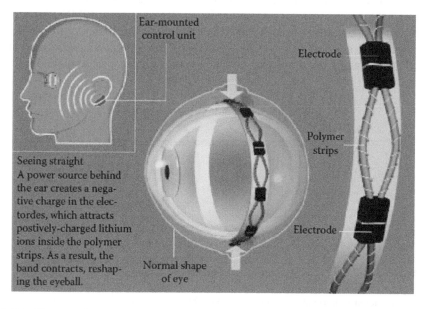

FIGURE 9.47   The essential operation of the active scleral band to create bionic vision.

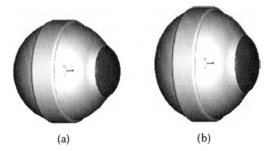

(a)   (b)

FIGURE 9.48 The eye in part (a) is myopic (long and thus shortsighted; image is formed inside the eye and does not reach the macula). The band expands the sclera outward to correct myopia (shortens the eye length and decreases corneal curvature) as in part (b).

by active scleral bands to create bionic eyes. The band has a built-in coil to be energized remotely by magnetic induction and thus provides power for the activation of IPCNC muscles. The active composite artificial muscle will deactivate on command, returning the axial length to its original position and vision back to normal (emmetropic vision).

Figure 9.48 depicts the general configuration for surgical correction of myopia and Figure 9.49 depicts the general configuration for surgical correction of hyperopia or presbyopia. The eye in Figure 9.48(a) is myopic, or long, and thus shortsighted; the image is formed inside the eye and does not reach the macula. The band expands the sclera outward to correct myopia (shorten the eye length and decrease corneal curvature) as in Figure 9.48(b). The eye in Figure 9.49(a) is hyperopic, or short, and thus far sighted; the image is formed outside and beyond the eye and does not reach the macula. The band contracts the sclera inward to correct hyperopia (increase the eye length and increase corneal curvature) as in Figure 9.49(b).

9.4 AEROSPACE APPLICATIONS

9.4.1 COMPOSITE WING FLAP

Figure 9.50 is a view showing multiple, IPMNC actuators in a stacked (sandwiched) configuration D that is designed to accommodate more power for specific actuations. IPMNC actuators A-a, b, c, d and e are each independent planar IPMNC actuators of different lengths (to provide different stiffness and therefore resonant frequency of the composite wing) manufactured according to the predescribed process and formed in a stacked configuration D, which as a whole comprises a top

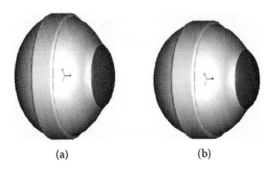

(a)   (b)

FIGURE 9.49 The eye in part (a) is hyperopic (short and thus farsighted; image is formed outside and beyond the eye and does not reach the macula). The band contracts the sclera inward to correct hyperopia (increases the eye length and increases corneal curvature) as in part (b).

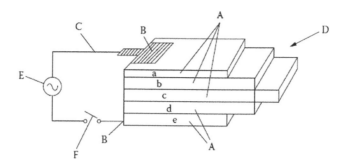

FIGURE 9.50   Composite wing flap made with IPMNCs.

surface and a bottom surface. Terminals *B* are connected to top and bottom surfaces, respectively, at first end of IPMNC actuator *D*. Terminals are also connected by electrical wires *C*, respectively, to a power source *E*. Electrical wire *C* contains an on-off switch *F*. Several of these IPMNC actuators *A* can be assembled in series and multiple amounts of voltage applied to increase power in the composite actuator. IPMNC actuators *A-a-e* act as series resistor elements, especially at higher frequencies.

The IPMNC actuator of Figure 9.50 is a resistive element by nature. Therefore, stacking several of the actuators in effect increases the overall resistance of the combined system. This in turn can allow for higher input voltages. The variation in length of each actuator is due to the desired stiffness of the wing as a whole. Since each actuator has conductivity through its thickness, there is no need to connect wires to faces. Just stacking them can produce a thicker and more powerful actuator that can handle higher loads. The only necessary terminal connections are on the top face of the top layer *A*-a and the bottom face of the bottom layer *A*-e to an alternating (oscillating) source of voltage.

9.4.2   RESONANT FLYING MACHINE

Figure 9.51 is a perspective view of the IPMNC actuator showing a flying machine *G* constructed from IPMNC actuator *B* formed in a single sheet having a top surface *A* and a central axis C. The terminal *D* attached to top surface extends along the central axis *C* of the membrane B.

Terminals *D* are connected at their ends to a power supply *H* by the electric wire *E*. As shown, wire *E* connecting terminal *D* to power supply *H* includes an on-off switch *F*. The IPMNC is packaged in this form for application as a resonant flying machine. In this configuration, the treated IPMNCs ("muscles") can flap like a pair of wings and create a flying machine. "Resonant" means excitation at the resonant frequency of the membrane, which causes the most violent vibration of the membrane. Each body of mass has a resonant frequency at which it will attain its maximum

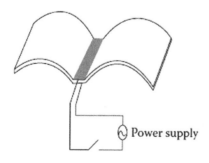

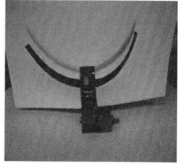

FIGURE 9.51   An illustrative view of the IPMNC actuator showing a flying machine.

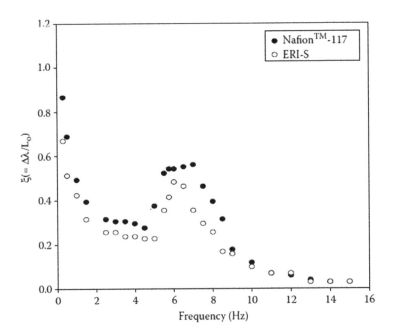

FIGURE 9.52 A fabricated IPMNC flying machine (top) and its resonance performance (bottom). Resonance was observed at about 7 Hz, where the associated displacement was observed approximately half of the cantilever length. Typical resonance characteristics of IPMNCs (Nafion™-117 and ERI-S). (Note: L_o=2", W= 0.5", Cation type=Li$^+$, E_{app}=1.5 volts (sine), vertically positioned.)

displacement when shaken by some input force or power. To obtain large displacements of the actuator, oscillating signals should be applied at a frequency close to its body resonant frequency.

Figure 9.52 shows a fabricated large IPMNC actuator strip with a pair of electrodes (terminals) in the middle fixed to the actuator surfaces of top and bottom. Connecting the circuit to an A.C.-power source (alternating current signal generator) can produce oscillating motion of the membrane actuator similar to a hummingbird's or insect's wing-flap motion. Furthermore, by applying the input voltage signal at or near the resonant frequency of the wing structure, large deformations can be obtained that will vibrate the wing structure in a resonant mode.

The wing assembly is preferably encapsulated in a thin elastic membrane to prevent dehydration of the IPMNC actuator. Also, solid-state polyelectrolytes can be incorporated.

In reality, the possible wake capture mechanism in typical flies is described in Figure 9.53, where nonlinear wing operation is necessary to mimic biological locomotion. Therefore, the IPMNCs should be controlled in a similar manner to carry out such locomotion either actively or passively. Figure 9.54 depicts a flapping wing system equipped with IPMNCs.

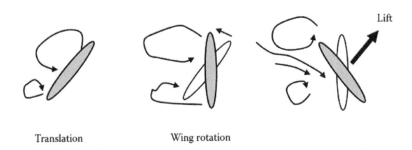

Translation   Wing rotation

FIGURE 9.53 Possible wake capture mechanism.

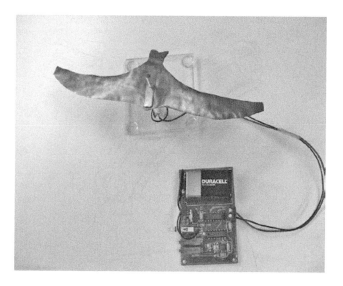

FIGURE 9.54   A flapping wing system equipped with IPMNCs and the electronic driver.

9.4.2.1   Artificial Coral Reefs for Underwater Mine and Moving Object Detection

Figures 9.55 and 9.56 depict large arrays of IPMNC strips acting like large colonies of coral reefs. These artificial coral reefs can act like large sensing arrays to detect any special movement of objects underwater – in particular, mines dropped from the surface – or even movement of surface objects.

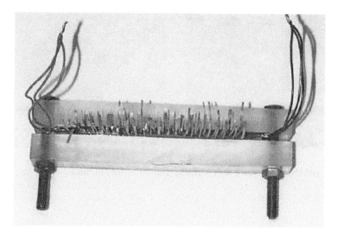

FIGURE 9.55   Large arrays of undulating IPMNC strips acting like artificial coral reefs.

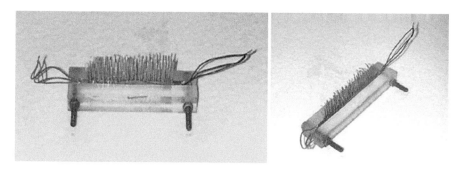

FIGURE 9.56   Another configuration of IPMNC artificial coral reefs.

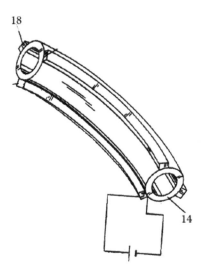

FIGURE 9.57   IPMNC bending muscles (18) equipped with undulating optical fiber (14) for optical switching and light moducation.

9.4.2.2   Other Uses

Figures 9.57 and 9.58 show additional applications of IPMNCs.

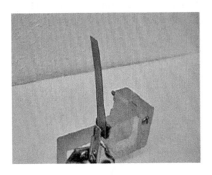

FIGURE 9.58   Using IPMNCs to create morphing and undulating antennas.

10 Epilogue and Conclusions

10.1 EPILOGUE

This book is the second edition of Artificial Muscles: Applications of Advanced Polymeric Nano Composites, which covers advanced ionic polymeric nanocomposites as distributed nanosensors, nano transducers, nanoactuators and artificial muscles. It is the second edition of the first book of its kind to cover such revolutionary and futuristic advanced nanocomposites and smart materials. This book is a result of more than 34 years of research and development on advanced ionic polymeric biomimetic nanocomposites. It essentially covers an introduction to ionic polymers, intelligent gels, and artificial muscles and goes on to cover:

- Ionic polymer-metal nanocomposite (IPMC) artificial muscles: fundamentals
- Manufacturing techniques for IPMC artificial muscles
- Ionic polyacrylonitrile (PAN) chemoelectromechanical artificial muscles/nanomuscles
- PAMPS gel artificial muscles
- Modeling and simulation for IPMCs as biomimetic distributed nanosensors, nanoactuators, nano transducers and artificial muscles
- Sensing, transduction, feedback control and robotic applications of artificial muscles
- Conductive polymers as artificial muscles
- Engineering, industrial and medical applications of IPMCs and smart materials

Furthermore, note that as the energy demand grows, more efficient energy conversion devices increase. One area of improvement is the use of direct energy conversion processes and devices. As discussed in this book, advanced ionic polymeric nanocomposites present tremendous potential for energy harvesting. They are the foundation of current state-of-the-art devices to convert chemical or electrical energy into mechanical energy to perform useful work. In sensing, devices equipped with IPMCs can efficiently convert mechanical energy into electrical or chemical forms.

The work done in this book summarizes efforts on a family of ionic polymeric nanocomposites that have proven to be a viable alternative to conventional means. The synthesis and application of these materials and their corresponding modeling show great promise as alternative smart and intelligent materials for soft robotics, engineering, medicine, biotechnology, and industrial applications.

10.2 CONCLUSION: PAN MUSCLES

Ionically and electrochemically activated muscles made with polyacrylonitrile (PAN) muscle fibers were reported. It was concluded that pH activation of these muscles in a fiber bundle form of 20 fibers could routinely lead to over 200% chemomechanical contraction and expansion, with strength exceeding that of biological muscles. Single fibers of PAN can undergo contractions and expansions of over 500%. But they can also be activated electrically by inducing electrolysis in an electrochemical bath. However, PAN fiber bundles' contraction and expansion reduce to about 45% for 100 fiber bundles in such a bath. The time of expansion and contraction increases to a few minutes instead of a few seconds in direct chemomechanical activation. Both the pH and dielectric constant of the saturating solution contribute to the PAN fibers' expansion or contraction. Whether the mechanisms by which they contribute to the fiber expansion and contraction are the same or not is a matter of further investigation.

DOI: 10.1201/9781003015239-10

A theoretical model for the expansion and contraction of cylindrical specimens of such muscles in an electric field was also presented based on a Poisson-Boltzmann equation. Numerical simulation appears to be consistent with experimental results reported in the literature on other polyelectrolyte gels.

Contraction of these gel fibers was likely due to the contracted PAN's hydrophobic properties and the resulting expulsion of water from the polymer matrix. Artificial muscles made from PAN gel fibers were shown to have a modulus of elasticity similar to and even greater than that of human tissue. The elastic properties and PAN gel fibers' unique ability to change length significantly indicate this material's potential use for linear actuators and artificial muscles. By combining PAN gel fibers with platinum or graphite fibers, the artificial muscles served as an electrode in an electrochemical cell. The artificial muscles contracted over 45% of their initial length through the electrolysis of water, thus indicating the feasibility of electrical activation. Although the contraction rate was significantly slower than that of muscle tissue, decreasing water and ion mass transfer limitations would dramatically enhance the contraction rate.

10.3   CONCLUSION: IPMC ACTUATORS

One of the most promising soft ionic actuators is this new class of polymeric and metal nanocomposites that shows remarkable displacement characteristics in the presence of small electrical voltages. Dramatic improvement has been made to original IPMCs by incorporating surfactant solutions to reduce the clustering of reduced metal nanoparticles within the polymer network. This unique technique improved the force output of the IPMNC by 100%. IPMCs have great potential in the aquatic robotic application and biotechnology because of their hydrophilic nature and compliance and elasticity in a viscous medium such as water. These IPMCs are virtually living, water-loving, advanced polymeric nanocomposites that will prove to be excellent soft robotic actuators for many medical and rehabilitation applications, including Braille cells for blind people.

10.4   CONCLUSION: IPMC SENSORS AND TRANSDUCERS

IPMCs generate output voltage upon deformation. Because deformation can be quite complex and dynamic, the sensing voltage generated can be quite dynamic and complex, showing the nature and the type of deformation. The hysteresis curve showed linear, followed by another linear, and, lastly, parabolic trends as the sensor was bent in a complete cycle. Finally, ionic polymeric metal nanocomposites as sensors and energy harvesting transducers will play a major role in sensor and energy harvesting technologies because they are self-powered and work in the air and polar fluids such as water. Thus, they are useful where simplicity and low cost are sought. One of the main conclusions regarding IPMC sensing and transduction capabilities is that they are capacitive under low frequencies and resistive under high frequencies, spanning the range of piezoelectric and piezoresistive sensors. Another remarkable advantage of IPMC sensors is that they are large-motion sensors.

References

Abe, Y., A. Mochizuki, T. Kawashima, S. Tamashita, K. Asaka, and K. Oguro. 1998. Effect on bending behavior of counter cation species in perfluorinated sulfonate membrane-platinum composite. *Polym. Adv. Technol.* 9:520–526.

Adolf, D., M. Shahinpoor, D. Segalman, and W. Witkowski. 1993. Electrically controlled polymeric gel actuators. U.S. patent no. 5,250,167.

Akazawa, K. 1991. Neuromuscular control system and hardware models. *Adv. Robot.* 5(1):75–85.

Akazawa, K., and K. Fuji. 1986. Theory of muscle contraction and motor control. *Adv. Robot.* 1(4):379–390.

Anquetil, P. A., H. Yu, J. D. Madden, P. G. Madden, T. M. Swager, and I. W. Hunter. 2002. Thiophene-based conducting polymers molecular actuation. In *Smart structures and materials EAPAD symposium proceedings, SPIE,* 4695.

Asahi Chemical Industry Co., Ltd. 1995. Aciphex formula. Public release.

Asaka, K., and K. Oguro. 2000. Bending of polyelectrolyte membrane platinum composites by electric stimuli. Part II. Response kinetics. *J. Electroanal. Chem.* 480:186–198.

Asaka, K., K. Oguro, Y. Nishimura, M. Misuhata, and H. Takenaka. 1995. Bending of polyelectrolyte membrane-platinum composites by electric stimuli. I. Response characteristics to various wave forms. *Polym. J.* 27:436–440.

Asaka, K., K. Oguro, and H. Takenaka. 1994. Bending response of polyelectrolyte membrane-Pt composite. In *Preprints Sapporo symposium on intelligent polymer gels, October 6–8, Sapporo, Japan,* 87–88.

Ashton-Miller, J. A., and M. G. Pandy. 1990. Issues in modeling and control of biomechanical systems. In *Proceedings of the winter annual meeting of the American Society of Mechanical Engineers, November 25–30, Dallas, TX.* ASME Publications, DSC-Vol. 25.

Atkins, P. W. 1982. *Physical chemistry,* 2nd ed. London: Oxford University Press.

Aviram, A. 1978. Mechanophotochemistry. *Macromolecules* 11:1275.

Bar-Cohen, Y., S. P. Leary, M. Shahinpoor, J. O. Simpson, and J. Smith. 1999a. Flexible low-mass devices and mechanisms actuated by electro-active polymers. In *Electroactive polymers SPIE publication no. 3669-38,* 51–56.

———. 1999b. Electro-active polymer (EAP) actuators for planetary applications. In *Electroactive polymers SPIE publication no. 3669-05,* 57–63.

Bar-Cohen, Y., T. Xue, B. Joffe, S.-S. Lih, M. Shahinpoor, J. Simpson, J. Smith, and P. Willis. 1997a. Electroactive polymers (IPMC) low mass muscle actuators. In *Proceedings of (1997) SPIE conference on smart materials and structures, March 5, San Diego, CA.*

Bar-Cohen, Y., T. Xue, B. Joffe, S.-S. Lih, P. Willis, J. Simpson, J. Smith, and M. Shahinpoor. 1997b. Electroactive polymers (EAP) low mass muscle actuators. In *Proceedings of SPIE 3041, smart structures and materials (1997) symposium, enabling technologies: Smart structures and integrated systems,* ed. M. E. Regelbrugge. Bellingham, WA: SPIE, 697–701. ISBN 0-8194-2454-4.

Bar-Cohen, Y., T. Xue, M. Shahinpoor, K. Salehpoor, J. Simpson, J. Smith, and P. Willis. 1998a. Low-mass muscle actuators using electroactive polymers (EAP). In *Proceedings of SPIE smart materials and structures conference, March 3–5, San Diego, CA, SPIE publication no. 3324-32.*

Bar-Cohen, Y., T. Xue, M. Shahinpoor, J. Simpson, and J. Smith. 1998b. Flexible low-mass robotic arm actuated by electroactive polymers (EAP). In *Proceedings of SPIE smart materials and structures conference March 3–5, San Diego, CA, SPIE publication no. SPIE 3329-07.*

———. 1998c. Flexible, low-mass robotic arm actuated by electroactive polymers and operated equivalently to human arm and hand. In *Proceedings of the space '98-robotics '98 conference, third international conference exposition on robotics for challenging environment,* eds. L. A. Demsetz, R. H. Byrne, and J. P. Weyzel. April 26–30, Albuquerque, NM: American Society of Civil Engineers.

Baughman, R. H. 1996. Conducting polymer artificial muscles. *Synth. Met.* 78:339–353.

Baughman, R. H., C. Cui, A. A. Zakhidov, Z. Iqbal, J. N. Basrisci, G. M. Spinks, G. G. Wallace, A. Mazzoldi, D. de Rossi, A. G. Rinzier, O. Jaschinski, S. Roth, and M. Kertesz. 1999. Carbon nanotube actuators. *Science* 284:1340–1344.

Baughman, R. H., L. W. Shacklette, R. L. Elsenbaumer, E. J. Plichta, and C. Becht. 1991. Microelectromechanical actuators based on conducting polymers. In *Molecular electronics,* ed. P. I. Lazarev. Dordrecht: Kluwer Academic Publishers.

Beatty, M. F. 1987. Topics in finite elasticity: Hyperelasticity of rubber, elastomers, and biological tissues—With examples. *Appl. Mech. Rev.* 40:1699–1733. (Reprinted with minor modifications as "Introduction to Nonlinear Elasticity." In *Nonlinear effects in fluids and solids*, eds. M. M. Carroll and M. A. Hayes. New York, NY: Plenum Press, 16–112, 1996.)

Bejczy, A. K., and J. K. Salisbury. 1983. Controlling remote manipulators through kinesthetic coupling. *Comput. Mech. Eng.* 2(1).

Bennet, M., and D. J. Leo. 2003. Manufacture and characterization of ionic polymer transducers with non-precious metal electrodes. *J. Smart Mater. Struct.* 12(3):424–436.

Bernardi, D. M., and M. W. Verbugge. 1992. A mathematical model of a solid polymer electrolyte fuel cell. *J. Electrochem. Soc.* 139(9):2477–2491.

Bernardi, P., G. D'Inzeo, and S. Pisa. 1994. A generalized ionic model of the neuronal membrane electrical activity. *IEEE Trans. Biomed. Eng.* 41(2):125–133.

Bone, Q., and N. B. Marshall. 1982. *Biology of fishes.* London: Blackie & Son.

Brand, H. R., and H. Finkelmann. 1998. Physical properties of liquid crystalline elastomers. In *Handbook of liquid crystals. Vol. 3: High molecular weight liquid crystals*, eds. D. Demus, J. Goodby, G. W. Gray, H.-W. Spiess, and V. Vill. Weinheim: Wiley–VCH, 277–289.

Brand, P. W., and A. Hollister. 1993. *Clinical mechanics of the hand*, 2nd ed. Mosby Year Book.

Brandt, H. R., and H. Pleiner. 1994. Electrohydrodynamics of nematic liquid crystalline elastomers. *Phys. A* 208:359.

Brock, D. 1991a. *Artificial muscle research review.* Cambridge, MA: MIT Artificial Intelligence Lab.

_____. 1991b. *Design and control of an artificial muscle.* Cambridge, MA: MIT Artificial Intelligence Lab.

_____. 1991c. *Review of artificial muscle based on contractile polymers.* MIT Artificial Intelligence Lab. AI Memo No. 1330.

Brock, D., W. Lee, D. Segalman, and W. Witkowski. 1994a. A dynamic model of a linear actuator based on polymer hydrogel. In *Proceedings of the international conference on intelligent materials*, 210–222.

_____. 1994b. A dynamic model of a linear actuator based on polymer hydrogel. *J. Intell. Mater. Syst. Struct.* 5(6):764–771.

Burgmayer, P., and R. W. Murray. 1982. An ion gate membrane: Electrochemical control of ion permeability through a membrane with an embedded electrode. *J. Am. Chem. Soc.* 104:6139.

Caldwell, D. 1993. Natural and artificial muscle elements as robot actuators. *Mechatronics* 3(3):269–283.

Caldwell, D. G., and P. M. Taylor. 1989. An artificial muscle actuator for robots. In *Proceedings of the fourth international conference on advanced robotics, June 13–15, Columbus, OH*, 244–258.

Caldwell, D., and P. Taylor. 1990. Chemically stimulated pseudo-muscular actuation. *Int. J. Eng. Sci.* 28(8):797–808.

Cao, Y., P. Smith, and A. J. Heeger. 1991. Mechanical and electrical properties of polyacetylene films oriented by tensile drawing. *Polymer* 32:1210.

Carlson, J. D. 1999a. Low-cost MR fluid sponge devices. *J. Intell. Syst. Struct.* 10:589–594.

_____. 1999b. Magnetorheological fluid actuators. In *Adaptronics and smart materials*, ed. H. Janocha. Berlin: Springer-Verlag, 180–195.

Carlson, J. D., and K. D. Weiss. 1994. A growing attraction to magnetic fluids. *Mach. Design* 8:61–66.

Chen, Z., X. Tan, and M. Shahinpoor. 2005. Quasistatic positioning of ionic polymer–metal composite (IPMC) actuators. In *Proceedings of the IEEE/ASME conference on advanced intelligent mechatronics (AIM 2005), July 24–28, Monterey, CA*, 60–65.

Chiang, C. K., A. A. Druy, S. C. Gau, A. J. Heeger, H. Shirakawa, E. J. Louis, S. C. Gau, A. G. Mac Diarmid, and W. Park. 1978. Synthesis of highly conducting films of derivatives of polyacetylene, $(CH)_x$. *J. Am. Chem. Soc.* 100:1013.

Chiang, C. K., C. R. Fischer, Y. W. Park, A. J. Heeger, H. Shirakawa, E. J. Louis, S. C. Gau, and A. G. Mac Diarmid. 1977. Electrical conductivity in doped polyacetylene. *Phys. Rev. Lett.* 39:1098.

Chiarelli, P., A. Della Santa, D. De Rossi, and A. Mazzoldi. 1994. Actuation properties of electrochemically driven polypyrrole free-standing films. In *Proceedings of the international conference on intelligent materials, Vol. ICIM-3, June 1996, Lyon, France*, 352–360.

Chiarelli, P., and D. De Rossi. 1988. Determination of mechanical parameters related to the kinetics of swelling in an electrically activated contractile gel. *Prog. Colloid Polym. Sci.* 78:4–8.

Chiarelli, P., D. De Rossi, and K. Umezawa. 1989. Progress in the design of an artificial urethral sphincter. In *Proceedings of the third Vienna international workshop on functional electrostimulation, September, Vienna, Austria.*

Chiarelli, P., K. Umezawa, and D. D. Rossi. 1991. A polymer composite showing electrocontractile response. *Polym. Gels* 195–201.

Choi, K., K. J. Kim, D. Kim, C. Manford, S. Heo, and M. Shahinpoor. 2006. Performance characteristics of electrochemically driven polyacrylonitrile fiber bundle actuators. *J. Intell. Mater. Syst. Struct.* 17(7):563–576.

Crowson, A. 1996. Smart materials technologies and biomimetics. In *Proceedings of the SPIE smart structures and materials conference, February, San Diego, CA.*

Dautzenberg, H., W. Jaeger, J. Kotz, B. Philipp, C. Seidel, and D. Stscherbina. 1994. *Polyelectrolytes, formation, characterization and application.* Munich: Hanser Publishers.

Davis, T. A., J. D. Genders, and D. Pletcher. 1997. *A first course in ion permeable membranes.* Romsey: Electrochemical Consultancy.

De, S. K., N. R. Aluru, B. Johnson, W. C. Crone, D. J. Beebe, and J. Moore. 2002. Equilibrium swelling and kinetics of pH-responsive hydrogel: Models, experiments, and simulation. *J. Microelectromech. Syst.* 11(5):544–555.

de Gennes, P. G., M. Hebert, and R. Kant. 1997. Artificial muscles based on nematic gels. *Macromol. Symp.* 113:39.

de Gennes, P. G., K. Okumura, M. Shahinpoor, and K. J. Kim. 2000. Mechanoelectric effects in ionic gels. *Europhys. Lett.* 50(4):513–518.

Della Santa, A., D. De Rossi, and A. Mazzoldi. 1997. Characterization and modeling of a conducting polymer muscle-like linear actuator. *Smart Mater. Struct.* 6:23–24.

Deole, U., R. Lumia, and M. Shahinpoor. 2004. Characterization of impedance properties of ionic polymer metal composite actuators. In *Proceedings of the second world congress on biomimetics and artificial muscle (biomimetics and nano-bio 2004), December 5–8, Albuquerque, NM.*

De Rossi, D., P. Charelli, G. Buzzigoli, C. Domenici, and L. Lazzeri. 1986. Contractile behavior of electrically activated mechanochemical polymer actuators. *Trans. Am. Soc. Artif. Intern. Organs* XXXII:157–162.

De Rossi, D., C. Domenici, and P. Charelli. 1988. Analogs of biological tissues for mechano transduction: Tactile sensors and muscle-like actuators. In *NATO ASI series, sensors sensory syst. adv. robots,* Vol. F43, 201–218.

De Rossi, D., D. Parrini, P. Chiarelli, and G. Buzzigoli. 1985. Electrically induced contractile phenomena in charged polymer networks: Preliminary study on the feasibility of muscle-like structures. *Trans. Am. Soc. Artif. Intern. Organs* XXXI:60–65.

De Rossi, D., M. Suzuki, Y. Osada, and P. Morasso. 1992. Pseudomuscular gel actuators for advanced robotics. *J. Intell. Mater. Syst. Struct.* 3:75–95.

Doi, M., M. Matsumoto, and Y. Hirose. 1992. Deformation of ionic polymer gels by electric fields. *Macromolecules* 25:5504–5511.

Dorfner, K. 1972. *Ion exchangers properties and applications,* 3rd ed. Ann Arbor, MI: Ann Arbor Science Publishers, Inc.

Dukhin, S. S., and V. N. Shilov. 1974. *Dielectric phenomena and the double layer in disperse systems and polyelectrolytes.* Jerusalem: Keter Publishing House Ltd.

Dusek, K., ed. 1993. *Advances in polymer science, vols. 109 and 110, responsive gels, transitions I and II.* Berlin: Springer-Verlag.

Dwyer-Joyce, R. S., W. A. Bullough, and S. Lingard. 1996. Elastohydrodynamic performance of unexcited electro-rheological fluids. *Int. J. Mod. Phys. B* 10(23–24):3181–3189.

Eckerle, J. S., G. B. Andeen, and R. D. Kornbluh. 1992. Exploring artificial muscle as robot actuators. *Rob. Int. Soc. Manuf. Eng.* 5(1):1–4.

Eguchi, M. 1925. Piezoelectric polymers. *Phil. Mag.* 49.

Eisenberg, A., and F. E. Bailey, eds. 1986. *Coulombic interactions in macromolecular systems,* Vol. 302. ACS Sym. Series.

Eisenberg, A., and M. King. 1977. *Ion-containing polymers.* New York, NY: Academic Press.

Eisenberg, A., and H. L. Yeager, eds. 1982. *Perfluorinated ionomer membranes,* Vol. 180. Washington, DC: American Chemical Society.

Ellington, C. P., and T. J. Pedley. 1995. Biological fluid dynamics. In *Proceedings of symposia of the society for experimental biology, no. XLIX, Society for Experimental Biology.*

Engdahl, G., and I. D. Mayergoyz. 2000. *Handbook of giant magnetostrictive materials,* 1st ed. New York, NY: Academic Press.

Escoubes, M., and M. Pineri. 1982. In *Perfluorinated ionomer membranes,* eds. A. Eisenberg and H. L. Yeager. Washington, DC: American Chemical Society.

Evans, R. B., and D. E. Thompson. 1993. The application of force to the healing tendon. *J. Hand Ther.* 266–284.

Ferrara, L., M. Shahinpoor, K. J. Kim, B. Schreyer, A. Keshavarzi, E. Benzel, and J. Lantz. 1999. Use of ionic polymer–metal composites (IPMCs) as a pressure transducer in the human spine. In *Electroactive polymer SPIE publication no. 3669-45*, 394–401.

Finkelmann, H., and H. R. Brand. 1994. Liquid crystalline elastomers—A class of materials with novel properties. *Trends. Polym. Sci.* 2:222.

Finkelmann, H., H. J. Kock, and G. Rehage. 1981. Investigations on liquid crystalline siloxanes: 3. Liquid crystalline elastomer—A new type of liquid crystalline material. *Makromol. Chem. Rapid. Commun.* 2:317–323.

Finkelmann, H., and M. Shahinpoor. 2002. Electrically controllable liquid crystal elastomer–graphite composite artificial muscles. In *Proceedings of SPIE 9th annual international symposium on smart structures and materials, March, San Diego, CA*, 4695–4653.

Firoozbakhsh, K., and M. Shahinpoor. 1998. Mathematical modeling of ionic interactions and deformation in ionic polymeric metal composite artificial muscles. In *Proceedings of SPIE smart materials and structures conference, March 3–5, San Diego, CA, publication no. SPIE 3323-66*.

Flory, P. J. 1941. Thermodynamics of high polymer solutions. *J. Chem. Phys.* 9(8):660.

Flory, P. J. 1953a. *Principles of polymer chemistry*. Ithaca, NY: Cornell University Press.

_____. 1953b. *Statistical mechanics of swelling network structures*. Ithaca, NY: Cornell University Press.

_____. 1969. *Statistical mechanics of polymer chains*. New York, NY: Interscience Publishers.

Fragala, A., J. Enos, A. LaConti, and J. Boyack. 1972. Electrochemical activation of synthetic artificial muscle membrane. *Electrochem. Acta* 17:1507–1522.

Full, R. J., and K. Meijer. 2001. Metrics of natural muscle. In *Electro active polymers (EAP) as artificial muscles, reality potential and challenges*, ed. Y. Bar-Cohen. SPIE & William Andrew/Noyes Publications, 67–83.

Full, R. J., T. Kubow, J. Schmitt, P. Holmes, and D. Koditschek. 2002. Quantifying dynamic stability and maneuverability in legged locomotion. *Int. Comp. Biol.* 42:149–157.

Furukawa, T., and N. Seo. 1990. Electrostriction as the origin of piezoelectricity in ferroelectrics polymers. *Jpn. J. Appl. Phys.* 29(4):675–680.

Furusha, J., and M. S. Sakaguchi. 1999. New actuators using ER fluids and their applications to force display devices in virtual reality and medical treatment. *Int. J. Mod. Phys., B* 13(14–16):2051–2059.

Gandhi, M. R., P. Murray, G. M. Spinks, and G. G. Wallace. 1995. Mechanisms of electromechanical actuation in polypyrrole. *Synth. Met.* 75:247–256.

Gandhi, M. V., B. S. Thompson, and S.-B. Choi. 1989a. A new generation of innovative ultraintelligent composite materials featuring electro-rheological: An experimental investigation. *J. Compos. Mater.* 23:1232–1255.

Gandhi, M. V., B. S. Thompson, S.-B. Choi, and S. Shakair. 1989b. Electro-rheological fluid-based articulating robotic systems. *J. Mech., Trans., Automation.* 111:328–336.

Gebel, G., P. Aldebert, and M. Pineri. 1987. Structure and related properties of solution-case perfluorosulfonate isonomer films. *Macromolecules* 20:1425–1428.

Genuini, G. et al. 1990. *Pseudomuscular linear actuators: Modeling and simulation experiences in the motion of articulated chains*. Maratea, Italy: NATO ACI Science.

Gierke, T. D., G. E. Munn, and F. C. Wilson. 1982. Morphology of perfluorosulfonated membrane products—Wide-angle and small-angle x-ray studies. *ACS Symp. Ser.* 180:195–216.

Gierke, T. D., and W. Y. Hsu. 1982. The cluster-network model of ion clustering in perfluorosulfonated membranes. In *Perfluorinated ionomer membranes*, eds. A. Eisenberg and H. L. Yeager. Washington, DC: ACS, 283–307.

Gobin, P. F., and J. Tatibouet, eds. 1996. In *Proceedings of the third international conference on intelligent materials, SPIE Vol. 2779, The International Society for Optical Engineering*.

Gong, J. P., T. Nitta, and Y. Osada. 1994a. Electrokinetic modeling of the contractile phenomena of polyelectrolyte gels—One-dimensional capillary model. *J. Phys. Chem.* 98:9583–9587.

_____. 1994b. Electrokinetic modeling of the contraction of polyelectrolyte gels. In *Proceedings of the international conference on intelligent materials*, 556–564.

Gong, J. P., and Y. Osada. 1994. Modeling and simulation of ionic polymer network. In *Preprints of the Sapporo symposium on intelligent polymer gels, October 6–8, Sapporo, Japan*, 21–22.

Grimshaw, P. E., J. H. Nussbaum, A. J. Grodzinsky, and M. L. Yarmush. 1990a. Kinetics of electrically and chemically induced swelling in polyelectrolyte gels. *J. Chem. Phys.* 93(6):4462–4472.

Grimshaw, P. E., J. H. Nussbaum, M. L. Yarmush, and S. R. Eisenberg. 1989. Dynamic membranes for protein transport: Chemical electrical control. *Chem. Eng. Sci.* 44(4):827–840.

_____. 1990b. Selective augmentation of macromolecular transport in gels by electrodiffusion and electro-kinetics. *Chem. Eng. Sci.* 45(9):2917–2929.

Grodzinsky, A. J. 1974. *Electromechanics of deformable polyelectrolyte membranes.* Sc.D. dissertation. MIT, Cambridge, MA.

Grodzinsky, A. J. 1975. Fields, forces and flows in biological tissues and membranes. Lecture notes.

Grodzinsky, A. J., and J. R. Melcher. 1974. Electromechanics of deformable, charged polyelectrolyte membranes. In *Proceedings of the 27th annual conference on engineering in medicine and biology,* 16, paper 53.2.

_____. 1976. Electromechanical transduction with charged polyelectrolyte membranes. *IEEE Trans. Biomed. Eng.* 23(6):421–433.

Guo, S., T. Fukuda, K. Kosuge, F. Arai, K. Oguro, and M. Negoro. 1994. *Micro catheter system with active guide wire structure, experimental results and characteristic evaluation of active guide wire catheter using ICPF actuator.* Osaka: Osaka National Research Institute, 191–197.

_____. 1995. *Micro catheter system with active guide wire.* Preprint, Osaka, Japan.

Hamlen, R. P., C. E. Kent, and S. N. Shafer. 1965. Electrolytically active polymers. *Nature* 206:1149–1150.

Happel, J., and H. Brenner. 1973. *Low Reynolds number hydrodynamics.* Englewood Cliffs, NJ: Prentice Hall Publisher, 393.

Hawkins, G. F., M. J. O'Brien, and T. S. Creasy. 2004. Nastic materials using internally generated pressure. In *International conference on adaptive structures and technologies, CAST 2004, October, Bar Harbor, ME.*

Hebert, M. R. K., and P. G. de Gennes. 1997. Dynamics and thermodynamics of artificial muscles based on nematic gels. *J. Phys. I France* 7:909–918.

Heeger, A. J. 2001. Semiconducting and metallic polymers: The fourth generation of polymeric materials. The Noble Prize 2000 lecture. *Curr. Appl. Phys.* 1:247–267.

Heitner-Wirguin, C. 1996. Recent advances in perfluorinated ionomer membranes: Structure, properties and applications. *J. Membr. Sci.* 120(1):1–33.

Helfferich, F. 1995. *Ion exchange.* New York, NY: Dover Publications, Inc.

Henderson, B., S. Lane, M. Shahinpoor, K. Kim, and D. Leo. 2001. Evaluation of ionic polymer-metal composites for use as near-DC mechanical sensors. In *AIAA space 2001—Conference and exposition, Albuquerque, NM, AIAA paper 2001–4600.*

Hess, C., and L. Li. 1989. Smart hands for EVA retriever. In *NASA conference on space telerobotics, January, Pasadena, CA.*

Higa, M., and A. Kira. 1994. A new equation of ion flux in a membrane: Inclusion of frictional force generated by the electric field. *J. Phys. Chem.* 98(25):6339–6342.

Hirai, M., T. Hirai, A. Sukumoda, H. Nemoto, Y. Amemiya, K. Kobayashi, and T. Ueki. 1995. Electrically induced reversible structural change of a highly swollen polymer gel network. *J. Chem. Soc. Faraday Trans.* 91:473–477.

Hirai, T., J. Zheng, and M. Watanabe. 1999. Solvent-drag bending motion of polymer gel induced by an electric field. In *Proceedings of the SPIE 6th annual international symposium on smart structures and materials,* Vol. 3669, ed. Y. Bar-Cohen, 209–217.

Hirokawa, Y., and T. Tanaka. 1984. Volume transition in nonionic gel. *J. Chem. Phys.* 81:6379.

Hirotsu, S., Y. Hirokawa, and T. Tanaka. 1987. Volume-phase transitions of ionized n-isopropylacrylamide gels. *J. Chem. Phys.* 87(2):1392–1395.

Hsu, W. Y., and T. D. Gierke. 1982. Ion clustering and transport in Nafion perfluorinated membranes. *J. Electrochem. Soc.* 129(3):C121–C129.

Hu, X. 1996. Molecular structure of polyacrylonitrile fibers. *J. Appl. Polym. Sci.* 62:1925–1932.

Huang, J. P., M. Karttunen, K. W. Yu, and L. Dong. 2003. Dielectrophoresis of charged colloidal suspensions. *Phys. Rev. E* 67:021403.

Huggins, M. L. 1941. Solutions of long chain compounds. *J. Chem. Phys.* 9(15):440.

Huxley, A. F. 1957. Muscle structure and theories of contraction. *Prog. Biophys. Chem.* 7:255.

Ichijo, H., O. Hirasa, R. Kishi, K. Oowada, K. Sahara, E. Kokufuta, and S. Kohno. 1995. Thermo-responsive gels. *Radiat. Phys. Chem.* 46:185–190.

Igawa, M., E. Koboyashi, A. Itakura, K. Kikuchi, and H. Okochi. 1994. Selective ion transport across monomeric or reversed micellar liquid membrane containing an open-chain polyether surfactant. *J. Phys. Chem.* 98(47):12447–12451.

Ito, S. 1989. Phase transition of aqueous solution of poly (N-alkylacrylamide) derivatives—Effects of side chain structure. *Kobunsha* 46(7):437.

_____. 1990. Phase transition of aqueous solution of poly (*N*-alkylacrylamide) derivatives—Effects of side chain structure. *Kobunsha* 47(6):467.

Itoh, Y., N. Okui, T. Matsumura, S. Umemoto, and T. Sakai. 1987. Contraction/elongation mechanism of acrylonitrile gel fibers. *Polym. Prepr.* 36(9):2897–2899.

Jacobsen, S. C. 1989. The state of the art in dexterous robotics. In *Conference dinner speaker, IEEE international conference on robotics and automation, May, Scottsdale, AZ.*

Jacobsen, S. C., E. K. Iverson, D. F. Knutti, R. T. Johnson, and K. B. Biggers. 1986. Design of the UTAH/M.I.T. dexterous hand. In *IEEE international conference on robotics and automation, April, San Francisco.*

Jang, B. Z., and Z. J. Zhang. 1994. Thermally and phase transformation-induced volume changes of polymers for actuator applications. *J. Intell. Mater. Syst. Struct.* 5:758–763.

Jolly, M. E., J. D. Carlson, and B. C. Muñoz. 1996. A model of the behavior of magnetorheological materials. *Smart Mater. Struct.* 5:607–614.

Jolly, M. R., and J. D. Carlson. 2000. Composites with field responsive rheology. In *Comprehensive composite materials*, Vol. 5. New York, NY: Elsevier Science Ltd., Chapter 27, 86.

Kabei, N., T. Miyazaki, H. Kurata, M. Ogasawara, T. Murayama, K. Nagatake, and K. Tsuchiya. 1992. Theoretical analysis of an electrostatic linear actuator developed as a biomimicking skeletal muscle. *JSME Int. J., Ser. III*, 35(3):400–405.

Kaga, Y., H. Okuzaki, H. Yasunaga, and Y. Osada. 1994. Electrically controlled insulin release from polyelectrolyte gel. In *Proceedings of the international conference on intelligent materials*, 711–718.

Kagami, Y., A. Matsuda, H. Yasunaga, and Y. Osada. 1994. Order–disorder transition of hydrogel with n-alkylacrylate and shape memory function as a smart diaphragm. In *Preprints of the Sapporo symposium on intelligent polymer gels, October 6–8, Sapporo, Japan*, 129–130.

Kajiwara, K., and S. B. Ross-Murphy. 1992. Synthetic gels on the move. *Nature* 355(6357):208–209.

Kajiyama, T., Y. Oishi, A. Takahara, and X. He. 1994. Higher order structures and thermoresponsive properties of polymeric gel with stearyl acrylate. In *Preprints of the Sapporo symposium on intelligent polymer gels, October 6–8, Sapporo, Japan*, 27–30.

Kaneto, K., M. Kaneko, Y. Min, and A. G. MacDiarmid. 1995. Artificial muscle: Electromechanical actuators using polyaniline films. *Synth. Met.* 71:2211.

Kanno, R., M. Hattori, S. Tadokoro, T. Takamori, and K. Oguro. 1994. *Modeling of electric characteristics of ICPF actuator.* Kobe: Kobe University Preprint.

Kanno, R., A. Kurata, M. Hattori, S. Tadokoro, T. Takamori, and K. Oguro. 1994a. Characteristics and modeling of ICPF actuators. In *Proceedings of Japan–USA symposium on flexible automation*, Vol. 2, 692–698.

_____. 1994b. Characteristics and modeling of ICPF actuator. In *Proceedings of the Japan-USA symposium on flexible automation*, Vol. 2, 691–698.

Kanno, R., S. Tadokoro, T. Takamori, and M. Hattori. 1996. Linear approximate dynamic model of ICPF actuator. In *Proceedings of the IEEE international conference on robotics and automation*, 219–225.

Katchalsky, A. 1949. Rapid swelling and deswelling of reversible gels of polymeric acids by ionization. *Experientia* V:319–320.

Keshavarzi, A., M. Shahinpoor, K. J. Kim, and J. Lantz. 1999. Blood pressure, pulse rate, and rhythm measurement using ionic polymer–metal composite sensors. In *Electroactive polymers SPIE publication no. 3669-36*, 369–376.

Kim, K. J., J. Caligiuri, K. Choi, M. Shahinpoor, I. D. Norris, and B. R. Mattes. 2002. Polyacrylonitrile nanofibers as artificial nano-muscles. In *Proceedings of the first world congress on biomimetics and artificial muscle (biomimetics 2002), December 9–11, Albuquerque, NM.*

Kim, K. J., J. Caligiuri, and M. Shahinpoor. 2003. Contraction/elongation behavior of cation-modified polyacrylonitrile fibers. In *Proceedings of SPIE 10th annual international symposium on smart structures and materials, March 2–6, San Diego, CA, SPIE publication no. 5051-23*, 207–213.

Kim, K., J. Detweiler, G. Lloyd, M. Shahinpoor, and A. Razani. 2002. Experimental and theoretical investigation of a metal hydride artificial muscle. In *Proceedings of the first world congress on biomimetics and artificial muscle (biomimetics 2002), December 9–11, Albuquerque, NM.*

Kim, D., and K. J. Kim. 2005. Self-oscillatory behavior of ionic polymer–metal composite (IPMC): A new finding. In *SPIE smart materials and structures conference, March, #5759-7.*

Kim, J., J.-Y. Kim, and S.-J. Choe. 2000. Electro-active paper: Its possibility as actuators. In *Proceedings of the SPIE EAPAD conference, part of the 7th annual international symposium on smart structures and materials, SPIE Proceedings*, Vol. 3987, ed. Y. Bar-Cohen, 203–209.

Kim, K. J., and M. Shahinpoor. 1999. Effect of the surface-electrode resistance on the actuation of the ionic polymer–metal composite (IPMC) artificial muscles. In *Electroactive polymers, SPIE publication no. 3669-43*, 308–319.

Kim, K. J., and M. Shahinpoor. 2001a. Development of three-dimensional polymeric artificial muscles. In *Proceedings of SPIE 8th annual international symposium on smart structures and materials, March, Newport Beach, CA*, 4329–(58).

_____. 2001b. The synthesis of nano-scale platinum particles—Their role in performance improvement of artificial muscles and fuel cells. In *Proceedings of SPIE 8th annual international symposium on smart structures and materials, March, Newport Beach, CA*, 4329–(26).

_____. 2002a. Application of polyelectrolytes in ionic polymeric sensors, actuators, and artificial muscles. Review chapter in *Handbook of polyelectrolytes and their applications*. In *Applications of polyelectrolytes and theoretical models*, Vol. 3, eds. Tripathy, S. K., J. Kumar, and H. S. Nalwa. Stevenson Ranch, CA: American Scientific Publishers.

_____. 2002b. Ionic polymer–metal nano-composites: Manufacturing techniques. In *Proceeding of SPIE 8th annual international symposium on smart structures and materials, March, San Diego, CA*, 4695, paper no. 26.

_____. 2002c. A novel method of manufacturing three-dimensional ionic polymer–metal composites (IPMCs) biomimetic sensors, actuators and artificial muscle. *Polymer* 43(3):797–802.

_____. 2002d. Electrical activation of contractile polyacrylonitrile (PAN)-conductor composite fiber bundles as artificial muscles. In *Proceedings of the first world congress on biomimetics and artificial muscle (biomimetics 2002), December 9–11, Albuquerque, NM*.

_____. 2003a. Ionic polymer–metal composites—II. Manufacturing techniques, smart materials and structures (SMS). *Inst. Phys. Publ.* 12(1):65–79.

_____. 2003b. Effective diffusivity of nanoscale ion–water clusters within ion-exchange membranes determined by a novel mechano-electrical technique. *Int. J. Hydr. Energy* 28(1):99–104.

Kim, K. J., M. Shahinpoor, and R. Razani. 1998. Solid polymer fuel cells for the next century. *Int. J. Environ. Conscious Des. Manuf.* 7(3):17–46.

_____. 1999. Electro-active polymer materials for solid polymer fuel cells. In *Electroactive polymers, SPIE publication no. 3669-42*, 385–393.

_____. 2000a. Preparation of IPMCs for use in fuel cells, electrolysis and hydrogen sensors. In *SPIE smart materials and structures, publication no. SPIE 3987-41*, 311–320.

_____. 2000b. Preparation of IPMCs for use in fuel cells, electrolysis, and hydrogen sensors. In *Proceedings of SPIE 7th international symposium on smart structures and materials*, Vol. 3687, 110–120.

Kishi, R., M. Hasebe, M. Hara, and Y. Osada. 1990. Mechanism and process of chemomechanical contraction of polyelectrolyte gels under electric field. *Polym. Adv. Technol.* 1:19–25.

Kishi, R., H. Ichijo, and O. Hirasa. 1993. Thermo-responsive devices using poly(vinyl methyl ether) hydrogels. *J. Intell. Mater. Syst. Struct.* 4: 533–537.

Kolde, J. A., B. Bahar, M. S. Wilson, T. A. Zawodzinski, and S. Gottesfeld. 1995. In *Proton conducting membranes fuel cells*, eds. S. Gottesfeld, G. Halpert, and A. Landgrebe. Pennington, NJ: The Electrochemical Society Proceedings Series, PV 95–23, p. 193.

Kolosov, O., M. Suzuki, and K. Yamanaka. 1993. Microscale evaluation of the viscoelastic properties of polymer gel for artificial muscles using transmission acoustic microscopy. *J. Appl. Phys.* 74(10):6407–6412.

Komoroski, R. A., and K. A. Mauritz. 1982. Nuclear magnetic resonance studies and the theory of ion pairing in perfluorinated isonomers. In *ACS symposium series*, Vol. 180. Washington, DC: The American Chemical Society.

Kottke, E. A., L. D. Partridge, and M. Shahinpoor. 2004. Bio-potential neural activation of artificial muscles. In *Proceedings of the second world congress on biomimetics and artificial muscle (biomimetics and nano-bio 2004), December 5–8, Albuquerque, NM*.

_____. 2007. Bio-potential activation of artificial muscles. *J. Intell. Mater. Syst. Struct.* 18(2).

Kuhn, W. 1949. Reversible Dehnung und Kontraktion bei Anderung der Ionisation eines Netzwerks Polyvalenter Fadenmolekulionen. *Experientia* V:318–319.

Kuhn, W., and B. Hargitay. 1951. Muskelahnliche Kontraktion und Dehnung von Netzwerken Polyvalenter Fadenmolekulionen. *Experientia* VII:1–11.

Kuhn, W., B. Hargitay, A. Katchalsky, and H. Eisenberg. 1950. Reversible dilation and contraction by changing the state of ionization of high-polymer acid networks. *Nature* 165:514–516.

Kuhn, W., O. Kunzle, and A. Katchalsky. 1948. Verhalten Polyvalenter Fademolekelionen in Losung. *Halvetica Chem. Acta* 31:1994–2037.

Kuntz, W. H., R. Larter, and C. E. Uhegbu. 1987. Enhancement of membrane transport of ions by spatially nonuniform electric fields. *J. Am. Chem. Soc.* 109(9):2582–2585.

Laverack, M. S. 1985. Physiological adaptations of marine animals. In *Symposia of the society for experimental biology*. No. XXXIX, Society for Experimental Biology.

Lavrov, A. N., S. Komiya, and Y. Ando. 2002. Antiferromagnets: Magnetic shape-memory effects in a crystal. *Nature* 418:385–386.

Leo, D. J., and J. Cuppoletti. 2004. High-energy density actuation based on biological transport mechanisms. In *International conference on adaptive structures and technologies (CAST 2004), October, Bar Harbor, ME.*

Li, J. Y., and S. Nemat-Nasser. 2000. Micromechanical analysis of ionic clustering in Nafion perfluorinated membrane. *Mech. Mater.* 32(5):303–314.

Li, Y., and T. Tanaka. 1989. Study of the universality class of the gel network system. *J. Chem. Phys.* 90(9):5161–5166.

Li, F. K., W. Zhu, X. Zhang, C. T. Zhao, and M. Xu. 1999. Shape memory effect of ethylene-vinyl acetate copolymers. *J. Appl. Polym. Sci.* 71(7):1063–1070.

Liang, C., and C. A. Rogers. 1990. One-dimensional thermomechanical constitutive relations for shape-memory materials. *J. Intell. Mater. Syst. Struct.* 1(2):54–59.

_____. 1992. Design of shape memory alloy actuators. *ASME J. Mech. Des.* 114:223–230.

Liang, C., C. A. Rogers, and E. Malafeew. 1991. Preliminary investigation of shape memory polymers and their hybrid composites. In *ASME conference on smart structures and materials*, Vol. 123, 97–105.

_____. 1997. Investigation of shape memory polymers and their hybrid composites. *J. Intell. Mater. Syst. Struct.* 8:380.

Lifson, S., and A. Katchalsky. 1954. The electrostatic free energy of polyelectrolyte solutions—II. Fully stretched macromolecules. *J. Polym. Sci.* 13:43–55.

Liu, Z., and P. Calvert. 2000. Multilayer hydrogens and muscle-like actuators. *Adv. Mater.* 12(4):288–291.

Lloyd, L. G., K. J. Kim, A. Razani, and M. Shahinpoor. 2002. Investigation of a solar-thermal bio-mimetic metal hydride actuator. In *Proc. Solar 2002 Conf. (Reno, NV); Proc. Solar Eng. 2002, SED 2002-1066*, 301–308.

_____. 2003. Investigation of a solar-thermal bio-mimetic metal hydride actuator. *ASME J. Sol. Energy Eng.* 125:95–100.

Lumia, R., and M. Shahinpoor. 1999. Microgripper design using electro-active polymers. In *Elactroactive polymer SPIE publication no. 3669-30*, 322–329.

_____. 2002. Artificial muscle micro-gripper. In *Proceedings of the first world congress on biomimetics and artificial muscle (biomimetics 2002), December 9–11, Albuquerque, NM.*

MacDiarmid, A. G. 2001. Synthetic metals: A novel role for organic polymers. The Noble Prize 2000 lecture. *Curr. Appl. Phys.* 1:269–279.

Maddock, L., Q. Bone, and J. M. V. Rayner. 1994. *Mechanics and physiology of animal swimming.* Cambridge: Cambridge University Press.

Mallavarapu, K., and D. J. Leo. 2001. Feedback control of the bending response of ionic polymer actuators. *J. Intell. Mater. Syst. Struct.* 12:143–155.

Margolis, J. M. 1989. *Conductive polymers and plastics.* London: Chapman & Hall.

Marieb, E. N. 1992. *Human anatomy and physiology*, 2nd ed. Redwood City, CA: Benjamin/Cummings Publishing.

Matencio, T., M. A. De Paoli, R. C. D. Peres, R. M. Torresi, and S. I. Cordoba de Torresi. 1995. Ionic exchanges in dodecylbenzene sulfonate doped polypyrrole. I—Optical beam deflection study. *Synth. Met.* 72:59.

Matsui, H., and K. Koboyashi. 1983a. Biomechanics VIII-A. International series on biomechanics. In *Proceedings of the 8th international congress of biomechanics*, Vol. 4A. Nagoya: Human Kinetics Publishers, Inc.

_____. 1983b. Biomechanics VIII-B. International series on biomechanics. In *Proceedings of the 8th international congress of biomechanics*, Vol. 4B. Nagoya: Human Kinetics Publishers, Inc.

Matsukata, M., T. Aoki, K. Sanui, N. Ogata, A. Kikuchi, Y. Sakurai, and M. Okano. 1994. New temperature-responsive biosystems by IPAA-modified enzymes. In *Preprints of the Sapporo symposium on intelligent polymer gels, October 6–8, Sapporo, Japan*, 37–38.

Matsuo, E. S., and T. Tanaka. 1988. Kinetics of discontinuous volume-phase transition of gels. *J. Chem. Phys.* 89(3):1695–1703.

McGehee, M. D., E. K. Miller, D. Moses, and A. J. Heeger. 1999. Twenty years of conductive polymers: From fundamental science to applications. In *Advances in synthetic metals: Twenty years of progress in science and technology*, eds. P. Bamier, S. Lefrant, and G. Bidan. New York, NY: Elsevier, 98–203.

Meghdari, A., M. Jafarian, M. Mojarrad, and M. Shahinpoor. 1993. Exploring artificial muscles as actuators for artificial hands. In *Intelligent structures, materials, and vibrations*, ASME publication DE-Vol. 58, eds. M. Shahinpoor and H. S. Tzou, 21–26.

Michopolous, J. G., and M. Shahinpoor. 2002a. Continuous electrodynamic estimation of impedance associated with multi-dimensional ionic polymeric artificial muscles. In *Proceedings of the first world congress on biomimetics and artificial muscle (biomimetics 2002), December 9–11, Albuquerque, NM.*

_____. 2002b. Towards a multiphysics formulation of electroactive large deflection plates made from ionic polymeric artificial muscles. In *Proceedings of the first world congress on biomimetics and artificial muscle (biomimetics 2002), December 9–11, Albuquerque, NM.*

_____. 2004. Experimental calibration of non-linear continuum multi-field ionic polymer plate *modeling. In Proceedings of the second world congress on biomimetics and artificial muscle (biomimetics and nanobio 2004), December 5–8, Albuquerque, NM.*

Millet, P., M. Pinneri, and R. Durand. 1989. New solid polymer electrolyte composites for water electrolysis. *J. Appl. Electrochem.* 19:162–166.

Mojarrad, M. 1997. Autonomous robotic swimming vehicle employing artificial muscle fin mimicking fish propulsion. In *AUSI conference on unmanned untethered submersible technology, September 7–10, Durham, NH.*

_____. 1999. Design of composite artificial muscle (CAM) fin for aquatic vehicle propulsion. In *AUSI conference on unmanned untethered submersible technology, August 23–25, Durham, NH.*

Mojarrad, M. 2001. *Study of ionic polymeric gels as smart materials and artificial muscles for biomimetic swimming robotic applications.* Ph.D. thesis. Department of Mechanical Engineering, University of New Mexico, Albuquerque, NM.

Mojarrad, M., and M. Shahinpoor. 1996a. Noiseless propulsion for swimming robotic structures using polyelectrolyte ion-exchange membranes. In *Proceedings of SPIE 1996 North American conference on smart structures and materials, February 26–29, San Diego, CA,* 2716, paper no. 27, 183–192.

_____. 1996b. Ion exchange membrane-platinum composites as electrically controllable *artificial muscles.* In *Proceedings of third international conference on intelligent materials, June 3–5, Lyon, France,* 1012–1016.

_____. 1997a. Ion-exchange-metal composite sensor films. In *Proceedings of SPIE conference on intelligent structures and materials, March 3–6, San Diego, CA.*

_____. 1997b. Ion-exchange-metal composite artificial muscle actuator load characterization *and modeling.* In *Proceedings of SPIE conference on intelligent structures and materials, March 3–6, San Diego, CA.*

_____. 1997c. Biomimetic robotic propulsion using polymeric artificial muscles. In *Proceedings of (1997) IEEE robotics and automation conference, April 20–25, Albuquerque, NM.*

Mojarrad, M., and D. Wilson. 2000. Electroactive polymeric composite fin design for propulsion of autonomous aquatic vehicles. In *Proceedings of the fourth international conference and exposition/demonstration on robotics for challenging situations and environments, February 27–March 2.*

Moneim, M. S., F. Keikhosrow, A.-A. Mustapha, K. Larsen, and M. Shahinpoor. 2002. Flexor tendon repair using shape memory alloy suture. *J. Clin. Orthop. Relat. Res.* 402:251–259.

Moor, W. 1972. *Physical chemistry,* 4th ed. Englewood Cliffs, NJ: Prentice-Hall.

Moor, R. B., K. M. Cable, and T. L. Croley. 1992. Barriers to flow in semicrystalline isonomer a procedure for preparing melt-processes perfluorosulfonate isonomer films and membranes. *J. Membr. Sci.* 75:7–14.

Motamedi, A. R., F. T. Blevins, M. C. Willis, T. P. McNally, and M. Shahinpoor. 2000. Biomechanics of the coracoclavicular ligament complex and augmentations used in its repair and reconstruction. *Am. J. Sports Med.* 28(3):380–384.

Naarmann, H. 1987. Conjugated polymers. In *Electronic properties of conjugated polymers,* eds. H. Kuzmany, M. Mehring, and S. Roth. New York, NY: Springer-Verlag.

Nagasawa, M. 1988. Molecular conformation and dynamics of macromolecules in condensed systems. In *Studies in polymer science.* New York, NY: Elsevier.

Narita, T. M., J. P. Gong, and Y. Osada. 1998. Kinetic study of surfactant binding into polymer gel— Experimental and theoretical analyses. *J. Phys. Chem. B* 102:4566–4572.

Nemat-Nasser, S. 2002. Micro-mechanics of actuation of ionic polymer–metal composites (IPMCs). *J. Appl. Phys.* 92(5):2899–2915.

Nemat-Nasser, S., and J. Y. Li. 2000. Electromechanical response of ionic polymer–metal composites. *J. Appl. Phys.* 87(7):3321–3331.

Nemat-Nasser, S., and Y. Wu. 2003. Comparative experimental study of ionic polymer–metal composites with different backbone ionomers and in various cation forms. *J. Appl. Phys.* 93:5255–5267.

Newbury, K. M. 2002. Characterization, modeling, and control of ionic polymer transducers. Ph.D. dissertation. Virginia Tech.

Newbury, K. M., and D. J. Leo. 2002. Electrically induced permanent strain in ionic polymer–metal composite actuators. In *Proceedings of SPIE (2002) North American conference on smart structures and materials, March, San Diego, CA,* Vol. 4695, 67–77.

_____. 2003a. Linear electromechanical model of ionic polymer transducers, Part I: Model *development*. *J. Intell. Mater. Syst. Struct.* 14(6):333–342.

_____. 2003b. Linear electromechanical model of ionic polymer transducers, Part II: Experimental validation. *J. Intell. Mater. Syst. Struct.* 14(6):343–358.

Nicoli, D. et al. 1983. Chemical modification of acrylamide gels: Verification of the role of ionization in phase transitions. *Macromolecules* 16:887–891.

Nitta, T., K. Akama, J. P. Gong, and Y. Osada. 1994. Ionic conduction and water transport in the charged polymer networks. In *Preprints of the Sapporo symposium on intelligent polymer gels, October 6–8, Sapporo, Japan*, 119–120.

Ogston, A. G., B. N. Preston, and J. D. Wells. 1973. On the transport of compact particles through solutions of chain-polymers. *Proc. R. Soc. Lond.* 333:297.

Oguro, K., K. Asaka, and H. Takenaka. 1993. Polymer film actuator driven by low voltage. In *Proceedings of the 4th international symposium of micro machines and human science, Nagoya, Japan*, 39–40.

Oguro, K., N. Fujiwara, K. Asaka, K. Onishi, and S. Sewa. 1999. Polymer electrolyte actuator with gold electrodes. In *Proceedings of the SPIE 6th annual international symposium on smart structures and materials, SPIE Proceedings 3*, Vol. 669, 64–71.

Oguro, K., Y. Kawami, and H. Takenaka. 1992. Bending of an ion-conducting polymer film-electrode composite by an electric stimulus at low voltage. *Trans. J. Micromach. Soc.* 5:27–30.

_____. 1993. Actuator element. U.S. patent no. 5,268,082.

Oguztoreli, M. N., and R. B. Stein. 1983. Optimal control of antagonistic muscles. *Biol. Cybern.* 48:1–99.

O'Handley, R. C. 1998. Model for strain and magnetostriction in magnetic shape memory alloys. *J. Appl. Phys.* 83:3263–3271.

O'Handley, R. C., K. Ullakko, J. K. Huang, and C. Kantner. 1996. Large magnetic-field-induced strains in Ni_2MnGa single crystals. *Appl. Phys. Lett.* 69:1966–1971.

Ohmine, I., and T. Tanaka. 1982. Salt effects on the phase transitions of polymer gels. *J. Chem. Phys.* 77:5725.

Okabayashi, K., F. Goto, K. Abe, and T. Yoshida. 1987. Electrochemical studies of polyaniline and its application. *Synth. Met.* 18:365.

Okuzaki, H., Y. Katsuyama, and Y. Osada. 1994. Electrically driven chemomechanical polymer gel based on molecular assembly reaction. In *Preprints of the Sapporo symposium on intelligent polymer gels, October 6–8, Sapporo, Japan*, 115–116.

Okuzaki, H., and Y. Osada. 1994a. Chemomechanical polymer gel with electrically driven motility. In *Proceedings of the international conference on intelligent materials*, 960–970.

_____. 1994b. Effects of hydrophobic interaction on the cooperative binding of surfactant to a polymer network. *Macromolecules* 27:502–506.

_____. 1994c. Electro-driven chemomechanical polymer gel as an intelligent soft material. *J. Biomater. Sci. Polym. Ed.* 5(5):485–496.

Onishi, K., S. Sewa, K. Asaka, N. Fujiwara, and K. Oguro. 2000. Bending response of polymer electrolyte actuator. In *Proc. smart structures and materials 2001: Electroactive polymer actuators and devices*.

Osada, Y. 1991. Chemical valves and gel actuators. *Adv. Mater.* 3(2):107–108.

_____. 1992a. A chemomechanical polymer gel with electrically driven motility. In *Proceedings of the conference on recent advances in adaptive and sensory materials and their applications*, 783–790.

_____. 1992b. Electro-stimulated chemomechanical system using polymer gels (an approach to intelligent artificial muscle system). In *Proceedings of the international conference on intelligent materials*, 155–161.

Osada, Y., and M. Hasebe. 1985. Electrically activated mechanochemical devices using polyelectrolyte gels. *Chem. Lett.* 4:1285–1288.

Osada, Y., and R. Kishi. 1989. Reversible volume change of microparticles in an electric field. *J. Chem. Soc.* 85:655–662.

Osada, Y., and A. Matsuda. 1995. Shape memory in hydrogels. *Nature* 376:219.

Osada, Y., H. Okuzaki, J. P. Gong, and T. Nitta. 1994. Electro-driven gel motility on the base of cooperative molecular assembly reaction. *Polym. Sci.* 36:340–351.

Osada, Y., H. Okuzaki, and H. Hori. 1992. A polymer gel with electrically driven motility. *Nature* 355:242–244.

Osada, Y., and S. B. Ross-Murphy. 1993. Intelligent gels. *Sci. Am.* 268:82–87.

Oster, G., and D. Auslander. 1971. Topological representations of thermodynamic systems—II. Some elemental subunits for irreversible thermodynamics. *J. Franklin Inst.* 292:77–92.

Otake, M., M. Inaba, and H. Inoue. 1999. Development of gel robots made of electro-active polymer PAMPS gel. In *Proceedings of IEEE International Conference on Systems, Man and Cybernetics*, WA20-2, CDROM (1999).

_____. 2000a. Development of electric environment to control mollusk-shaped gel robots made of electro-active polymer PAMPS Gel. In *Proceedings of the SPIE electroactive polymer actuators and devices (EAPAD)*, Vol. 3987, 321–330.

_____. 2002a. Kinematics of gel robots made of electro-active polymer PAMPS gel. In *Proceedings of the IEEE international conference on robotics and automation, San Francisco*, 488–493.

Otake, M., Y. Kagami, M. Inaba, and H. Inoue. 2000b. Behavior of a mollusk-type robot made of electro-active polymer gel under spatially varying electric fields. In *Intelligent autonomous systems*, Vol. 6, eds., E. Pagello et al. IOS Press, 686–691.

Otake, M., Y. Kagami, M. Inaba, and H. Inoue. 2002b. Motion design of a starfish-shaped gel robot made of electro-active polymer gel. *Rob. Auton. Syst.* 40:185–191.

_____. 2002c. Starfish-shaped gel robots made of EAP. *WW-EAP Newsletter* 4(2):7–8.

Otero, T. F. 1997. Artificial muscles, electrodissolution and REDOX processes in conducting polymers. In *Handbook of organic conductive molecules and polymers*, ed. H. S. Nalwa. New York, NY: John Wiley & Sons Ltd.

Otero, T. F., E. Angulo, J. Rodríguez, and C. Santamaría. 1992a. Spanish patent. E.P. 9200095, EP-9202628.

_____. 1992b. Electrochemomechanical properties from a bilayer: Polypyrrole/non-conducting and flexible material artificial muscle. *J. Electroanal. Chem.* 341:369–375.

Otero, T. F., H. Grande, and J. Rodriguez. 1995. A new model for electrochemical oxidation of polypyrrole under conformational relaxation control. *J. Electrical. Chem.* 394:211–216.

_____. 1996a. Reversible electrochemical reactions in conducting polymers: A molecular approach to artificial muscles. *J. Phys. Org. Chem.* 9:381.

_____. 1996b. Influence of the counterion size on the rate of electrochemical relaxation in polypyrrole. *Synth. Met.* 9:285.

Otero, T. F., and J. Rodríguez. 1992. Electrochemomechanical and electrochemopositioning devices: Artificial muscles. In *Intrinsically conducting polymers: An emerging technology*, ed. M. Aldissi. The Netherlands: Kluwer Academic Publishers, 40.

Otero, T. F., J. Rodriguez, E. Angulo, and C. Santamaria. 1993. Artificial muscles from bilayer structures. *Synth. Met.* 55–57:3713–3717.

Otero, T. F., J. Rodríguez, and C. Santamaría. 1994. Conductive polymers. *Mater. Res. Soc. Symp. Proc.* 330:333.

Otero, T. F., and J. M. Sansinena. 1995. Artificial muscles based on conducting polymers. *Bioelectrochem. Bioenerg.* 38:411–414.

Otero, T. F., and J. M. Sansifiena. 1998. Soft and wet conducting polymers for artificial muscles. *Adv. Mater.* 10(6):491–494.

Paquette, J., and K. J. Kim. 2002. An electric circuit model for ionic polymer-metal composites. In *Proceedings of the first world congress on biomimetics and artificial muscles, December 9–11, Albuquerque, NM*.

Park, Y. H., and M. H. Han. 1992. Preparation of conducting polyacrylonitrile/polypyrrole composite films by electrochemical synthesis and their electrical properties. *J. Appl. Polym. Sci.* 45:1973–1982.

Park, I.-S., K. J. Kim, and D. Kim. 2006. Multi-fields responsive ionic polymer–metal composites. In *Proceedings of SPIE smart materials and structures conference, March*, 6168–6136.

Peace, G. S. 1993. *Taguchi methods: Hands-on approach*. New York, NY: Addison–Wesley Publishing Company, Inc.

Pei, Q., and O. Inganäs. 1993a. Electrochemical muscles: Bending strips built from conjugated polymers. *Synth. Met.* 55–57:3718–3723.

_____. 1993b. Electrochemical applications of the bending beam method. 2. Electroshrinking and slow relaxation in polypyrrole. *J. Phys. Chem.* 97:6034.

Pei, Q., O. Inganas, and I. Lundstrom. 1993. Bending bilayer strips built from polyaniline for artificial electrochemical muscles. *Smart Mater. Struct.* 2(1):1–5.

Perline, R., R. Kornbluh, and J. P. Joseph. 1998. Electrostriction of polymer dielectrics with compliant electrodes as a means of actuation. *Sens. Actuators* 64:77–85.

Perline, R., R. Kornbluh, Q. Pei, and J. Joseph. 2000. High speed electrically actuated elastomers with strain greater than 100%. *Science* 287:836–839.

Phadke, M. S. 1989. *Quality engineering using robust design*. Englewood Cliffs, NJ: Prentice Hall.

Pollack, G. H. 2001. *Cells, gels and the engines of life*. Seattle, WA: Ebner and Sons.

Price, W. E., and G. G. Wallace. 1994. Conducting polymer membranes as intelligent separation systems. In *Proceedings of the international conference on intelligent materials*, 994–1002.

Ramirez, P., H. J. Rapp, S. Reichle, H. Strathmann, and S. Mafe. 1992. Current-voltage curves of bipolar membranes. *J. Appl. Phys.* 72(1):259–264.

Rashid, T., and M. Shahinpoor. 1999. Force optimization of ionic polymeric platinum composite artificial muscles by means of an orthogonal array manufacturing method. In *Electroactive polymers, SPIE publication no. 3669-28*, 289–298.

Ratna, B. R., J. V. Selinger, and H. G. Jeon. 2002. Isotropic-nematic transition in liquid-crystalline elastomers. *Phys. Rev. Lett.* 89:225701.

Rieder, W. G., and H. R. Busby. 1990. *Introductory engineering modeling emphasizing differential models and computer simulations*. Malabar, FL: Robert E. Krieger Publishing Company.

Roentgen, W. C. 1880. About the changes in shape and volume of dielectrics caused by electricity. *Ann. Phys. Chem.* I(1):771–786.

Rohen, I. W. et al. *Color atlas of anatomy: A photographic study of the human body*, 5th ed. Philadelphia, PA: Lippincott Williams & Wilkins, 208–209.

Rosen, M. W. 1959. Water flow about a swimming fish. Master's thesis. U.S. Naval Ordinance Test Station.

Sacerdote, M. P. 1899. Strains in polymers due to electricity. *J. Phys.* 3:31, series t, VIII.

Sadeghipour, K., R. Salomon, and S. Neogi. 1992. Development of a novel electrochemically active membrane and "smart" material based vibration sensor/damper. *Smart Mater. Struct.* 1:172–179.

Sakurada, I. et al. 1959. Vapor pressures of polymer solutions. II. Vapor pressure of the poly(vinyl alcohol)-water system. *J. Polym. Sci.* 35:497–505.

Salehpoor, K., M. Shahinpoor, and M. Mojarrad. 1996a. Electrically controllable ionic polymeric gels as adaptive optical lenses. In *Proceedings of SPIE conference on intelligent structures and materials, February 26–29, San Diego, CA*, Vol. 2716, 36–45.

———. 1996b. Electrically controllable artificial PAN muscles. In *Proceedings of SPIE conference on intelligent structures and materials, February 26–29, San Diego, CA*, Vol. 2716, 116–124.

Salisbury, J. K. 1984. Design and control of an articulated hand. In *International symposium of design synthesis, July, Tokyo*.

Samec, Z., A. Trojanek, and E. Samcova. 1994. Evaluation of ion transport parameters in a Nafion membrane from ion-exchange measurements. *J. Phys. Chem.* 98(25):6352–6358.

Schmitz, K. S. 1994. Macro-ion characterization from dilute solutions to complex fluids. In *ACS symposium series 548*. New York, NY: American Chemical Society.

Schreyer, H. B., N. Gebhart, K. J. Kim, and M. Shahinpoor. 2000. Electric activation of artificial muscles containing polyacrylonitrite gel fibers. *Biomacromol. J.* 1:42–647.

Schreyer, H. B., M. Shahinpoor, and K. J. Kim. 1999. Electrical activation of PAN-Pt artificial muscles. In *Proceedings of SPIE/smart structures and materials/electroactive polymer actuators and devices, March, Newport Beach, CA*, Vol. 3669, 192–198.

Segalman, D. J., and W. R. Witkowski. 1994. Two-dimensional finite element analysis of a polymer gel drug delivery system. In *Proceedings of the international conference on intelligent materials*, 1055–1066.

Segalman, D., W. Witkowski, D. Adolf, and M. Shahinpoor. 1991. Electrically controlled polymeric gels as active materials in adaptive structures. In *Proceedings of the ADPA/AIAA/ASME/SPIE conference on active materials and adaptive structures*, 335–345.

———. 1992a. Theory of electrically controlled polymeric muscles as active materials in adaptive structures. *Int. J. Smart Mater. Struct.* 1(1):44–54.

———. 1992b. Numerical simulation of dynamic behavior of polymeric gels. In *Proceedings of the 1st international conference on intelligent materials, ICIM'92, July, Tsukube, Japan, Technomic Publishing Co.*, 310–313.

———. 1992c. Theory and application of electrically controlled polymeric gels. *Smart Mater. Struct.* 1:95–100.

Segalman, D., W. Witkowski, R. Rao, D. Adolf, and M. Shahinpoor. 1993. Finite element simulation of the 2D collapse of a polyelectrolyte gel disk. In *Proceedings (1993) of SPIE North American conference on smart structures and materials, February, Albuquerque, NM*, Vol. 1916, 14–22.

Sen, A., J. I. Scheinbeim, and B. A. Newman. 1984. The effect of plasticizer on the polarization of poly(vinylidene fluoride) films. *J. Appl. Phys.* 56(9):2433–2439.

Senador, A. E., M. T. Shaw, and P. T. Mather. 2001. *MRS symposium proceedings*. Warrendale, PA: Materials Research Society, 661, KK5.9.1.

Sessler, G. M., and J. Hillenbrand. 1999. Novel polymer electrets. *MRS Symp. Proc. Electroact. Polym. (EAP)* 600:143–158.

Shahinpoor, M. 1991. Conceptual design, kinematics and dynamics of swimming robotic structures using active polymer gels. In *Proceedings of ADPA/AIAA/ASME/SPIE conference on active materials & adaptive structures, November, Alexandria, VA*.

_____. 1992. Conceptual design, kinematics and dynamics of swimming robotic structures using ionic polymeric gel muscles. *Smart Mater. Struct.* 1(1):91–94.

_____. 1993a. Novel applications of ionic polymeric gels as smart materials and artificial muscles. In *Interdisciplinary research in smart materials*, eds. Crowson, A., and J. A. Bailey. U.S. Army Research Office Publications, 78–89.

_____. 1993b. Microelectro-mechanics of ionic polymeric gels as artificial muscles for robotic applications. In *Proceedings (1993) IEEE international conference on robotics & automation, May, Atlanta, GA*, Vol. 2, 380–385.

_____. 1993c. Nonhomogenous large deformation theory of ionic polymeric gels in electric and pH fields. In *Proceedings of the SPIE conference on smart structures and materials, February 1–4, Albuquerque, NM*, Vol. 1916, 40–55.

_____. 1993d. Electro-mechanics of bending of ionic polymeric gels as synthetic muscles for adaptive structures. In *Adaptive structures and material systems*, ed. G. P. Carman and E. Garcia. ASME Publication AD-35, 11–22.

_____. 1994a. Micro-electro-mechanics of ionic polymeric gels as electrically controllable *artificial muscles*. In *Proceedings of the international conference on intelligent materials*, 1095–1104.

_____. 1994b. Electro-mechanics of resilient contractile fiber bundles with applications to ionic polymeric gel and SMA robotic actuators. In *Proceedings (1994) IEEE international conference on robotics & automation, May, San Diego, CA*, Vol. 2, 1502–1508.

_____. 1994c. Design and modeling of a novel spring-loaded ionic polymeric gel actuator. In *Proceedings SPIE (1994) North American conference on smart structures and materials, February, Orlando, FL*, Vol. 2189, paper no. 26, 255–264.

_____. 1994d. Microelectro-mechanics of ionic polymeric gel as synthetic robotic muscles. In *Proceedings SPIE (1994) North American conference on smart structures and materials, February, Orlando, FL*, Vol. 2189, paper no. 27, 265–274.

_____. 1994e. Continuum electromechanics of ionic polymeric gels as artificial muscles for robotic applications. *Smart Mater. Struct. Int. J.* 3:367–372.

_____. 1994f. Micro-electro-mechanics of ionic polymeric gels as electrically controlled synthetic muscles. In *Biomedical engineering recent advances*, Vol. 1, ed. Vossoughi, J. Washington, DC: University of District of Columbia Press, 756–759.

_____. 1995a. A new effect in ionic polymeric gels: The ionic flexogelectric effect. In *Proceedings SPIE (1995) North American conference on smart structures materials, February 28–March 2, San Diego, CA*, paper no. 05, 2441.

_____. 1995b. Active polyelectrolyte gels as electrically controllable artificial muscles an intelligent network structures. In *Active structures, devices and systems*, eds. H. S. Tzou, G. L. Anderson, and M. C. Natori. Lexington, KY: World Science Publishing.

_____. 1995c. Spring-loaded ionic polymeric gel linear actuator. U.S. patent no. 5,389,222. Owner: U.S. Department of Energy, Sandia National Laboratories.

_____. 1995d. Design, modeling and fabrication of micro-robotic actuators with ionic *polymeric gel and SMA micro-muscles*. In *Proceedings (1995) ASME design engineering technical conference, September, Boston*.

_____. 1995e. Micro-electro-mechanics of ionic polymeric gels as electrically controllable artificial muscles. *Int. J. Intell. Mater. Syst.* 6(3):307–314.

_____. 1995f. Design and development of micro-actuators using ionic polymeric micromuscles. In *Proceedings of the ASME design engineering technical conference, September, Boston*.

_____. 1996a. Intelligent materials and structures revisited. In *Proceedings of SPIE conference on intelligent structures and materials, February 26–29, San Diego, CA*, 238–250.

_____. 1996b. Ionic polymeric gels as artificial muscles for robotic and medical applications. *Int. J. Sci. Technol.* 20(1): 89–136, transaction B.

_____. 1996c. The ionic flexogelectric effect. In *Proceedings of (1996) third international conference on intelligent materials, ICIM'96, and third European conference on smart structures and materials, June, Lyon, France*, 1006–1011.

_____. 1998. Active polyelectrolyte gels as electrically controllable artificial muscles and intelligent network structure. In *Structronic systems, Part II*, eds. H. S., Tzou, A. Guran, U. Gabbert, J. Tani, and E. Breitbach. London: World Scientific Publishers, 31–85.

_____. 1999a. Ionic polymer metal nano-composites as biomimetic sensors and actuators. In *Evolving and revolutionary technologies for the new millennium*, eds. L. J. Cohen, J. L. Bauer, and W. E. Davis, Vol. 44, book 2. Covina, CA: SAMPE, 1950–1960.

_____. 1999b. Electro-mechanics of iono-elastic beams as electrically controllable artificial muscles. In *Elactroactive polymer publication no. 3669-12*, 109–121.

_____. 2000a. Ion-exchange membrane-metal composite as biomimetic sensors and actuators. In *Polymer sensors and actuators*, eds. Y. Osada, and D. De Rossi. Heidelberg: Springer-Verlag, Chapter 12, 325–359.

_____. 2000b. Electromechanical modeling of ionic polymer-conductive composite (IPCC) artificial muscles. In *SPIE smart materials and structures, publication no. 3984-37*, 310–320.

_____. 2000c. Potential applications of electroactive polymer sensors and actuators in MEMS technologies. In *SPIE smart materials & MEMS, publication no. 4234-40*, 450–459.

_____. 2000d. Electrically activated artificial muscles made with liquid crystal elastomers. In *Proceedings of the SPIE 7th annual international symposium on smart structures and materials*, ed. Y. Bar-Cohen. *EAPAD conference, 3987*, 187–192.

_____. 2002a. Metal hydride artificial muscle systems. In *Proceedings of the first world congress on biomimetics and artificial muscle (biomimetics 2002), December 9–11, Albuquerque, NM.*

_____. 2002b. Applications of ionic polymer conductor composites (IPCCs) to nanotechnology and nano robots. In *Proceedings of the first world congress on biomimetics and artificial muscle (biomimetics 2002), December 9–11, Albuquerque, NM.*

_____. 2002c. Metal hydride artificial muscle systems. In *Proceedings of the first world congress on biomimetics and artificial muscle (biomimetics 2002), December 9–11, Albuquerque, NM.*

_____. 2002d. Fundamentals of ionic polymer conductor nano-composites as biomimetic sensors, soft actuators and artificial muscles—A review of recent findings. In *Proceedings of 14th U.S. national congress on theoretical and applied mechanics.* Keynote presentation, soft actuators and sensors, June 23–28. Blacksburg, VA: Virginia Polytech.

_____. 2002e. Mechnoelectric effects in ionic polymers. In *Proceedings of 14th U.S. national congress on theoretical and applied mechanics.* Special presentation in honor of Professor Millard F. Beatty, contemporary issues in mechanics special symposium, June 23–28. Blacksburg, VA: Virginia Polytech.

_____. 2002f. Electrically controllable deformation in ionic polymer metal nano-composite actuators. In *Proceedings of ASME 2002 international mechanical engineering congress & exhibition, November 17–22, New Orleans, MECE2002-39037.*

_____. 2002g. Ionic polymer-conductor composites (IPCCs) as biomimetic robotic actuators, sensors and artificial muscles. In *Proceedings of 8th international conference on new actuators, June 10–12, Bremen, Germany.*

_____, ed. 2003a. *Proceedings of the first world congress on biomimetics and artificial muscles, December 8–11, Albuquerque, NM.* Albuquerque, NM: ERI Press.

_____. 2003b. Ionic polymer conductor nano-composites as distributed nanosensors, nanoactuators and artificial muscles—A review of recent findings. In *Proceedings of the international conference on advanced materials and nanotechnology, AMN-1, Mac-Diarmid Institute for Advanced Materials and Nanotechnology, February 9–11, Wellington, NZ*, 14–22.

_____. 2003c. Ionic polymer-conductor composites as biomimetic sensors, robotic actuators and artificial muscles—A review. *Electrochim. Acta* 48(14–16):2343–2353.

_____. 2003d. Mechanoelectrical phenomena in ionic polymers. *J. Math. Mech. Solids* 8(3):281–288, special issue in honor of Professor Millard F. Beatty.

_____. 2004a. Electroactive ion-containing polymers. In *Handbook of smart systems*. London: Institute of Physics (IOP).

_____. 2004b. Artificial muscles. In *Encyclopedia of biomaterials and biomedical engineering*, eds. G. Wnek and G. Bowlin. New York, NY: Marcel Dekker Publishers.

_____. 2004c. Smart thin sheet batteries made with ionic polymer–metal composites (IPMCs). In *IMECE2004-60954, Proceedings of ASME-IMECE 2004, ASME international mechanical engineering congress and RD&D exposition, November 13–19, Anaheim, CA.*

_____. 2004d. Ionic polymer conductor composite materials as distributed nanosensors, nanoactuators and artificial muscles—A review. In *Proceedings of the second world congress on biomimetics and artificial muscle (biomimetics and nano-bio 2004), December 5–8, Albuquerque, NM.*

_____. 2005a. Smart ionic polymer conductor composite materials as multifunctional distributed nanosensors, nanoactuators and artificial muscles. In *IMECE2005-79394, Proceedings of ASME-IMECE 2005, ASME international mechanical engineering congress and RD&D exposition, November 5–11, Orlando, FL.*

_____. 2005b. Recent advances in ionic polymer conductor composite materials as distributed nanosensors, nanoactuators and artificial muscles. In *Proceedings of SPIE 12th annual international symposium on smart structures and materials, March 7–10, San Diego, CA, publication no. 5759*, 49–63.

Shahinpoor, M., and M. Ahghar. 2002. Modeling of electrochemical deformation in polyacrylonitrile (PAN) artificial muscles. In *Proceedings of the first world congress on biomimetics and artificial muscle (biomimetics 2002), December 9–11, Albuquerque, NM.*

Shahinpoor, M., M. Ahghar, K. J. Kim, and L. O. Sillerud. 2002. Industrial and medical applications of ionic polymer–metal composites as biomimetic sensors, actuators, and artificial muscles. In *Proceedings of SPIE 9th annual international symposium on smart structures and materials, March, San Diego, CA, publication no. 4695-43.*

Shahinpoor, M., and R. Alvarez. 2002. Simulation and control of iono-elastic beam dynamic deformation model. In *Proceedings of SPIE 9th annual international symposium on smart structures and materials, March, San Diego, CA, publication no. 4695-40.*

Shahinpoor, M., Y. Bar-Cohen, J. Simpson, and J. Smith. 1998a. Ionic polymer–metal composites (IPMCs) as biomimetic sensors, actuators and artificial muscles—A review. *Smart Mater. Struct. J.* 7:R15–R30.

———. 1999. Ionic polymer-metal composites (IPMCs) as biomimetic sensors and actuators. In *Field-responsive polymers,* eds. I. M. Khan, and J. S. Harrison. American Chemical Society Publication, ACS-FRP, ACS Symposium Series 726. Oxford: Oxford University Press, 25–50.

Shahinpoor, M., Y. Bar-Cohen, T. Xue, J. Simpson, and J. Smith. 1998b. Some experimental results on ion-exchange polymer–metal composites as biomimetic sensors and actuators. In *Proceedings of SPIE smart materials and structures conference, March 3–5, San Diego, CA, publication no. 3324-37.*

Shahinpoor, M., and A. Guran. 2003. Ionic polymer-conductor composites (IPCCs) as biomimetic sensors, actuators and artificial muscles. In *Selected topics in structronics and mechatronic systems,* eds. A. Belyaev, and A. Guran. London: World Scientific Publishers, 417–436.

Shahinpoor, M., and K. J. Kim. 2000a. The effect of surface-electrode resistance on the performance of ionic polymer–metal composite (IPMC) artificial muscles. *Smart Mater. Struct.* 9:543–551.

———. 2000b. Artificial sarcomere and muscle made with conductive polyacrylonitrile (C-PAN) fiber bundles. In *SPIE smart materials and structures, publication no. 3987-34,* 243–251.

———. 2000c. Effects of counter-ions on the performance of IPMCs. In *SPIE smart materials and structures, publication no. 3987-18,* 110–120.

———. 2001a. Fully dry solid-state artificial muscles exhibiting giant electromechanical *effect.* In *Proceedings of SPIE 8th annual international symposium on smart structures and materials, March, Newport Beach, CA, publication no. 4329-58.*

———. 2001b. Novel ionic polymeric hydraulic actuators. In *Proceedings of SPIE 8th annual international symposium on smart structures and materials, March: Newport Beach, CA, publication no. 4329-23.*

———. 2001c. A novel physically loaded and interlocked electrode developed for ionic polymer-metal composites (IPMCs). In *Proceedings of SPIE 8th annual international symposium on smart structures and materials, March, Newport Beach, CA, publication no. 4329-24.*

———. 2001d. A mega-power metal hydride anthroform biorobotic actuator. In *Proceedings of SPIE 8th annual international symposium on smart structures and materials, March, Newport Beach, CA, publication no. 4327-18.*

———. 2001e. Design, development, and testing of a multi-fingered mammalian heart compression/assist device equipped with IPMC. In *Proceedings of SPIE 8th annual international symposium on smart structures and materials, March, Newport Beach, CA, publication no. 4329-53.*

———. 2001f. Nano and micro sensors, actuators and artificial muscles made with ionic polymeric conductive composites. In *NanoSpace (2001), the international conference, March, Galveston, TX.*

———. 2001g. Ionic polymer-metal composites—I. Fundamentals. *Smart Mater. Struct. Int. J.* 10:819–833.

———. 2002a. A solid-state soft actuator exhibiting large electromechanical effect. *Appl. Phys. Lett. (APL)* 80(18):3445–3447.

———. 2002b. Novel ionic polymer–metal composites equipped with physically loaded particulate electrode as biomimetic sensors, actuators and artificial muscles. *Actuators Sens. A, Phys.* 3163:1–8.

———. 2002c. Electric deformation memory effects in ionic polymers. In *Proceedings of SPIE 8th annual international symposium on smart structures and materials, March, San Diego, CA, paper no. 13, 4695.*

———. 2002d. Ionic polymer-metal composites: Fundamentals and phenomenological modeling. In *Proceedings of SPIE 8th annual international symposium on smart structures and materials, March, San Diego, CA, paper no. 36, 4695.*

———. 2002e. A novel physically loaded and interlocked electrode developed for ionic polymer-metal composites (IPMCs). *Actuator Sens. A. Phys.* 96(2/3):125–132.

———. 2002f. Metal hydride artificial muscles. U.S. patent 6,405,532.

_____. 2002g. Electrically controllable deformation memory effects in ionic polymers. In *Proceedings of SPIE 8th annual international symposium on smart structures and materials, March, San Diego, CA, publication no. 4695-13*, 85–94.

_____. 2002h. Experimental study of ionic polymer-metal composites in various cation forms. *Actuator Behav., Sci. Eng. Compos. Mater.* 10(6):423–436.

_____. 2002i. A solid-state soft actuator exhibiting large electromechanical effect. *Appl. Phys. Lett. (APL)* 80(18):3445–3447.

_____. 2002j. Mass transfer induced hydraulic actuation in ionic polymer–metal composites. *J. Intell. Mater. Syst. Struct. (JIMSS)* 13(6):369–376.

_____. 2004. Ionic polymer–metal composites. III. Modeling and simulation as biomimetic sensors, actuators, transducers and artificial muscles (review paper). *Smart Mater. Struct. Int. J.* 13(4):1362–1388.

_____. 2005. Ionic polymer–metal composites—IV. Industrial and medical applications (review paper. *Smart Mater. Struct. Int. J.* 14(1):197–214.

Shahinpoor, M., K. J. Kim, S. Griffin, and D. Leo. 2001. Sensing capabilities of ionic polymer–metal composites. In *Proceedings of SPIE 8th annual international symposium on smart structures and materials, March, Newport Beach, CA, publication no. 4329-28*.

Shahinpoor, M., K. J. Kim, and D. Leo. 2003. Ionic polymer–metal composites as multifunctional materials. *Polym. Compos.* 24(1):24–33.

Shahinpoor, M., K. J. Kim, and M. Mojarrad. 2005. *Ionic polymeric conductor composite artificial muscles*, 2nd ed. Albuquerque, NM: ERI/AMRI Press.

Shahinpoor, M., K. J. Kim, L. O. Sillerud, I. D. Norris, and B. R. Mattes. 2002. Electroactive polyacrylonitrile nanofibers as artificial nanomuscles. In *Proceedings of SPIE 8th annual international symposium on smart structures and materials, March, San Diego, CA, publication no. 4695-42*.

Shahinpoor, M., and M. Mojarrad. 1994. Active musculoskeletal structures equipped with a circulatory system and a network of ionic polymeric gel muscles. In *Proceedings of the international conference on intelligent materials, 1079–1085. Also preprints of the Sapporo symposium on intelligent polymer gels, October 6–8, Sapporo, Japan*, 127–128.

_____. 1996. Ion-exchange membrane-platinum composites as electrically controllable artificial muscles. In *Proceedings (1996) of third international conference on intelligent materials, ICIM '96, and third European conference on smart structures and materials, June, Lyon, France*, 1012–1017.

_____. 1997a. Biomimetic robotic propulsion using ion-exchange membrane metal composite artificial muscles. In *Proceedings (1997) of IEEE robotic automation conference, April, Albuquerque, NM*.

_____. 1997b. Ion-exchange-metal composite sensor films. In *Proceedings (1997) of SPIE smart materials and structures conference, March, San Diego, CA, publication no. 3042-10*.

_____. 1997c. Ion-exchange-metal composite artificial muscle actuator load characterization and modeling. *Smart Mater. Technol.* 3040:294–301.

_____. 1997d. Electrically induced large amplitude vibration and resonance characteristics of ionic polymeric membrane-metal composites. In *Proceedings of SPIE conference on intelligent structures and materials, March 3–6, San Diego, CA*.

_____. 2000. Soft actuators and artificial muscles. U.S. patent # 6,109,852.

Shahinpoor, M., I. D. Norris, B. R. Mattes, K. J. Kim, and L. O. Sillerud. 2002. Electroactive polyacrylonitrile nanofibers as artificial nano-muscles. In *Proceedings of the 2002 SPIE conference on smart materials and structures, March, San Diego, CA, publication no. 4695-42*, 169–173.

Shahinpoor, M., and Y. Osada. 1995a. Electrically induced dynamic contraction of ionic polymeric gels. In *Preprints of the Sapporo symposium on intelligent polymer gels, October 6–8, 1994, Sapporo, Japan. Also in Proceedings of SPIE conference on smart materials and structures, February 28–March 2, San Diego, CA*.

_____. 1995b. Heart tissue replacement with ionic polymeric gels. In *Proceedings (1996) of ASME winter annual meeting, November 12–18, San Francisco*.

Shahinpoor, M., and S. G. Popa. 2002. Recent technology development in contactless monitoring of the electromagnetic activities of the human heart and brain. In *Proceedings of SPIE 9th annual international symposium on smart structures and materials, March, San Diego, CA, publication no. 4694-44*.

Shahinpoor, M., K. Salehpoor, and M. Mojarrad. 1996. Electrically controllable artificial PAN muscles. In *Proceedings of the 1996 SPIE North American conference on smart structures and materials, February, San Diego, CA*, Vol. 2716, 16–124.

_____. 1997a. Some experimental results on the dynamic performance of PAN muscles. *Smart Mater. Technol.* 3040:169–173.

_____. 1997b. Linear and platform type robotic actuators made from ion-exchange-membrane-*metal composites*. In *Proceedings of SPIE conference on intelligent structures and materials, March 3–6, San Diego, CA.*

Shahinpoor, M., K. Salehpoor, and A. Razani. 1998. Role of ion transport in dynamic sensing and actuation of ionic polymeric platinum composite artificial muscles. In *Proceedings of SPIE smart materials and structures conference, March 3–5, San Diego, CA, publication no. 3330-09.*

Shahinpoor, M., L. O. Sillerud, and S. G. Popa. 2002. Smart fiber optic magnetometer. In *Proceedings of the first world congress on biomimetics and artificial muscle (biomimetics 2002), December 9–11, Albuquerque, NM.*

Shahinpoor, M., and M. S. Thompson. 1994. The Venus fly trap and the waterwheel plant as smart carnivorous botanical structures with built-in sensors and actuators. In *Proceedings (1994) of the international conference on intelligent materials, ICIM '94, June, Williamsburg, VA*, 1086–1094.

_____. 1995. The Venus flytrap as a model for biomimetic material with built-in sensors and actuators. *J. Mater. Sci. Eng.* C2:229–233.

Shahinpoor, M., and H. S. Tzou. 1993. Intelligent structures, materials, and vibrations. In *Proceedings of the 14th biennial conference on mechanical vibration and noise, September 19–22*. Albuquerque, NM: ASME Publications.

Shahinpoor, M., and G. Wang. 1995. Design, modeling and performance evaluation of a novel large motion shape memory alloy actuator. In *Proceedings of SPIE (1995) North American conference on smart structures and materials, February 28–March 2, San Diego, CA, paper no. 31*, 2447.

Shahinpoor, M., G. Wang, and M. Mojarrad. 1994. Electro-thermo-mechanics of spring-loaded contractile fiber bundles with applications to ionic polymeric gel and SMA actuators. In *Proceedings of the international conference on intelligent materials*, 1105–1116.

Shiga, T. 1997. Deformation and viscoelastic behavior of polymer gels in electric fields. *Adv. Polym. Sci.* 134:131–163.

Shiga, T., Y. Hirose, A. Okada, and T. Kurauchi. 1993. Bending of ionic polymer gel caused by swelling under sinusoidally varying electric fields. *J. Appl. Polym. Sci.* 47:113–119.

Shiga, T., and T. Kurauchi. 1990. Deformation of polyelectrolyte gels under the influence of electric field. *J. Appl. Polym. Sci.* 39:2305–2320.

Shinohara, S., N. Tajima, and K. Yanagisawa. 1994. Photo fabrication for small polymer gels and the properties. In *Preprints of the Sapporo symposium on intelligent polymer gels, October 6–8, Sapporo, Japan*, 91–92.

Shirakawa, H. 2001. The discovery of polyacetylene films—The dawning of an era of conductive polymers. The Noble Prize 2000 lecture. *Curr. Appl. Phys.* 1:281–286.

Shirakawa, H., E. J. Louis, A. MacDiarmid, C. K. Chiang, and A. J. Heeger. 1977. Synthesis of electrically conducting organic polymers: Halogen derivatives of polyacetylene, $(CH)x$ *J. C. S. Chem. Commun.* 285:578.

Singh, D., D. M. Lu, and N. Djilali. 1999. A two-dimensional analysis of mass transport in proton exchange membrane fuel cells. *Int. J. Eng. Sci.* 37:431–452.

Skalak, R., and S. Chien. 1987. *Handbook of bioengineering.* New York, NY: McGraw–Hill.

Smela, E., and N. Gadegaard. 2001. Volume change in polypyrrole studied by atomic force microscopy. *J. Phys. Chem. B* 105:9395.

Smela, E., O. Inganäs, and I. Lundström. 1995. Controlled folding of micrometer-size structures. *Science* 268:1735–1738.

Smela, E., O. Inganas, Q. Pei, and I. Lundstrom. 1993. Electrochemical muscles: Micromachining fingers and corkscrews. *Adv. Mater.* 5(9):630–632.

Soltanpour, D., and M. Shahinpoor. 2002. Development of a synthetic-muscle-based miniature diaphragm pump for medical applications. In *Proceedings of the first world congress on biomimetics and artificial muscle (biomimetics 2002), December 9–11, Albuquerque, NM.*

Soltanpour, D., P. Shahinpoor, and M. Shahinpoor. 2002. Development of an artificial muscle-based smart band to correct refractive errors of the human eye. In *Proceedings of the first world congress on biomimetics and artificial muscle (biomimetics 2002), December 9–11, Albuquerque, NM.*

Springer, T. E., M. S. Wilson, and S. Gottesfeld. 1993. Modeling and experimental diagnostics in polymer electrolyte fuel cells. *J. Electrochem. Soc.* 140(12).

Springer, T. E., T. A. Zawodzinski, and S. Gottesfeld. 1991. Polymer electrolyte fuel cell model. *J. Electrochem. Soc.* 138(8).

Stein, J. L., J. A. Ashton-Miller, and M. G. Pandy. 1989. Issues in modeling and control of biomechanical systems. In *Proceedings of the winter annual meeting of the American Society of Mechanical Engineers, December 10–15*. San Francisco, CA: ASME Publications, DSC-Vol. 17.

Steinberg, I. Z., A. Oplatka, and A. Katchalsky. 1966. Mechanochemical engines. *Nature* 210:568–571.

Steinmetz, C. G., and R. Larter. 1988. Electrostatic model for enhancement of membrane transport by a non-uniform electric field. *J. Phys. Chem.* 92(21):6113–6120.

Stone, R. J., and J. A. Stone. 1990. *Atlas of the skeletal muscles.* Dubuque, IA: Wm. C. Brown Publishers.

Strobel, M., C. S. Lyons, and K. L. Mittal. 1994. *Plasma surface modification of polymers: Relevance to adhesion.* Utrecht: VSP Publication.

Tadokoro, S. 2000. An actuator model of ICPF for robotic applications on the basis of physicochemical hypotheses. In *Proceedings of the IEEE, ICRA*, 1340–1346.

Tadokoro, S., T. Takamori, and K. Oguro. 2001. Application of the Nafion–platinum composite actuator. In *Proceedings of the SPIE*, Vol. 4329, 28–42.

Tadokoro, S., S. Yamagami, T. Takamori, and K. Oguro. 2000, Modeling of Nafion-Pt composite actuators (ICPF) by ionic motion. In *Proceedings of the SPIE*, Vol. 3987, 92–102.

Takenaka, H., E. Torikai, Y. Kawami, and N. Wakabayshi. 1982. Solid polymer electrolyte water electrolysis. *Int. J. Hydrogen Energy* 7:397–403.

Tal, J. 1994. *Step-by-step design of motion control systems.* Ontario: Electromate Industries Publication.

Tanaka, T. 1978. Collapse of gels and the critical endpoint. *Phys. Rev. Lett.* 40:820.

_____. 1981. Gels. *Sci. Am.* 124–138.

_____. 1987. Gels. In *Structure and dynamics of biopolymers*, ed. C. Nicolini. Boston, MA: Martinus Nijhoff Publishers, 237–257.

Tanaka, T. et al. 1987. Mechanical instability of gels at the phase transition. *Nature* 325(6107):796–798.

Tanaka, T., D. Fillmore, S. T. Sun, I. Nishio, G. Swislow, and A. Shah. 1980. Phase transitions in ionic gels. *Phys. Rev. Lett.* 45:1636–1639.

Tanaka, T., I. Nishio, and S. T. Sun. 1982. Collapse of gels in an electric field. *Science* 218:467–469.

Tobushi, H., S. Hayashi, and S. Kojima. 1992. Mechanical properties of shape memory polymer of polyurethane series. *JSAE Int. J., Ser. 1* 35(3):296–302.

Tortora, G., and S. R. Grabowski. 1993. *Principles of anatomy and physiology*, 7th ed. New York, NY: Harper-Collins College Publishers.

Tourillon, G., and F. Garnier. 1984. Structural effect on the electrochemical properties of polythiophene and derivatives. *J. Electroanal. Chem.* 161:51.

_____. 1994. Effect of dopant on conjugated polymers. *J. Polym. Sci., Polym. Phys. Ed.* 22:33.

Tozzi, P., M. Shahinpoor, D. Hayoz, and L. von Segesser. 2004. Electroactive polymers to assist failing heart: The future is now. In *Proceedings of the second world congress on biomimetics and artificial muscle (biomimetics and nano-bio 2004), December 5–8, Albuquerque, NM.*

Triantafyllou, M. S., and G. S. Triantafyllou. 1995. An efficient swimming machine. *Sci. Am.* 64–70.

Truesdell, C. A., and W. Noll. 1965. The nonlinear field theories of mechanics. In *Handbuch der physik*, ed. S. Flügge. Berlin: Springer-Verlag, III/3.

Tyldesley, B., and J. I. Grieve. 1989. *Muscles, nerves and movement.* Glasgow: Blackwell Scientific Publications.

Tyson, J., T. Schmidt, K. Galanulis, and M. Shahinpoor. 2002. Full-field deformation and strain measurement in biomechanics and biomimetics. In *Proceedings of the first world congress on biomimetics and artificial muscle (biomimetics 2002), December 9–11, Albuquerque, NM.*

Tyson, J., T. Schmidt, M. Shahinpoor, and K. Galanulis. 2003. Biomechanics deformation and strain measurements with 3D image correlation photogrammetry. *J. Exp. Tech., Soc. Exp. Mech.* 27(4):32–46.

Tzou, H. S., and T. Fukuda. 1992. *Precision sensors, actuators and systems.* Dordrecht: Kluwer Academic Publishers.

Ueoka, Y., N. Isogai, and T. Nitta. 1994. Eel-like gel actuator with biomimetic movement. In *Preprints of the Sapporo symposium on intelligent polymer gels, October 6–8, Sapporo, Japan*, 117–118.

Ugural, A. C., and S. K. Fenster. 1987. *Advanced strength and applied elasticity.* New York, NY: Elsevier.

Ullakko, K. 1996. Magnetically controlled shape memory alloys: A new class of actuator materials. *J. Mater. Eng. Perform.* 5:405–409.

Umemoto, S., N. Okui, and T. Sakai. 1991. Contraction behavior of poly(acrylonitrile) gel fibers. In *Polymer gels*, eds. D. DeRossi et al. New York, NY: Plenum Press, 257–270.

Urayama, K., T. Takaigawa, T. Masuda, and S. Kohjiya. 1994. Stress relaxation and creep of polymer gels in solvent under various deformation modes. In *Preprints of the Sapporo symposium on intelligent polymer gels, October 6–8, Sapporo, Japan*, 89–90.

Vander, A. J., J. H. Sherman, and D. S. Luciano. 1985. *Human physiology, the mechanism of body function*, 4th ed. New York, NY: McGraw-Hill, 255–298.

van der Veen, G., and W. Prins. 1971. Photomechanics directed bending of a polymer film by light. *Phys. Sci.* 230:70.

Veltink, P. H., H. J. Chizeck, P. E. Crago, and A. El-Bialy. 1992. Nonlinear joint angle control for artificially stimulated muscle. *IEEE Trans. Biomed. Eng.* 39(4):368–380.

Verbrugge, M. W., and R. F. Hill. 1993. Ion and solvent transport in ion-exchange membranes. *J. Electrochem. Soc.* 140(5):1218–1225.

Wang, G. 1998. A general design of bias force shape memory alloy (BFSMA) actuators and an electrically controlled SMA knee and leg muscle exerciser for paraplegics and quadriplegics. Ph.D. dissertation. Department of Mechanical Engineering, University of New Mexico, Albuquerque.

Wang, Q., X. Du, B. Xu, and L. E. Cross. 1999. Electromechanical coupling and output efficiency of piezoelectric bending actuators. *IEEE Trans. Ultrason. Ferroelect. Freq. Control* 46(3):638–646.

Wang, T. T., J. M. Herbert, and A. M. Glass, eds. 1988. *The applications of ferroelectric polymers*. New York, NY: Chapman & Hall.

Wang, G., and M. Shahinpoor. 1997a. Design for shape memory alloy rotatory joint actuators using shape memory effect and pseudoelastic effect. *Smart Mater. Technol.* 3040:23–30.

———. 1997b. Design, prototyping and computer simulation of a novel large bending actuator made with a shape memory alloy contractile wire. *Smart Mater. Struct. Int. J.* 6(2):214–221.

———. 1997c. A new design for a rotatory joint actuator made with shape memory alloy *contractile wire. Int. J. Intell. Mater. Syst. Struct.* 8(3):191–279.

———. 1998. Design of a knee and leg muscle exerciser using a shape memory alloy rotary actuator. In *Proceedings of SPIE smart materials and structures conference, March 3–5, San Diego, CA, publication no. 3324-29.*

Webb, P. W., and D. Weihs. 1983. *Fish biomechanics*. New York, NY: Praeger.

Weiss, K. D., J. D. Carlson, and D. A. Nixon. 1994. Viscoelastic properties of magneto- and electro-rheological fluids. *J. Intell. Mater. Syst. Struct.* 5:772–775.

Whiting, C. J., A. M. Voice, P. D. Olmsted, and T. C. B. McLeish. 2001. Shear modulus of polyelectrolyte gels under electric field. *J. Phys.: Condens. Matter.* 13:1381–1393.

Williams, W., and D. Bannister. *Gray's anatomy*, 37th ed. Edinburgh: Churchill Livingstone, 614–616.

Winter, D. A. 1990. *Biomechanics and motor control of human movement.* New York, NY: Wiley-Interscience.

Winter, D., R. W. Norman, R. P. Wells, K. C. Hayes, and A. E. Patla. 1985a. Biomechanics IX—A, international series on biomechanics. In *Proceedings of ninth international congress of biomechanics*, Vol. 5A. Waterloo: Human Kinetics Publishers, Inc.

———. 1985b. Biomechanics IX—B, international series on biomechanics. In *Proceedings of ninth international congress of biomechanics*, Vol. 5B. Waterloo: Human Kinetics Publishers, Inc.

Wood, J. E., S. G. Meek, and S. C. Jacobsen. 1989a. Quantitation of human shoulder anatomy for prosthetic arm control—I. Surface modeling. *J. Biomech.* 22(3):273–292.

———. 1989b. Quantitation of human shoulder anatomy for prosthetic arm control—II. Anatomy matrices. *J. Biomech.* 22(4):309–325.

Woojin, L. 1996. Polymer gel-based actuator: Dynamic model of gel for real-time control. Ph.D. thesis. Massachusetts Institute of Technology.

Wu, Y. T., C. J. Brokaw, and C. Brennen. 1975. *Swimming and flying in nature*, Vols. I and II. New York, NY: Plenum Press.

Yamada, H. 1970. *Strength of biological materials*. Baltimore, MD: The Williams & Wilkins Company.

Yannas, I. V., and A. J. Grodzinsky. 1973. Electromechanical energy conversion with collagen fibers in an aqueous medium. *J. Mechanochem. Cell Motil.* 2:113–125.

Yasuda, K., P. Gong, Y. Katsuwama, A. Nakavama, Y. Tanabe, E. Kondo, M. Ueno, and Y. Osada. 2005. Biomechanical properties of high-toughness double network hydrogels. *Biomater. J.* 26(21):4468–4475.

Yeager, H. L. 1982. Perfluorinated ionomer membranes. In *Perfluorinated ionomer membranes*, eds., A. Eisenberg, and H. L. Yeager. American Chemical Society Symposium Series, 180.

Yen, C. C., and T. C. Chang. 1989. Preparation of electric-conducting film from polyacrylonitrile-silver nitrate complex. *J. Appl. Polym. Sci.* 40:267–277.

Yoshida, H., and Y. Miura. 1992. Behavior of water in perfluorinated ionomer membranes having various monovalent cations. *J. Membr. Sci.* 68:1–10.

Yoshida, R., T. Yamaguchi, and H. Ichijo. 1996. Novel oscillation, swelling-deswelling dynamic behavior for pH-polymer gels. *Mater. Sci. Eng. C* 4:107–113.

Young, R. J., and P. A. Lovell. 1991. *Introduction to polymers*, 2nd ed. London: Chapman & Hall, 353.

Yuh, J. 1994. Underwater robotic vehicle: Design and control. In *NSF workshop on future research directions in underwater robotics, Maui, HI.*

Zhang, Q. M., V. Bharti, and X. Zhao. 1998. Giant electrostriction and relaxor ferroelectric behavior in electron-irradiated poly(vinylidene fluoride-trifluoroethylene) copolymer. *Science* 280:2101–2104.

Zhang, Q. M., T. Furukawa, Y. Bar-Cohen, and J. Scheinbeim, eds. 1999. *Proceedings of the fall MRS symposium on electroactive polymers (EAP)*, Vol. 600. Warrendale, PA: Materials Research Society, 1–336.

Zrinyi, M., L. Barsi, D. Szabo, and H. G. Kilian. 1997. Direct observation of abrupt shape transition in ferrogels induced by nonuniform magnetic field. *J. Chem. Phys.* 106(13):5685–5692.

Zrinyi, M., D. Szabo, and J. Feher. 1999. Comparative studies of electro- and magnetic field sensitive polymer gels. In *Proc. SPIE 6th Annu. Int. Symp. Smart Struct. Mater. EAPAD conf.*, Vol. 3669, ed. Y. Bar-Cohen. 406–413.

Author Index

Subject Index

Note: Locators in *italics* represent figures and **bold** indicate tables in the text.

Printed and bound by CPI Group (UK) Ltd, Croydon, CR0 4YY

17/10/2024

01775700-0005